Eurekas
and euphorias

Walter Gratzer is at the Randall Centre for Molecular
Mechanisms of Cell Function, King's College, London.
He is the author of *The Undergrowth of Science; Delusion,
Self-Deception, and Human Frailty* (OUP, 2000). Professor
Gratzer is known to a wide readership through his book
reviews, many of them in *Nature*.

Eurekas and euphorias

The Oxford book of scientific anecdotes

WALTER GRATZER

OXFORD
UNIVERSITY PRESS

OXFORD
UNIVERSITY PRESS

Great Clarendon Street, Oxford OX2 6DP

Oxford University Press is a department of the University of Oxford.
It furthers the University's objective of excellence in research, scholarship,
and education by publishing worldwide in

Oxford New York

Auckland Bangkok Buenos Aires Cape Town Chennai
Dar es Salaam Delhi Hong Kong Istanbul Karachi Kolkata
Kuala Lumpur Madrid Melbourne Mexico City Mumbai Nairobi
São Paulo Shanghai Taipei Tokyo Toronto

Oxford is a registered trade mark of Oxford University Press
in the UK and in certain other countries

Published in the United States
by Oxford University Press Inc., New York

First published 2002

First published as an Oxford University Press Paperback (with corrections) 2004

British Library Cataloguing in Publication Data

Data available

ISBN 0-19-860940-X
1

Typeset in Minion
by Footnote Graphics Limited, Warminster, Wiltshire
Printed in Great Britain by
Clays Ltd., St Ives plc

Contents

Introduction

There was once a cleric in rural England, who used to enliven his sermons with dramatic asides, prefaced by the following appeal to the Almighty: 'And if, oh Lord, this lesson be not clear, permit thy servant to illustrate it with an anecdote . . .' A better-known divine, the genial Reverend Sydney Smith, was once heard to conclude his evening prayer, which he always spoke out loud, with 'Now Lord, I'll tell you an anecdote'. Well, if the Lord could be supposed to relish a ripe reminiscence, why not the rest of us? An American diplomat, John Hay, moreover, opined that true history could best be sought 'in the personal anecdotes and private letters of those who make history'. Dr Johnson, too, proclaimed the value of the revealing biographical morsel. Writing in his magazine, *The Rambler*, he observed: 'More knowledge may be gained of a man's real character by a short conversation with one of his servants than from a formal and studied narrative, begun with his pedigree and ended with his funeral.'

So what constitutes an anecdote? The *Oxford English Dictionary* gives as its first definition: 'Secret, private, or hitherto unpublished narratives or details of history', and, borrowing from the Chambers dictionary of 1727, it continues: 'a term used by some authors, for the titles of Secret Histories; that is, of such as relate to the secret affairs and transactions of princes; speaking with too much freedom, or too much sincerity, of the manner and conduct of persons in authority, to allow of their being made public'. The implication is that the revelations must be in some measure indiscreet or scurrilous if they are to count. But the *OED*'s next definition broadens the scope: 'The narrative of a detached incident or of a single event, told as being in itself interesting or striking. (*At first* an item of gossip).' For illustration it offers a trope from a novel by Benjamin Disraeli: 'a companion who knows everything, full of wit and anecdote'. An anecdote, then, should be both entertaining and thought-provoking. To Disraeli's bibliophiliac father, Isaac, anecdotes were 'minute notices of human nature and of human learning'; and he went on: 'Some people exclaim, "Give me no anecdotes of an author, but give me his works"; and yet I have often found that the anecdotes are more interesting than the works.'

In science this is all too often true: to all but a few, stories of Einstein's private life are more captivating, or at all events rather more accessible, than his works. It would be preposterous, of course, to pretend that these gobbets of the past will set the reader on a painless path to scientific knowledge, but I hope they may cast at least a flickering light on the sociology and the history of science.

Science differs from other realms of human endeavour in that its substance does not derive from the activity of those who practise it; the nature of the atom or the structure of DNA would have been discovered if Bohr and Rutherford, and Watson and Crick had not lived. It would merely have taken longer. Science is above all a collective activity. *L'art c'est moi, la science c'est nous*, was how Claude Bernard, the father of modern physiology, put it. In this sense individuals are of marginal importance. Science, though, is not deficient in flamboyant or eccentric personalities, from Tycho Brahe with his silver nose and attendant dwarf, or Henry Cavendish with his morbid fear of human contact, on into our own era. Consider, for instance, the literary industry that has grown up around the physicist Richard Feynman, or the personality cult that enveloped the extravagant persona of the great theoretician Wolfgang Pauli, whose aphoristic utterances have passed into the currency of everyday scientific discourse.

It was Pauli who allegedly declared of a colleague's learned effusion: 'This paper isn't right, it isn't even wrong.' And then there is his oft-quoted observation after he had endured a seminar by an aspiring lecturer: 'So young and already so unknown.' When, during a discussion, Pauli was interrupted by a lesser physicist, Eugene Guth, with a pedantic observation, he listened for a moment and then interjected, 'Guth, whatever you know I know'. It seemed wholly in keeping with Pauli's remarkable career that he should have ended his life in a Swiss hospital room with the number 137, a 'magic number' thrown up by quantum theory and relating to the fine-structure of the hydrogen spectrum, a subject that had preoccupied Pauli through much of his life. (The cosmologist, Arthur Eddington, made a fetish of always seeking out a cloakroom peg with this same number on which to hang his hat.) The malign coincidence disturbed Pauli and overshadowed his last days.

Assembled here is a collection of historical incidents and not of quotations, epigrams, or witticisms by or about scientists. Many such circulate through the scientific fraternity, pass from student to student, and serve their teachers as leavening in boring lectures. Every scientist remembers, I would guess, Niels Bohr's reported rejoinder, when taxed with the absurdity of displaying a good-luck horseshoe on the door of his country cottage,

'Of course it's nonsense, but I am told it works even when you don't believe it'. And everyone surely relishes the judgement of Einstein's schoolteacher that the boy would never amount to anything, or indeed the story that Albert's first words, uttered when he was already three and a half, were the vociferous complaint that the milk was too hot; 'But you can talk!', his startled and delighted parents were supposed to have exclaimed, 'why have you never spoken before?' 'Because', came the answer (or so legend has it), 'previously everything was in order'.* And all students of chemistry have been entertained by the fable of the voluminous beard, sported by Professor Kipping of Cambridge (or was it Oxford's Nevil Sidgwick, or even Adolf von Baeyer in Munich?): it was said to harbour a crystal of every known organic chemical compound; so when the substance you had synthesized failed to crystallize you had only to ask the professor for his advice, for then the beard, hovering inquiringly over the test tube, would shed into it the microscopic crystal that would seed the process of crystallization.

All of us who have spent our lives in the laboratory cherish recollections of comic or outlandish episodes, born of ingenuity or ham-handedness, misfortune or immoderate luck, well polished generally in the telling. Most often they are too slight to bear the burden of print, but I treasure, for instance, the incident, related to me by a friend, which momentarily shook his faith in his vocation: his sublimely maladroit high-school physics teacher, while explaining the nature of gravity, holds a piece of chalk up high and asks the class to consider what happens when it falls. He releases the chalk, which catches on his cuff and vanishes up his sleeve. Similarly Charles Daubeny, Oxford's first Professor of Chemistry, once held up before his lecture audience two bottles, the contents of which, he dramatically announced, would, if mixed, destroy the lecture hall in an awesome explosion. Daubeny turned, tripped, and dropped both bottles, the audience shrank back—but nothing happened, for a prudent technician had

---

*That Einstein was a late developer seems indisputable, and he himself conjectured that this may have contributed in no small way to the magnitude of his scientific achievement. Here is how he put it in a letter to his friend, the physicist, James Franck:

> I sometimes ask myself how it came about that I was the one to develop the theory of relativity. The reason, I think, is that a normal adult never stops to think about problems of space and time. These are things which he has thought of as a child. But my intellectual development was retarded, as a result of which I began to wonder about space and time only when I had already grown up.

replaced the contents before the lecture. (What the liquids might have been is not revealed.) Or consider the reminiscence of the celebrated theoretical physicist Abdus Salam, of his schooldays between the wars in what is now Pakistan: 'Our teacher', Salam relates, 'spoke of gravitational force. Of course, gravity was well known and Newton's name had penetrated even to a place like Jhang. Our teacher then went on to speak of magnetism. He showed us a magnet. Then he said, "Electricity! Ah, that is a force which does not live in Jhang. It lives only in Lahore, a hundred miles east! And the nuclear force? That is a force which lives only in Europe! It does not live in India and we were not to worry about it".' (Salam's contribution, for which he received the Nobel Prize in 1979, was to link electromagnetism and the weak nuclear force.) Such tiny, if heart-warming, vignettes do not altogether qualify as anecdotes.

The themes that pervade scientific anecdotage are coloured commonly by the character traits of scientists, the great much more than the inconsequent. One is the intensity that so often marks their approach to their work—a capacity to exclude their surroundings, which sometimes reveals itself in a preternatural absence of mind. One of his friends told of a visit to Einstein, who had been left by his wife to look after their newborn son in the small family flat in Berne: Einstein was writing equations with one hand, while mechanically rocking the cradle with the other, oblivious to rending screams issuing from its depths. There are many such stories about other scientific luminaries, as will appear. Niels Bohr was one who could not be deflected from his purpose: he once cornered the Austrian theoretician Erwin Schrödinger, after a lecture, determined to get to the bottom of a thorny point. Schrödinger was weary and in the grip of a bad cold and wanted only to get to bed, but Bohr pursued him mercilessly into the bedroom and continued the one-sided debate through the night. Bohr's running argument with Einstein over quantum theory was terminated only by death and is touched on in some of the reminiscences recorded here.

Such single-minded pursuit of a scientific quarry has been known to lead to a truly suicidal (or homicidal) detachment. Bertrand Russell remarked of his colleague, the mathematician G. H. Hardy, that, if he had calculated that Russell would die in five minutes, his regret at losing a friend, should his prediction be borne out, would be outweighed by his satisfaction at being proved right. It is this kind of passionate commitment that has led to many a death by self-experimentation and occasionally experimentation on others as well. The noted evolutionary biologist W. D. (Bill) Hamilton, who met an untimely death after a collecting expedition to the Congo in 2000, made it a practise to plunge his hand down any hole in the jungle floor to discover what might be lurking there. (He had indeed lost the tips of several fingers on his right hand, but this was the result of a childhood experiment with explosives.) Whenever he came upon an insect nest he would beat at it with a stick to see what emerged. His obituarist told that he saw Hamilton run only once—after he had disturbed a hive of killer bees.

Ambitions and jealousies form another recurring theme. Niels Bohr was a rare example of a man, seemingly devoid of personal ambition and driven by an unremitting and disinterested desire to grasp the truth, whether by his own revelations or those of others. More in keeping with the run of human nature was the confession of an eminent American mathematician that he had rather a theorem remained undiscovered than that he not be the one to think of it. The biographer (Paul Hoffman) of the remarkable Hungarian mathematician Paul Erdös relates that after Erdös and a colleague, Atle Selberg, had discovered a proof of an ancient puzzle, the Prime Number Theorem, Selberg overheard a mathematician whom he did not know remark to a colleague: 'Have you heard? Erdös and what's-his-name have an elementary proof of the Prime Number Theorem.' Selberg was so mortified by his implied insignificance that he published the work on his own and was duly rewarded with the Fields medal, the mathematicians' Nobel Prize.

There also comes to mind the Japanese physicist, who missed by a hairs-breadth discovering the neutron; at the mention of the word (a sport to his cruel students) tears would start and course down his cheeks. And what of Philipp Lenard, the paranoiac German physicist, Nobel Laureate, Nazi, and passionate enemy of Einstein? Robbed, as he felt, of credit for various discoveries, he wrote to a confrère, James Franck, then serving on the Western front during the First World War, to inspire him with martial ardour; for Lenard desired, according to Franck, that especially the English should be beaten because they had never cited his works properly. These may be extreme cases, but fragile or over-inflated egos have always been a rich source of piquant reminiscence.

Incautious predictions by eminent savants have engendered many moments of farce. It was the British Astronomer Royal who delivered himself of the opinion that 'space travel is utter bunk', not many years before the first manned satellite went into orbit; and one of the greatest physicists of the twentieth century, Ernest Rutherford, held that notions of commercial exploitation of atomic energy was 'moonshine'. But, as Niels Bohr was supposed to have remarked, 'prediction is very difficult, especially of the future'. And, finally, because science in general proceeds slowly and is short on day-to-day drama, it is the moments of sudden startling illumination, and especially of those rarest of all gifts from nature, chance revelations, that live in the memory—a drip from a rheumy nose falls on a Petri dish and dissolves the colonies of bacteria, a thermometer breaks in a reaction vessel and mercury is revealed as the catalyst for the reaction that starts a new industry. More often, of course, the ineluctable law—that anything that can go wrong will do just that—holds sway. You may unsuccessfully consume an entire box of matches in trying to set light to a heap of dry twigs and newspaper in the grate, yet toss a match out of a car window and start a forest fire. This sort of implausible conjunction, too, is the stuff of anecdote.

It remains only to concede that authenticity is an ever-present bugbear in a collection of this nature. Some stories have attached themselves to more than one protagonist; where this is the case, I have noted as much. Nor can I, with hand on heart, swear that none is apocryphal, but I have taken the view that if they deserve to be true then they are worth including. I may, therefore, have erred in leaning sometimes on the editor's friend, *ben trovato*.

I am conscious, moreover, of a paucity of scholarly apparatus: more than literary or historical anecdotes, of which there are compilations without number, those of science often exist only in the tribal consciousness and pass through the generations by word of mouth. Therefore it is sometimes difficult to give sources for reference, but I have tried to do so whenever possible. In many instances secondary sources have been the best I could find. One has, of course, to draw the line somewhere. There is a story that has circulated around Oxford for about a century about an oral examination in physics. 'What is electricity?' the examiner asks. The panic-stricken student stammers: 'Ah, sir, I am sure I knew that, but I seem to have forgotten.' The examiner sighs: 'What a pity, *what* a pity! Only two persons have ever known what electricity is, yourself and the Author of Nature, and now one of them has forgotten.' I wonder whether this was somehow derived from Lord Palmerston's famous remark that only three people had ever understood the Schleswig-Holstein affair (the dispute between Denmark and Germany): one of them had gone mad, the second was dead, and the third was Palmerston himself, and he had forgotten. At all events, I have tried to trace the Oxford story (which has several times appeared in print) to its origins, but without success, and I have to infer that it is apocryphal. This is one example of many that I have in all conscience felt bound to omit.

I have, where necessary, sketched in briefly the nature of any underlying scientific issues to give the stories an otherwise elusive point. Cross-references to topics or persons discussed elsewhere are denoted by entry numbers in bold in square brackets. Sources, original or secondary, appear at the end of each entry. I hope that the odd reader who finds diversion in this collection of what Winston Churchill called 'the gleaming toys of history' may be lured deeper into the thickets of science, where lurk so many surprises and rewards.

I am indebted for many suggestions and corrections to Dr Bernard Dixon and the Revd Dr John Polkinghorne, and especially to Dr Michael Rodgers of Oxford University Press. This book was his idea, and without his encouragement, alternating with admonitions, it would not have got started, let alone finished. Sarah Bunney brought her meticulous skills to bear on the arduous process of editing the manuscript. Her keen eye and exceptional scientific and historical grasp eliminated many errors of fact and infelicities of style. I am grateful to several kind readers who have drawn my attention to errors, which I have tried to correct in this paperback edition.

1 The great stench

Chemistry is commonly associated by outsiders with offensive smells, and there are undoubtedly chemical bouquets that cling tenaciously to hands and clothes. Professor W. H. Perkin, Jr was supposedly once turned off a bus in Manchester while returning home from the laboratory, where he had been working on odoriferous amines. But there are few stories to match the following, told by John Read [23], Professor of Chemistry at the University of St Andrews.

Read was working at the time in the laboratory of Sir William Jackson Pope (1870–1939) in Cambridge. Pope was one of the founders of stereo-chemistry and made the fundamental discovery that compounds in which a carbon atom is bonded to four different chemical atoms or groups can take two forms. Suppose that the four groups, **a**, **b**, **c**, and **d**, lie at the corners of a tetrahedron [120], then if any two of them are interchanged, the resulting tetrahedron cannot be superimposed on the first, but is its mirror image. Such a structure has an intrinsic asymmetry, which is detected by its ability to rotate the plane of polarized light (light with its wave motion made to oscillate in a plane, rather than three dimensions) to the left or to the right.

Pope and his colleagues had prepared several such optically active (as they are called) compounds, based also on other elements than carbon. These included some sulphur compounds, many (like the simplest of all, hydrogen sulphide) highly malodorous. Now Pope wanted to see whether the optical activity would be altered by changing the central atom from sulphur to its close relative, selenium. (Hydrogen selenide is similar to hydrogen sulphide, the gas famous for its bad-egg smell, but even more disagreeable: when the great chemist Berzelius [19] was working with the substance his landlady accused him of gorging on garlic.) Chemists will want to know that Pope's compound was methylethylselenetine bromide, written [Me.Et.Se.CH$_3$.COOH]Br. Read relates:

> In our Cambridge investigations, a stick of selenium, contained in a long test-tube of hard glass, was heated in a current of hydrogen with a vigorous Bunsen flame. The selenium slowly disappeared through conversion into hydrogen selenide, and the issuing mixture of that gas with hydrogen was passed into alcoholic sodium hydroxide. The resulting solution of sodium

hydrogen selenide was warmed firstly with an equivalent of ethyl iodide, and secondly with equivalents of methyl iodide and sodium ethoxide. The methylethyl selenide had then to be warmed with bromoacetic acid, in order to yield the inoffensive methylethylselenetine bromide.

The initial operation was comfortably accomplished in the private laboratory; but for the succeeding stages, on account of the sensational odour, it was found necessary to work on the open roof of the building, with the wind behind the operator. The selenide was used only in small quantities of a few grams, with strict precautions against its escape into the air; but nevertheless the incidents that followed were worthy of the imaginative powers of a Wells and the graphic pen of a Defoe. It is said of some perfumes that their full strength becomes apparent only at great dilutions. The same rule seemed to apply to the alkyl selenides; the odour grew increasingly unbearable with dispersion; indeed it seemed to pass out of the realm of odours into that of a creepy sensation of a nightmare. Defying the restraining action of alkaline permanganate traps, the demoralising whiffs of vapour swept down upon defenceless Cambridge.

It was particularly unfortunate that the roof-experiments chanced to coincide with the Darwin centenary celebrations, held at Cambridge in June 1909. Open-air tea-parties in the houses of gardens bordering the distant Parker's Piece were interrupted, and when the guests retired indoors the objectionable odour pursued them and haunted their tea-cups. On the following afternoon a garden party at Christ's College unhappily lay also to leeward, and similar discomforts ensued. At street corners, in college combination rooms, in taverns and barbers' shops, in the old horse-drawn trams—wherever men gathered in Cambridge—the dominant subject of conversation and debate was The Smell.

Vigorous protests poured in to the local authorities, letters from indignant ratepayers appeared in the press, business men in Petty Cury and elsewhere to leeward closed their offices and sent their staffs away on hurried holidays, until the discomfiture should be overpast: in brief, a general unease sat in the usually serene air of Cambridge. At last, the seat of the disturbance was traced to the University Chemical Laboratory, and the *Cambridge Daily News* came out with the clarifying headlines: 'WHAT WAS IT? SUSPECTED DRAINS EXONERATED. SCIENCE THE SINNER.'

At this juncture it was decided to continue the work in the open country of the fens.

So Pope, Read, and another colleague sought permission from a farmer in remote Waterbeach to create 'a thunderous smell' on his land. The farmer invited the chemists to 'come and smell my muck heap', which was indeed potent, but the farmer was in for a surprise. Pope and his friends set out up the Cam by motorboat, carrying two large packing cases with apparatus and chemicals. They set up spirit burners and a Primus stove for heating the reaction mixtures. It did not take long for the farmer to take flight, but, Read continues,

> A large herd of cows formed into a semicircle to leeward and provided a silent, but appreciative audience. A few hundred yards downstream, the river took a bend to the right, and it was just before reaching this point that boats and barges, coming upstream from the direction of Ely, entered the odoriferous belt; the confusion among the occupants of these craft, as the strange invisible perfume hit them one after another, was most entertaining. Soon we began to experience the reaction of the smaller fauna: creeping and flying insects of many kinds swarmed over the apparatus, some of them even making determined attempts to force a way past the stoppers into the flasks. Their whole behaviour indicated that they felt they were missing something really good.

At this stage, sad to relate, the experiment was abandoned, partly because of the terrors of the stench, but more especially because Pope was now onto a new and more exciting project.

From *Humour and Humanism in Chemistry* by John Read (George Bell, London, 1947).

~

2 Culture clash

Sir Nevill Mott (1905–98) was a distinguished theoretical physicist and Cavendish Professor of Physics in Cambridge, best remembered for his contributions to solid-state physics. It was rumoured that, attending a party to celebrate Alexander Todd's Nobel Prize [54], Mott became aware, on looking round the room, that everyone present, except only himself, was also a Nobel Laureate; chagrined, he reached for his hat and abruptly left. He made good the mortifying deficiency some years later, and was perhaps unique in being thus honoured for research begun only after his retirement.

Mott was noted for his vague, distracted manner. Francis Crick recollects trying to introduce him to James Watson, two years after Watson and Crick's celebrated discovery of the DNA structure [88]:

'I'd like to introduce you to Watson', I said, 'since he's working in your lab.' He looked at me in surprise. 'Watson?', he said. 'Watson? I thought your name was Watson-Crick.'

Before his appointment to the Cavendish chair, Mott was for many years Professor of Physics at Bristol University and thereto attaches a story that encapsulates his preternatural absence of mind:

> Mott was travelling on the Paddington to Bristol train when three thoughts occurred to him. First, he was no longer at the physics department in Bristol but Cavendish Professor at Cambridge; second, he had travelled to London earlier that day by car; and, third, he had been accompanied by his wife.

The following episode dates from Mott's time in Bristol and was recalled by Gilbert Beaven, at the time a junior laboratory assistant in the Chemistry Department. This department in the 1930s sheltered a well-known physical chemist, Morris W. Travers, known, by reason of his work on adsorption of gases, as 'Rare-Gas' Travers. (He was also for some years Director of the All-India Institute of Sciences in Bangalore.)

Travers had invited the newly appointed Professor of Physics to visit his laboratory and gave him a tour and a comprehensive commentary on the research in progress. Mott listened in silence and when Travers's hour-long dissertation had drawn to a welcome close, he courteously thanked his host and headed for the door. Then he turned and, as an afterthought, asked: 'By the way, Travers, what *is* ethylene?'

The point of the story, of course, is that it was as though a church historian had been asked at the end of a lecture what he meant by 'the Reformation'. A very similar story attaches to another distinguished theoretician, Arnold Sommerfeld, professor at the University of Munich for much of the first half of the twentieth century. After a lecture on the assimilation of carbon dioxide by plants, he walked home with the lecturer, his faculty colleague, the famous organic chemist, Richard Willstätter. What actually was this carbonic acid about which Willstätter had waxed so eloquent? Willstätter was evidently dumbstruck by the question.

Francis Crick's reminiscence is from his book *What Mad Pursuit* (Basic Books, New York, 1988); the story about Nevill Mott on the Bristol train is told by M. Rodgers, *Nature* 383, 381 (1996).

3 The reveries of Kekulé

August Kekulé von Stradonitz was one of the founders of structural organic chemistry. Born in Darmstadt in 1829, he studied at the University of Giessen with Justus von Liebig, greatest of all organic chemists, then in France, and finally in England. He later took the chair of chemistry at the University of Ghent and then in 1865 he moved to Bonn, where he remained for the rest of his life. Kekulé was a noted teacher, but is now mainly remembered for his celebrated dreams, in which came to him the two inspirations that changed the face of chemistry.

It happened twice, the first time while he was in London. Kekulé was living in lodgings in Clapham and was accustomed to spend frequent evenings with a friend, another German chemist, Hugo Mueller. They talked about chemistry and, most of all, the structure of molecules, Kekulé's special preoccupation: how were the atoms arranged within the molecule, and how did it come about that two molecules with the same atomic composition—containing, suppose, 5 carbon atoms and 12 hydrogens—could be different substances? After one such congenial evening, Kekulé caught the last bus home. It was a fine summer evening and he took his seat on the open top deck of the horse-drawn conveyance. Here is how, many years later, he described his experience:

> I fell into a reverie, and lo, the atoms were gambolling before my eyes. Whenever, hitherto, these diminutive beings had appeared to me, they had always been in motion. Now, however, I saw how, frequently, two smaller atoms united to form a pair; how a larger one embraced the two smaller ones; how still larger ones kept hold of three or even four of the smaller; whilst the whole kept whirling in a giddy dance. I saw how the larger ones formed a chain, dragging the smaller ones after them, but only at the ends of the chain.

Kekulé, roused by the conductor's cry of 'Clapham Road', returned to his room and spent the rest of the night sketching the formulae on which his theory of structure came to be based. It was known that carbon has a valency of four; each carbon atom, in other words, can attach itself to four other atoms in forming a compound. In the simple example given above, then, the molecule C_5H_{12}, pentane, exists in three forms, where CH_3 and CH_2 represent carbon atoms linked to three and to two hydrogen atoms:

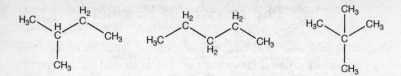

While in Ghent, Kekulé experienced a similar epiphany. This time the object of his reverie was the molecule benzene, which has the composition C_6H_6. This is the archetype of the aromatic compounds, a class to which a large proportion of the most interesting synthetic and naturally occurring substances belong. Here again is Kekulé's reminiscence:

> I was sitting writing on my textbook, but the work did not progress; my thoughts were elsewhere. I turned my chair to the fire and dozed. Again the atoms were gambolling before my eyes. This time the smaller groups kept modestly in the background. My mental eye, rendered more acute by repeated visions of the kind, could now distinguish larger structures of manifold conformation: long rows sometimes more closely fitted together all twining and twisting in snake-like motion. But look! What was that? One of the snakes had seized hold of its own tail, and the form whirled mockingly before my eyes. As if by a flash of lightening I awoke; and this time also I spent the rest of the night working out the consequences of the hypothesis.

The hypothesis, of course, was that benzene was a cyclic molecule, the six carbon atoms forming a hexagon, one hydrogen attached at each corner. Adolf von Baeyer [84], the great organic chemist, said that he would have exchanged his lifetime's accomplishments for this one insight of Kekulé's. It is no surprise that during the heyday of Freudian interpretation of dreams Kekulé's vision of snakes was given a sexual connotation, for he was living in bachelor quarters, far removed from his wife, whom he would have seen rather seldom. But then there is little that has not at some time been interpreted in such terms.

See, for example, O. T. Benfey, *Journal of Chemical Education*, **35**, 21 (1958).

~

4 Röntgen's rays

Wilhelm Conrad Röntgen was a respected, if not outstanding, experimental physicist, who in 1888 at the age of 43 was appointed professor and head of the physics institute at the University of Würzburg in Bavaria. This was at the time something of an academic backwater, but Röntgen, though essentially a loner, who preferred to work by himself, seems to have been an effective laboratory director and exerted himself to raise support and build up a good department. He was interested in electromagnetic radiation and set out to resolve the hotly contested question whether the recently discovered radiations, in particular the negatively charged cathode rays, should be regarded as particles or waves. (We now know that they partake of the character of both.) Here is how he came by his discovery, one of the most startling in the history of physics.

Röntgen was working alone in his personal laboratory on the evening of Friday 8 November 1895. To observe the trajectory of the cathode rays generated in a vacuum tube, a fluorescent screen was placed in their path. The pale-green glow of the screen where the radiation fell on it could be difficult to see, so the laboratory was meticulously blacked-out and the cathode-ray tube was itself shielded with a black cardboard screen to retain light from the spark discharge used to generate the cathode rays. In the darkness Röntgen became aware of a flickering spot of light some distance away on the bench. Was there a leak in the black-out curtains? There was none.

Closer inspection revealed that what was twinkling was a letter painted with phosphorescent material on a card. Röntgen knew that the cathode rays could not penetrate several feet of air outside the vacuum tube. A secondary radiation must therefore have passed through the cardboard screen unimpeded. Röntgen now put a playing card in the way, which proved transparent to the rays, as indeed did the whole pack. A book cast only a weak shadow on the illuminated screen, which revealed to Röntgen that the rays travelled in a straight line. When he next held a small piece of lead sheet in the beam he was astonished to observe beside the shadow of the lead the outline of his fingers and the image of the bones within.

Röntgen must undoubtedly have instantly recognized that he had made a discovery which would splinter the smooth surface of nineteenth-century physics. Around this time, the great theoretical physicist Max Planck, then a student, had been counselled by his physics professor, Philipp von Joly, to enter another field, for there was little more to learn about the nature of matter.

Röntgen showed on that same momentous evening that the radiation, which he called X-rays, originated at the point at which the cathode rays struck the wall of the tube and that (unlike cathode rays) they were not deflected by an magnetic field and were devoid of electric charge. Röntgen scarcely emerged from his laboratory during the weeks that followed. He obtained images of various objects, including (to her consternation) his wife's hand, showing the bone structure and the outline of her rings. His first report was published early in the New Year and caused a sensation. Lord Kelvin [10], one of the leading physicists of the time, believed that Röntgen's paper was a hoax until persuaded otherwise by confirmations from around the world. In the next few years thousands of papers were published on the radiation, most often referred to (as it still is in Germany) as Röntgen rays, and it did not take long for the medical profession to perceive its potential. Commercial interests followed, including such novelties as 'X-ray-proof' underwear, offered by an English company. The hazards also soon became apparent.

Röntgen was never at ease with his rays; his devotion was to classical physics and he was pained by the emergence of new phenomena that did not fit into the classical scheme. His most distinguished student, Rudolf Ladenburg, who was to become a professor at Princeton University, arrived in Würzburg some years after the X-ray episode and was allocated a problem in viscosity. The speed with which a ball falls through a liquid is determined by the viscosity of the liquid, according to an equation derived in the mid-nineteenth century by G. G. Stokes in Cambridge; but what, Röntgen wanted to know, would be the effect of confining the ball and the liquid in a narrow tube, in which there would be viscous drag at the wall? A long tube was set up, running from the top of the building to the basement, and filled with castor oil. According to Ladenburg, nothing ever gave Röntgen so much pleasure as to see the ball arrive in the basement at precisely the calculated instant.

For his discovery of X-rays Röntgen was awarded the first Nobel Prize for Physics in 1901. Several physicists, who had worked with cathode rays, lamented their failure to spot X-rays. Frederick Smith in Oxford had noticed that photographic plates stored in the vicinity of a cathode-ray tube tended to become fogged; so he had them moved further away, for it did not occur to him to look into the cause of the phenomenon. The most distraught was Philipp Lenard (later to win a Nobel Prize for his studies on radiation), who could never bring himself to pronounce Röntgen's name (nor could he accept the theoretical upheavals that shook physics in the first two decades of the twentieth century: he became a savage and intemperate opponent of Albert Einstein and a passionate Nazi).

See, for example, Otto Glasser, *Dr W. C. Röntgen*, 2nd edn (Charles C. Thomas, Springfield, Ill., 1958). Rudolf Ladenburg's reminiscence was recorded by E. U. Condon in an article called '60 years of quantum physics', in *History of Physics*, ed. Spencer R. Weart and Melba Phillips (American Institute of Physics, New York, 1985).

~

5 Light on sweetness: the discovery of aspartame

Aspartame, or Nutrasweet, must have relieved a corpulent population of many tons of excess weight. It is a sweetener devoid of the disagreeable aftertaste of saccharine and seems to have no pathological side-effects. Its discovery arose by pure accident, like indeed those of most other 'artificial' sweeteners, including the very first: saccharine was synthesized in 1879 by Constantine Fahlberg, a student of the foremost American organic chemist of the time, Ira Remsen, at Johns Hopkins University in Baltimore. Fahlberg, struck by a curious sweet taste on his fingers while eating his dinner, realized that it came from something he had spilled on his hand during the day. It was *ortho*-sulphobenzoic acid imide. He took out a patent on the substance, from which he excluded his mentor, and became rich, a lapse that Remsen never forgave him.

Then in 1937 a delinquent American research student, with the now unthinkable habit of smoking in the laboratory, was attempting to synthe-size an antipyretic agent; he took a drag from the cigarette, which he had left standing on end on the bench, and experienced a sensation of sweetness. This was the genesis of the cyclamate sweeteners. Yet another sweetener, acesulfame, came to light when a laboratory worker licked his finger to pick up a piece of paper.

James Schlatter was an organic chemist working in the laboratories of a pharmaceutical company, G. D. Searle, in search of a treatment for gastric ulcers. He was synthesizing a peptide (a chain of linked amino acids, such as makes up a protein) corresponding to a part of a hormone, gastrin. With a colleague he had prepared a simple compound of this kind (aspartyl-phenylalanine methyl ester) and was purifying it, as organic chemists do, by recrystallization. It was December 1965. Here is how he described what happened:

I was heating the aspartame in a flask with methanol when the mixture bumped [boiled abruptly] onto the outside of the flask. As a result some of the powder got onto my fingers. At a slightly later stage, when licking my fingers to pick up a piece of paper, I noticed a very strong, sweet taste. Initially, I thought that I must have had some sugar on my hands from earlier in the day. However, I quickly realised this could not be so, since I had washed my hands in the meantime. I therefore traced the powder on my hands back to the container into which I had placed the crystallised aspartylphenylalanine methyl ester. I felt that this dipeptide ester was not likely to be toxic and I therefore tasted a little of it and found that it was the substance which I had previously tasted on my finger.

Today the wearing of gloves is considered obligatory in chemical laboratories and the discovery would never be made, so the sweetness of the dipeptide would probably remain hidden from us for ever.

The oddest manner in which a new sweetener came to light was when, one day in 1976, a foreign research student at King's College in London misheard the instructions of his supervisor, Professor L. Hough. Hough was searching for possible synthetic industrial applications of sucrose, the common sugar of cane and beet, and several derivatives had been produced in the laboratory. One of these was a trichlorosucrose (sucrose into which three atoms of chlorine had been introduced). Hough asked Shashikant Phadnis to 'test' the substance, but, his ear being imperfectly attuned to the language, Phadnis instead tasted it. Sucralose, as it became known, is one of the sweetest of all substances and can replace sucrose at less than one-thousandth of the concentration.

See J. M. Schlatter in *Aspartame: Physiology and Biochemistry*, ed. L. D. Steginik and L. J. Filer (Marcel Dekker, New York, 1984).

~

6 Otto Stern's sulphurous cigars

Otto Stern (1888–1969) used to describe himself as an experimental theoretician. Isidore Rabi, the great American physicist [21], took Stern, with whom he had worked as a young man, for his model of all a scientist should be. He was genial, unpretentious, and generous, and had pre-eminently, Rabi wrote, 'good taste' in his physics: he infallibly sought out problems of the first importance, revelled in ingenuity, and his experimental approach was marked by style and wit. Stern was drawn initially to theoretical physics and he served for some years as assistant to Albert Einstein, to whom he was related. Stern told his assistant, Otto Frisch [20], that he and Einstein would visit the local brothels together, for these were quiet, relaxing places in which to discuss physics. One of Stern's theoretical studies, which solved a longstanding problem in statistical mechanics, was published during the First World War under the dateline 'Lomsha, Russian Poland'; this was a dingy provincial town to which Stern was posted to tend a weather station, and where he occupied his spare time with these arduous calculations.

Later, as Professor of Physical Chemistry in Hamburg, he built up a thriving department, dedicated mainly to his particular passion for atomic and molecular beams. These are streams of atoms or molecules, which travel in straight lines through a high vacuum and offered, as Stern showed, the opportunity for many kinds of fundamental measurements.

Stern's method was to settle, after deep thought, on the next experiment. He would then design the apparatus, which he would invite his assistants or students and technicians to construct. He then withdrew from the laboratory until told that the equipment was ready, when he would reappear, preceded by a cloud of cigar smoke, and begin the measurements himself. The experiment finished, the paper would be written and the apparatus dismantled in preparation for the next campaign.

Stern perceived that atomic beams afforded a means of detecting effects demanded by quantum theory, then still in its infancy and fiercely debated. Quantum theory predicted that some atoms, such as silver, should possess a magnetic moment—to behave as magnets—resulting from the circulating charge of the single electron remotest from the nucleus. Stern's plan was to observe the deflection of a beam of atoms in the silver vapour emerging from a furnace when it passed through a powerful magnetic field in a vacuum; in this way the magnetic moment could be measured. The separation of the beams, with the external magnetic field on and off, would be minute and extremely difficult, if not impossible, to measure. Stern discussed the prospects with his colleague, Walther Gerlach. 'Shall we do it?', he demanded excitedly, 'Let's do it!'

Gerlach encountered every kind of technical obstacle, but after many setbacks he fancied he could see the deflection in faint deposits of silver, barely visible on a glass plate. He took the plate to Stern for his opinion, and as the two men peered closely at it the hairline deposit turned gradually black and two lines emerged, minutely separated. Stern realized that it was the essence of a cheap cigar on his breath that had brought forth the image; for Stern, who was rich, was at this time unaccountably hard up and had to forgo his favourite brand for something much cheaper and full of sulphur compounds, potent enough to convert the silver into black silver sulphide. But this was not the end of the story, for closer inspection revealed that the deflected beam was itself split into two lines, separated by a hairsbreadth. It was only later that the full explanation emerged, and wrought a mighty change in the understanding of quantum theory: the magnetic moment, reflecting in effect the rate of rotation of the electron, could not take just any value, but was quantized—could assume only specified (and minutely different) values. The resulting populations of atoms with electrons in the different spin states then respond distinguishably to the magnetic field and splitting results. This result was the birth of 'space quantization', a new and at the time startling dimension of quantum theory. Isidor Rabi called it 'the glorious Stern-Gerlach experiment'. Stern was delighted not only with the result, but with the manner in which it had been made manifest.

Otto Stern received the Nobel Prize for his work on atomic and molecular beams in 1943. Ten years before, he had been driven out of Germany and had settled in the United States. There he received poor support during the war years and after, and, while still in his fifties, he retired to California, where, by now somewhat reclusive, he dedicated himself to the pleasures of good food and the cinema, to which he was much addicted. He died at 81, while watching a film.

For reminiscences of Otto Stern see Otto Frisch *What Little I Remember* (Cambridge University Press, Cambridge, 1979); John S. Rigden, *Rabi: Scientist and Citizen* (Basic Books, New York, 1987).

~

7 Metchnikoff restored to life

Ilya (later Élie) Metchnikoff was a Russian, born in the Ukraine in 1845, who spent most of his working life at the Pasteur Institute in Paris. During a stay in Italy he discovered the phenomenon of phagocytosis, when he observed amoeba-like cells surging towards a foreign body (a thorn) thrust into a translucent starfish larva. In the human body the phagocytes—white blood cells—are the first line of defence against invaders, such as bacteria, which they engulf and destroy. In later years Metchnikoff was captivated by the notion that our phagocytes run riot when we age and that all symptoms of senescence result from their destructive action. He believed this was fomented by waste products formed in the gut; these, he proclaimed, could be eliminated by promoting the proliferation of benign bacteria. All we needed to do was to ingest massive quantities of 'kephir' or yogurt, which would ensure a thriving population of *Lactobacillus bulgaricus* in the gut. Metchnikoff presumably followed his own prescription to avert the ravages of age, but died nevertheless at seventy-one.

Metchnikoff was a depressive, who twice attempted suicide. His second wife, Olga, related in her biography of her husband how science saved him from despair. He had taken a large dose of morphine, so large in fact that he had vomited and so merely sunk into a state of inertia.

> He fell into a sort of torpor, of extraordinary comfort and absolute rest; in spite of this comatose state he remained conscious and felt no fear of death. When he became himself again, it was with a feeling of dismay. He said to himself that only a grave illness could save him, either by ending in death or by awakening the vital instinct in him. In order to attain his object, he took a very hot bath and then exposed himself to cold. As he was coming back by the Rhone bridge, he suddenly saw a cloud of winged insects flying around the flame of a lantern. They were Phryganidae, but in the distance he took them for Ephemeridae, and the sight of them suggested the following reflection: 'How can the theory of natural selection be applied to these insects? They do not feed and only live a few hours; they are therefore not subject to the struggle for existence, they do not have time to adapt themselves to surrounding conditions.'
>
> His thoughts turned towards Science; he was saved; the link with life was re-established.

(Metchnikoff was not blessed with a knowledge of today's genetics: it is the mutations that promote survival up to the time of reproduction that give the insects, as much as ourselves, an evolutionary advantage.)

See Olga Metchnikoff, *Life of Élie Metchnikoff 1845–1916* (Constable, London, 1921).

8 An ill wind

At 7.30 in the evening of 3 December 1943, German bombers converged on the port of Bari in the heel of Italy. Their target was the harbour, which was full of ships unloading supplies for the Allied armies, fighting their way up Italy. The air-raid warnings had failed to sound and few people had taken cover. One bomb struck the USS *Liberty*, which was loaded with high explosives and 100 tons of mustard gas. Although neither side deployed chemical weapons during the war, both came close and neither trusted the other to abstain from their use. At all events, the Liberty exploded and a cloud of mustard gas engulfed the harbour area. The gas alarm was sounded, but it was too late for many soldiers and civilians. Stationed in Bari as a medical officer with the US Army was Dr Cornelius Rhoads, who had already made a name for himself in medical research before his country entered the war. He was now called upon to treat the victims of mustard-gas poisoning.

Rhoads was struck by the effect of the gas on the blood cells: immediately after exposure the white-cell count rose, but over the next days first the lymphocytes (on which the body's immune response in large measure depends) and then the other types of white blood cell fell practically to zero. Soon immature cells began to appear in the blood, indicating that the body was reacting to the insult that it had suffered. Mildly affected patients recovered within days or weeks; those who had suffered severe exposure died or were sometimes saved by massive transfusions. But, Rhoads observed with interest, infections were rare even in seriously affected patients and there was no evidence of any other tissue damage. Was the toxicity of the mustard gas specific, then, for white blood cells? And, if so, could it perhaps be useful for treating leukaemias—conditions characterized by production of excessive numbers of white blood cells? Rhoads's inspiration marked the start of chemotherapy in cancer research and treatment. Within months an oncologist in Chicago had successfully used nitrogen mustards—mustard gases and related compounds—to treat patients with leukaemia and Hodgkin's disease.

See, for example, M. L. Podolsky, *Cures out of Chaos* (Harwood, Amsterdam, 1998).

9 Marie Curie and the Immortals

The discovery of radium, for which Marie and Pierre Curie shared the Nobel Prize in 1903 with Henri Becquerel (who had made the accidental discovery of radioactivity [36]), was the most notable achievement of its time for French science. Marie Curie, born Maria Skłodowska in Poland in 1867, attracted opprobrium from the xenophobic right-wing press, inflamed by her well-publicized affair, years after Pierre's death in a street accident, with the celebrated physicist Paul Langevin, trapped in an unhappy marriage to a termagent wife. Marie Curie craved recognition by her adopted country and sought election to the Académie des Sciences. She was supported by several of the leading luminaries of French science, including the greatest of all, Henri Poincaré. But the Académie was an exclusively male body, which had repulsed earlier assaults from female aspirants. One of the five academies created after the Revolution, it was steeped in tradition and ritual. Still today the members wear a gold-embroidered green uniform and on their election honour their friends by inviting them to contribute to the (considerable) cost of their ceremonial sword.

In 1911, there were 68 members and death had liberated one vacancy. Three times a year the five academies met in plenary session, and, by chance, it was before one of these that Mme Curie's candidature went forward. The resurgence of the female membership issue caused a stir in the press and in high society, and the meeting was attended by 163 academicians, about twice the usual number. The proceedings began with speeches by Mme Curie's opponents and supporters. The President of the Académie des Sciences Morales et Politiques began by adverting to the intentions of the academies' founders, who, he assured his listeners, had never desired the incursion of women, and he urged the members 'not to breach the unity of this elite body, which is the Institute of France'. His appeal was met with tumultuous cheers.

Poincaré objected that the autonomy of the separate academies was just as hallowed a tradition and that members of other academies had no business interfering in the decisions of his own. This was countered by a lawyer from the Académie des Sciences Morales et Politiques, who suggested that the question of the election of women to the Institut de France (within which the five academies were subsumed) was of concern to all. The consequences would be too alarming to contemplate, for if the Institut was so imprudent as to admit women, might not one of their number eventually become President? After more such exchanges the demand for a vote could be heard and a clamour then arose from members impatient to speak. The chairman rang for order and was forced to climb on the presidential throne in an effort to control the proceedings. The vote by the members of the Académie des Sciences went in favour of protecting the 'immutable trad-itions of the Institut' by 85 to 60. The press was vociferous, though, of course, divided.

Yet this was not the end of the matter, for some days later the committee of the Académie des Sciences met behind closed doors to consider the nom-ination. Disregarding the vote at the plenary session they proposed Marie Curie as their first recommendation, with six other names as worthy second choices. The vote followed a week later and was preceded by fiercely articu-lated opinions from members. The underhand argument was voiced, not for the first time, that Marie Curie's great work had been done in associ-ation with her husband, whom perhaps she had merely assisted, and after his death with other men (her assistants), who were scientists of stature. Moreover, had not Mme Curie received a sufficiency of honours? Was it not now someone else's turn? A movement was launched to elect one of the other choices, an electrical engineer, who had made important contri-butions to wireless telegraphy, Édouard Branly. The extreme right-wing chauvinistic and xenophobic paper, *L'Action Française* represented Marie Curie's candidature as a plot by a group of left-wing Dreyfusards, seeking to block the election of the devoutly Catholic and thoroughly French Branly. Whatever the arguments that swayed the illustrious academicians, it was the ageing Branly who was duly elected.

For a full account see the fine biography *Marie Curie* by Susan Quinn (Simon and Schuster, New York, 1995).

~

10 'Every integer his personal friend': Hardy visits Ramanujan

The sublime Indian mathematician Srinivasa Aayengar Ramanujan (1887–1920) was 'discovered' by G. H. Hardy, Professor of Mathematics at Cambridge. Hardy was passionate about his subject and expounded his credo in his book, *A Mathematician's Apology* (1940). Bertrand Russell recorded a remark of Hardy's to the effect that 'if he could find a proof that I [Russell] was going to die in five minutes he would of course be sorry to lose me, but this sorrow would be quite outweighed by pleasure in the proof'. 'I entirely sympathized with him', Russell continued, 'and was not at all offended.'

Ramanujan was a young man, living with his parents in the post office of a small town in India. He discovered mathematics from an English text-book and proceeded to explore many areas of the subject, recording his cogitations in lined school exercise books. These he sent to several mathematicians in Britain; only Hardy took notice, realizing that he was in the presence of an unlettered genius. He brought Ramanujan to Cambridge at his own expense and became his mentor and collaborator. Hardy later wrote that he did not rate his own contributions very highly, far at any rate below Ramanujan's, but he had been able to speak to Ramanujan and Littlewood [144] (his famous Cambridge colleague) on equal terms, and this was reward enough.

Ramanujan became a Fellow of Trinity College and of the Royal Society, but he was miserable in Cambridge. A devout Brahmin, he adhered to a strict diet and could not accept English food or procure most of what he was accustomed to eat. He was perpetually chilled in his rooms in Trinity, heated only by a coal fire, and tormented by colds. In due course he developed tuberculosis, which all too quickly carried him off. The story of Hardy's visit to Ramanujan on his sickbed in a London hospital has been often told, here by C. P. Snow, who knew Hardy well.

Hardy had gone out to Putney by taxi, as usual his chosen method of conveyance. He went into the room where Ramanujan was lying. Hardy, always inept about introducing a conversation, said, probably without a greeting, and certainly as his first remark: 'The number of my taxi-cab was 1729. It seemed to me a rather dull number.' To which Ramanujan replied: 'No, Hardy! No, Hardy! It is a very interesting number. It is the smallest number expressible as the sum of two cubes in two different ways.'

That is the exchange as Hardy recorded it. It must be substantially accurate. He was the most honest of men; and further, no one could possibly have invented it.

Such prodigious numerical facility seems to be not uncommon amongst the best mathematicians. Consider, for example, the following:

Someone once asked A. C. Aitken, professor at Edinburgh University, to make 4 divided by 47 into a decimal. After four seconds he started and gave another digit every three-quarters of a second: 'point 085106382978723404 25531914'. He stopped, discussed the problem for one minute, and then restarted: '191489'—five-second pause—'361702127659574468. Now that's the repeating point. It starts again with 085. So if that's forty-six places I'm right.' To many of us such a man is from another planet, particularly in his final comment.

Here is an example of a different kind. Lord Kelvin (1824–1907), though a physicist, was also no mean mathematician. He had been (to his disappointment) Second Wrangler—the student placed second in the final honours school of mathematics—at Cambridge. (Indeed, it was related that on the morning the examination results were to be announced he sent his servant to find out 'who is Second Wrangler?', and was devastated when he was told, 'You, sir!'). One of Kelvin's mathematical heroes was the Frenchman Joseph Liouville, and, when lecturing in Glasgow one day, he asked his class: 'Do you know what a mathematician is?' He then wrote on the blackboard the equation

$$\int_{-\infty}^{+\infty} e^{-x^2}\,dx = \sqrt{\pi}$$

'A mathematician', he said, pointing to the board, 'is one to whom *that* is as obvious as that twice two are four is to you. Liouville was a mathematician.'

The quote about Hardy and Ramanujan is from C. P. Snow, *Variety of Men* (Macmillan, London, 1967; Penguin Books, London, 1969) and the example of Aitken's facility with numbers is from Anthony Smith, *The Mind* (Viking, New York, 1984; Penguin Books, London, 1985). For the Kelvin stories, see Sylvanus P. Thompson's biography, *The Life of William Thomson Lord Kelvin of Largs*, Vol. 2 (Macmillan, London, 1910) and E. T. Bell, *Men of Mathematics* (Gollancz, London, 1937).

11 David Hilbert's eulogy

David Hilbert (1862–1943) was a renowned German mathematician, head of the Mathematical Institute at the University of Göttingen, which then harboured the most remarkable galaxy of talent in that subject in the world. An old man when the Nazis rose to power, he forthrightly opposed the dismissal of his Jewish colleagues.

Hilbert's absence of mind was legendary. One of his students cited an example: one evening when Hilbert and his wife were preparing to receive guests for a dinner party, she instructed him to change the deplorable tie he was wearing. The guests arrived but Hilbert did not reappear. Eventually a search was made and he was found in bed asleep. Taking off his tie, he had merely continued with his usual sequence of actions, terminating in night-shirt and bed.

At some time in the 1920s one of Hilbert's brilliant students had written a paper, purporting to prove the Riemann hypothesis—a longstanding challenge to mathematicians, concerned with an important aspect of number theory. The student had shown the paper to Hilbert, who

studied it carefully and was really impressed by the depth of the argument; but unfortunately he found an error in it which even he could not eliminate. The following year the student died. Hilbert asked the grieving parents if he might be permitted to make a funeral oration. While the student's relatives and friends were weeping beside the grave in the rain, Hilbert came forward. He began by saying what a tragedy it was that such a gifted young man had died before he had an opportunity to show what he could accomplish. But, he continued, in spite of the fact that this young man's proof of the Riemann hypothesis contained an error, it was still possible that some day a proof of the famous problem would be obtained along the lines which the deceased had indicated. 'In fact,' he continued with enthusiasm, standing there in the rain by the dead student's grave, 'let us consider a function of a complex variable . . .'

See Constance Reid's biography, *Hilbert* (Copernicus, Springer-Verlag, New York, 1996).

~

12　Rabi meets his match

Isidore Rabi [21], when Chairman of the Physics Department at Columbia University and a mandarin of the American physics establishment, gave the following account of his first encounter with a youthful prodigy. It was 1935 and Rabi was pondering a controversial paper, just published by

Einstein, Podolsky, and Rosen, which sought to undermine, with a paradox, the foundations of quantum theory.

'I was reading the paper, and my way of reading a paper was to bring in a student and explain it to him. In this case the student was Lloyd Motz, who's now a professor of astronomy at Columbia. We were arguing about something, and after a while Motz says there was someone waiting outside the office, and asked if he could bring him in. He brought in this kid.' Schwinger was then sixteen. 'So I told him to sit down someplace, and he sat down. Motz and I were arguing, and this kid pipes up and settles the argument by the use of the completeness theorem'—an important mathematical theorem used frequently in quantum theory. 'And I said, "Who the hell is this?" Well, it turned out he was a sophomore at City College, and he was doing very badly—flunking his courses, not in physics, but doing very badly. I talked to him for a while and was deeply impressed. He had already written a paper on quantum electrodynamics. So I asked him if he wanted to transfer, and he said yes.'

Rabi managed, with great difficulty and with the help of a letter of support from another great physicist, Hans Bethe [62], to get Julian Schwinger admitted to Columbia.

Schwinger became one of the most celebrated theoreticians of the twentieth century. During the Second World War he worked in the Radiation Laboratory, set up at MIT, the Massachusetts Institute of Technology, to develop radar and other techniques. Rabi was associate director and recalled Schwinger's habit of working by night and sleeping during the day:

'At five o'clock, when everybody was leaving, you'd see Schwinger coming in', Rabi said. I was once told that people would leave unsolved problems on their desks and blackboards, and find when they returned the next morning that Schwinger had solved them. 'The problems he solved were just fantastic', Rabi continued. 'He lectured twice a week on his current work. As soon as Schwinger would make an advance, guys all around—Dicke and Ed Purcell [two prominent experimental physicists, famous especially for their work on nuclear magnetism]—would invent things like mad. All sorts of things.'

In 1965, while a professor at Harvard, Julian Schwinger was awarded the Nobel Prize and became a legend for his astonishing ability in lectures to develop any line of theoretical argument at the blackboard, with no apparent effort and no notes.

MIT's Radiation Laboratory gave birth to a series of important inventions and discoveries, and radar, in the event, helped much more than the atom bomb to defeat Germany and Japan. The single most critical achievement was the construction of the cavity magnetron by John Randall and Harry Boot in England. This instrument, the design of which appeared to defy logic, was the first source of high-intensity radiation in the centimetre wavelength range, required for air- and seaborne radar. Its beam could ignite a cigarette and cause car headlights in the far distance to glow. When the device was brought to MIT and examined by the assembled talent of the American physics establishment,

> the group included some of the best nuclear physicists in the country. They knew something about high-frequency radiation from their work on cyclotrons, but the magnetron confounded even them at first.
>
> 'It's simple', Rabi told the theorists who were seated around a table staring at the disassembled parts of the tube. 'It's just a kind of whistle.'
>
> 'Okay, Rabi,' Edward U. Condon asked, 'how does a whistle work?'
>
> Rabi was at a loss for a satisfactory explanation.

The stories about Schwinger and Rabi are taken from Jeremy Bernstein's book, *Experiencing Science* (Dutton, New York, 1978), and that of the magnetron episode is told in *The Physicists: The History of a Scientific Community in Modern America* by Daniel J. Kevles (Harvard University Press, Cambridge, Mass., 1971).

~

13 The Bucklands explode a miracle

William Buckland (1784–1856) was the first incumbent of the Chair of Zoology at Oxford, who passed his trait of extreme eccentricity to his son, Francis, a zoologist—author of *Curiosities of Natural History* and for some years Inspector of Salmon Fisheries. The Bucklands made it a custom to eat, in a spirit of scientific inquiry, any animal that crossed their path. Francis arranged with the London Zoo to receive a cut from anything that died there. Visitors to the Buckland ménage, as well as suffering the overtures of the pet donkey and other creatures not generally found in the drawing room, were apt to be offered such delicacies as mice *en croûte* or sliced head of porpoise. William maintained that roast mole had been the most unpleasant thing he had eaten until he tried the stewed bluebottles. When he was shown by his friend, the Archbishop of York, a snuffbox containing the embalmed heart of Louis XVI, which the prelate had bought in Paris at the time of the Revolution, William Buckland announced that he had never eaten the heart of a king and before he could be prevented had seized and swallowed it.

Nothing in the natural world was alien to the Bucklands. When a local clergyman, who was also a keen naturalist, excitedly brought William Buckland a fossilized bone that he had unearthed, Buckland held it out to his seven-year-old son: 'What is it Frankie?' 'It is a vertebra of an ichthyosaurus', the child replied without hesitation. Mrs Buckland shared the family's enthusiasms. When her husband awoke one night with the words on his lips: 'My dear, I believe that *Cheirotherium*'s footsteps are undoubtedly testitudinal', she at once accompanied him downstairs and prepared some flour paste in the kitchen while William retrieved a tortoise from the garden; and, indeed, to their delight, the impression in the paste proved almost identical with the fossil's footprint.

Frank Buckland recalled an embarrassing moment when returning to England in a coach with a stranger. Both dozed. Buckland had collected some red slugs in Germany (whether for his dinner is not recorded) and, awakening, was alarmed to see a procession of the creatures making their stately way across the bald pate of his sleeping companion. Rather than explain and apologize, Buckland felt it prudent to leave the coach at the first stop.

The ever-curious Bucklands, on a visit to Italy, were shown a stain on the floor of a church on the spot on which a saint had been martyred. Every morning, they were told, the fresh blood miraculously renewed itself. William at once kneeled on the floor and applied his tongue to the moist patch. Not blood, he informed his hosts: he knew exactly what it was—nothing more than bats' urine.

See, for example, *The Curious World of Francis Buckland* by G. H. O. Burgess (John Baker, London, 1967).

~

14 Farmyard thermodynamics

Walther Nernst, born in 1864, was one of the German moguls of physics and physical chemistry. His most famous contributions were in thermodynamics (of which he formulated the Third Law) and in electrochemistry. In 1920, he acquired Zibelle, an extensive estate in East Prussia. There were cows, pigs, a pond with carp, and a thousand acres of land, which included fields of cereals and other crops. Nernst pursued his new interest in farming with characteristic single-mindedness.

It is related that on a tour of inspection on a cold winter's morning he entered the cowshed and was astonished to discover how warm it was. Why was it heated, he asked? The reply came that the heat was generated only by the cows, the result of metabolic activity. Nernst was dumbstruck and immediately resolved to sell his cows and invest instead in carp: a thinking man, he said, cultivates animals that are in thermodynamic equilibrium with their surroundings and does not waste his money in heating the universe. So the old system of ponds on the estate was stocked with carp, which did not noticeably heat the water of their pond.

From Kurt Mendelssohn, *The World of Walther Nernst: The Rise and Fall of German Science* (Macmillan, London, 1973).

~

15 Newton ponders

Sir Isaac Newton (1642–1727) attracted many legends. His nature in maturity was sour and ungenerous; he was jealous of contemporaries and ferociously competitive. In his interminable dispute with his Hanoverian contemporary Gottfried Leibniz over who had first hit on the differential calculus, he was ruthless to the point of dishonesty. At the end he exulted that he had finally 'broken Leibniz's heart'. John Flamsteed, the first Astronomer Royal (1646–1719), and himself no easy character, was once heard to sigh: 'I dreamt Newton was dead.'

Newton's only close attachment was to his niece, not counting his small dog, Diamond. (When the dog knocked over a candle and caused a conflagration in which books and manuscripts were consumed, his master's only comment was 'Oh, Diamond, Diamond, thou little knowest what thou hast done!') But despite a fatherless, alienated childhood, the schoolboy at Grantham Grammar School in Lincolnshire was not altogether a stranger to schoolboy pranks: 'He first made lanterns of paper crimpled, which he used to go to school by, in winter mornings, with a candle, and tied them to the tails of the kites in a dark night, which at first affrighted the country people exceedingly, thinking they were comets' (for comets were regarded at the time as portents of alarming events).

The story of the falling apple at Woolsthorpe Manor may have had some foundation in truth, or at least stemmed from Newton himself, and his admirer, Voltaire [168], had it from Newton's niece, Catherine Conduitt. Newton was remarkable for the intensity of his concentration when working. His mind was often elsewhere while life went on around him. It was related, for instance, that he was found in his kitchen by the maid one day, standing before a saucepan of boiling water in which reposed his watch, while he gazed in bafflement at the egg in his hand. His nephew, Humphrey, wrote after Isaac's death in 1727:

> At some seldom times when he designed to dine in hall [at Trinity College, Cambridge], would turn to the left hand and go out into the street, when making a stop he found his mistake, and then sometimes instead of going into hall, return to his chamber again.

And in Thomas Moore's diary is to be found the following:

> Anecdote of Newton showing his extreme absence; inviting a friend [it was Dr Stukely] at dinner and forgetting it: the friend arriving, and finding the philosopher in a fit of abstraction. Dinner brought up for one: the friend (without disturbing Newton) sitting down and dispatching it, and Newton, after recovering from his reverie, looking at the empty dishes and saying, 'Well really, if it wasn't for the proof before my eyes, I could have sworn that I had not yet dined.'

Newton's genius retains, even now that his many achievements have been assimilated into the common currency of knowledge and usage, the capacity to provoke awe and astonishment. Contemplating his crowning opus, William Whewell, the Victorian scholar, declared: 'As we read the *Principia* we feel as we might when we are in an ancient armoury where the weapons are of a gigantic size; and as we look at them we marvel what manner of man was he who could use as a weapon what we can scarcely lift as a burden.'

Mark Kac, the Polish-American mathematician, recognized two types of genius: there are, he said, on the one hand the 'ordinary geniuses'—those who are just as you and I would be, were we only half as bright as they are—and, on the other, the 'magicians', those with whose minds we can never connect. Kac identified Richard Feynman [89] as a magician. When Feynman's contemporary and rival, Murray Gell-Mann, was asked how Feynman set about tackling a problem he answered: 'Dick goes like this', imitating a man in deep thought, hand to forehead; 'and then he writes down the answer.' (The remark may not have been altogether kindly meant.) Our foremost Newton scholar, Richard Westfall wrote as follows:

> The more I have studied him, the more Newton has receded from me. It has been my privilege at various times to know a number of brilliant men, men whom I acknowledge without hesitation to be my intellectual superiors. I have never, however, met one against whom I was unwilling to measure myself, so that it seemed reasonable to say that I was half as able as the person in question, or a third or a fourth, but in every case a finite fraction. The end result of my study of Newton has served to convince me that with him there is no measure. He has become for me wholly other, one of the tiny handful of supreme geniuses who have shaped the categories of the human intellect, a man not finally reducible to the criteria by which we comprehend our fellow beings.

Two biographies of Newton regarded as definitive are: *Never at Rest: A Biography of Isaac Newton* by Richard S. Westfall (Cambridge University Press, Cambridge, 1980) and *Isaac Newton: Adventurer in Thought* by A. Rupert Hall (Cambridge University Press, Cambridge, 1992).

~

16 Rutherford finds a solution

Ernest Rutherford, later Lord Rutherford of Nelson, was born in 1871 on a sheep farm in New Zealand. 'Always', one of his contemporaries greeted him, 'on the crest of the wave, eh Rutherford', he was responsible for a remarkable series of discoveries in atomic structure, which made him arguably the pre-eminent experimental physicist of his generation. 'Well, I made the wave, didn't I?' had been his response. A. V. Hill [115], the physiologist, recalled how Rutherford had remarked to him one day, without preamble: 'I've just been reading some of my early papers; and when I'd read them for a bit I said to myself, "Ernest my boy, you used to be a damned clever fellow".' Modesty was never a weakness of Rutherford's: he was a boisterous extrovert, and so good-natured that he seems never to have incurred envy or malice.

Among many other consequences, the discovery of radioactivity unlocked a puzzle, which had tormented Charles Darwin [43] in the last decades of his life: the age of our planet, inferred from the fossil record, vastly exceeded the calculated time required for Earth to have cooled from its temperature (that of the Sun) when formed. This calculation, performed by the uncontradictable Victorian physicist William Thomson (later Lord Kelvin of Largs) [10], seemed to jeopardize the whole of Darwin's theory. But the Sun is an atomic furnace and Earth is full of radioactivity; and radioactive decay of elements generates, as Rutherford had shown, an abundance of energy to account with ease for the shortfall.

In 1904, Rutherford was invited to lecture on the new revelations before a distinguished audience, which included the formidable Lord Kelvin, then 80 years old. His presence caused Rutherford some unease. Here, in his own words, is how he dealt with the ticklish situation:

> To my relief Kelvin fell asleep, but as I came to the important point, I saw the old bird sit up, open an eye and cock a baleful glance at me! Then a sudden inspiration came, and I said, 'Lord Kelvin had limited the age of the earth, *provided no new source of heat was discovered*. That prophetic utterance refers to what we are now considering tonight, radium'. Behold! The old boy beamed at me.

Kelvin, in fact, voiced doubts even two years later about whether radioactivity could really account for the extra energy. Another great physicist, Lord Rayleigh, invited Kelvin to lay a bet of five shillings that before six months had passed he would declare Rutherford to have been right. Within the allotted time Kelvin indeed came round, confessed as much in public before the British Association for the Advancement of Science, and paid up his five shillings.

See, for example, David Wilson's definitive biography, *Rutherford: Simple Genius* (Hodder and Stoughton, London, 1983).

~

17 The vanishing blackboard

André Marie Ampère (1775–1836), the remarkable French man of science, whose name is commemorated in the unit of electrical current, broke open the field of electromagnetism (or, as he called it, electrodynamics). He was an infant prodigy and was said in early childhood to have memorized all 20 volumes of the *Encyclopédie* edited by Diderot and d'Alembert. Long before Esperanto was thought of, he invented a universal language with a complete new vocabulary and syntax.

Ampère was capable of the intense concentration, common to many geniuses, that often reveals itself in an eccentric absence of mind. When thinking about physical or mathematical problems, he would draw a chalk stub from his pocket and use any convenient surface as a blackboard. It is related that, while walking in Paris, a thought came to him and he searched eagerly for a surface on which to work out its consequences. What he found was the back of a hansom cab and in a while he had covered it in equations. As his cogitations reached a denouement, he was startled to see his blackboard recede, gather speed, and, before he could react, vanish into the distance, bearing away the solution to his problem.

Thomas Hobbes (1588–1679), the combative philosopher, dabbled to some effect in science, and formulated, for instance, a clever (though incorrect) theory of light propagation. But he did not trouble himself with mathematics until one day, as John Aubrey tells it in *Brief Lives*, his eye fell on a volume of Euclid, open in the library. On the page was a proposition that Hobbes instantly decided was impossible. Reading on, he was referred to another proposition, which in turn directed him to another, and so on, until he was convinced that the first proposition was after all correct. 'This', says Aubrey, 'made him in love with geometry', and

> I have often heard Mr Hobbes say that he was wont to draw lines on his thigh
> and on the sheets abed, and also multiply and divide.

(It has to be conceded that Hobbes was no great mathematician, or at least the valuation he put on his abilities bore little relation to the results. He persuaded himself that he had solved the problem, still then teasing the mathematical intelligences of Europe, of squaring the circle [151], and plunged too eagerly into an acrimonious dispute on the matter with the equally factitious but vastly superior Oxford mathematics professor, John Wallis. In his rage, Hobbes unloosed the full horsepower of his eloquence on Wallis and his Oxford colleague, Seth Ward, in an extended philippic: 'So go your ways', he instructed them, 'you Uncivil Ecclesiastics, Inhuman Divines, Dedoctors of morality, Unasinous Colleagues, Egregious pair of

Issachars, most wretched Vindices and Indices Academiatrum . . . ' The sense hardly matters, nor did it win Hobbes the argument.)

There are many other recorded instances of scientific work written on surfaces designed for other purposes. The remarkable mathematical school, which sprang from a seemingly arid soil in the city of Lwów (now Lviv) in what was then the Polish Ukraine in the 1920s, had as its standing venue for discussion the Café Szkocka. This was chosen for its marble tabletops, which were so receptive to pencil and could be wiped clean at the end of an arduous day.

An attested case of abstraction concerns Niels Bohr [79]. Niels and his brother Harald, a distinguished mathematician, were both in their youth notable athletes. Harald played football for the Danish national team and won a silver medal in the Olympic Games of 1908. Niels played for his club as goalkeeper.

> Among his memorable exploits on the field was a game against a German club during which most of the action took place in the German half of the field. Suddenly, however, 'the ball came rolling toward the Danish goal and everyone was waiting for Niels Bohr to run out and grab it. But astonishingly he kept standing in the goal, devoting his attention rather to the goal post. The ball would certainly have gone in if the shouts of a resolute spectator had not awakened Bohr. After the match he gave the embarrassed excuse that a mathematical problem had suddenly occurred to him that absorbed him so strongly that he had carried out some calculations on the goal post.'

Ampère and the blackboard is told by Ralph E. Oesper in *The Human Side of Scientists* (University Publications, University of Cincinnati, Cincinnati, Ohio, 1975). The story about Bohr is related by Abraham Pais in his biography, *Neils Bohr's Times* (Oxford University Press, Oxford, 1991). For an account of Thomas Hobbes's vendettas with the mathematians, see, for example, *Great Feuds in Science* by Hal Hellman (Wiley, New York, 1998).

~

18 Cats and dogmas

Like animals on the theatrical stage, those in behavioural experiments all too often get the better of their trainers. Lewis Thomas [160], researcher and essayist, tells of such an embarrassing episode in one of his amiable essays. He alludes first to Clever Hans, the arithmetically gifted German horse, which its master, Herr von Osten, displayed as a prodigy in 1903.

The animal appeared to perform mental calculations, giving the answer to questions by tapping its hoof on the ground the correct number of times. A decade or so before, a similar phenomenon had been reported in England: a horse called Mahomet could tell the time from a watch displayed before its eyes. When a commission of psychologists examined Clever Hans it emerged that the horse was responding to minute, and as agreed by all, unconscious, movements by its master, which indicated to the sagacious animal when it was time to stop tapping.

Here is Thomas's account of the cat story:

The mind of a cat is an inscrutable mystery, beyond human reach, the least human of all creatures and at the same time, as any cat owner will attest, the most intelligent. In 1979, a paper was published in *Science* by B. R. Moore and S. Stuttard entitled 'Dr Guthrie and *Felis domesticus* or: tripping over the cat', a wonderful account of the kind of scientific mischief native to this species. Thirty-five years ago, E. R. Guthrie and G. P. Horton described an experiment in which cats were placed in a glass-fronted puzzle box and trained to find their way out by jostling a slender vertical rod at the front of the box, thereby causing a door to open. What interested these investigators was not so much that the cats could learn to bump into the vertical rod, but that before doing so each animal performed a long ritual of highly stereotyped movements, rubbing their heads and backs against the front of the box, turning circles and finally touching the rod. The experiment has ranked as something of a classic in experimental psychology, even raising in some minds the notion of a ceremony of superstition on the part of the cats: before the rod will open the door, it is necessary to go through a magical sequence of motions.

Moore and Stuttard repeated the Guthrie experiment, observed the same complex of 'learning' behaviour, but then discovered that it occurred only when a human being was visible to the cat. If no one was in the room with the box, the cat did nothing but take naps. The sight of a human being was all that was needed to launch the animal on a series of sinuous movements, rod or no rod, door or no door. It was not a learned pattern of behaviour, it was a cat greeting a person.

See Lewis Thomas, *Late Night Thoughts* (Oxford University Press, Oxford, 1984; the US edition, published by Viking Press in 1983, is called *Late Night Thoughts on Listening to Mahler's Ninth Symphony*).

19 But what use is it?

When Chancellor of the Exchequer William Gladstone, having witnessed Michael Faraday's demonstration of the newly discovered phenomenon of electromagnetic induction, asked: 'But what use is it?' Faraday famously replied, 'I don't know, but one day, sir, you may be able to tax it'. (According to another version, however, or perhaps even on a different occasion, his riposte was supposed to have been, 'What use is a newborn baby?')

Jöns Jacob Berzelius (1779–1848), one of the founders of modern chemistry, encountered similar incomprehension in Sweden. It is related that the servant, whom he engaged as laboratory assistant, was confronted one day by a group of stolid Stockholm burghers, demanding to know what went on in Berzelius's house. What were the assistant's duties, they asked. 'In the morning I go to the cupboard and shelves and bring the master all manner of things, powders, crystals, liquids of different colours and smells.' 'And then what?' 'He looks them over, takes some of each and puts these amounts in a big pot.' 'And then what?' 'Then he heats the pot and puts the whole affair into smaller pots after the contents of the large pot have boiled for an hour or two.' 'And then what does he do?' 'Then he puts the whole lot into a pail. Then the next morning I carry the pail outside and empty it down the drain.'

See Ralph E. Oesper, *The Human Side of Scientists* (University Publications, University of Cincinnati, Cincinnati, Ohio, 1975).

~

20 Unlocking the chains

The concept of a chain reaction—a process that gathers pace by multiplying the reactive entities as it progresses—entered chemistry in 1913 and physics some 20 years later. It is a characteristic of such reactions that they start slowly, sometimes after a marked lag, and accelerate explosively. The most familiar example is nuclear fission: when a neutron strikes a uranium (^{235}U) atom and is captured the nucleus breaks in two and releases two or three neutrons. These in turn act on surrounding uranium atoms and the process of fission accelerates sharply. In chemistry autocatalytic reactions with similar characteristics had been observed in the nineteenth century and had exercised such luminaries as Robert Bunsen, the great German chemist.

Max Bodenstein, a physical chemist, did notable work in Germany on the mechanisms of chemical processes. In 1913, he was puzzling over a light-induced reaction between hydrogen and chlorine, which exhibited just such a lag after irradiation of the reaction vessel, accelerated, and then unaccountably stopped. Bodenstein's assistant, Walter Dux, described what happened. As the two men pondered the meaning of the phenomenon, Bodenstein undid his gold watch-chain. He asked Dux to hold one end, while he himself twirled the other. 'If we give this chain an impulse', he ruminated, 'it will propagate through its whole length, but if we were to hold one link or take it out, the movement will be halted.' 'Perhaps', Dux rejoined, 'this is what happens in our reaction.' 'Good idea. Perhaps we might call it a *chain reaction*; let's test it.'

The idea caught on quickly and began to permeate research in chemical kinetics, notably in the formation of high polymers, the giant molecules of fibres and plastics. After Bodenstein's death in 1942, Dux asked his family for the watch-chain, but it turned out that Bodenstein had patriotically donated it to the German war effort soon after the experiment and had replaced it with a chain of steel. Dux had a replica gold chain fashioned and donated it to the University of Hanover as a memento.

Leo Szilard (1898–1964), the peripatetic Hungarian physicist who lived most of his life in hotel rooms and whose worldly goods were throughout this time contained in two suitcases, departed from Berlin with the advent of Hitler:

In the fall of 1933, I found myself in London. I kept myself busy trying to find positions for German colleagues who lost their university positions with the advent of the Nazi regime. One morning I read in the newspaper about the annual meeting of the British Association where Lord Rutherford was reported to have said that whoever talks about the liberation of atomic energy on the industrial scale is talking moonshine. Pronouncements of experts to the effect that something cannot be done have always irritated me. That day I was walking down Southampton Row [in Bloomsbury, where his hotel was] and was stopped for a traffic light. I was pondering whether Lord Rutherford might not prove to be wrong. As the light changed to green and I crossed the street it suddenly occurred to me that if we could find an element which is split by neutrons and which would emit two neutrons when it absorbed one neutron, such an element if assembled in sufficiently large mass, could sustain a nuclear chain reaction, liberate energy on an industrial scale, and construct atomic bombs. The thought that this might be possible became an obsession with me. It led me to go into nuclear physics, a field in which I had not worked before.

Szilard found a laboratory in London and tried to put his idea to the test, but the time was not ripe: none of the elements he tried to bombard with neutrons emitted secondary neutrons. Szilard nevertheless thought the scheme sufficiently realistic that he was prompted a few months later to take out a patent, and to prevent the dangerous notion from becoming public, he assigned it to the Admiralty.

About this time, Szilard was the victim of an innocent hoax, which succeeded beyond its perpetrators' expectations. They were two young physicists—Carl Bosch, a German, and R. V. Jones [106], then at Oxford. Jones, introducing himself as the editor of the *Daily Express*, telephoned Szilard and asked whether he could confirm that he had invented a radio-active death-ray. Szilard reacted explosively, for he had just then registered his patent on the nuclear chain reaction, and his panic at what he must have supposed was a leaked, if garbled, disclosure can be imagined.

It was another five years before Szilard's vision (of which he had in the meantime despaired) became reality: Lise Meitner (1878–1968) was a physicist, who had worked with the chemists Otto Hahn and Fritz Strassmann in Berlin on the identification of the products of nuclear transformations. Though Jewish, she had not, as an Austrian citizen, fallen victim to the Nazi employment laws, but the *Anschluss*, when Austria was conjoined with Germany, in 1938 deprived her of this refuge and she was hastily spirited out of the country before she could be arrested. Washed up in an inhospitable laboratory in Sweden, she could do no more than keep in touch by mail with her friend Otto Hahn. In December 1938, she received a visit from her physicist nephew, Otto Frisch (1904–70), who had found a position in England but was then working in Niels Bohr's famous institute in Copenhagen. It was the custom of nephew and aunt to spend Christmas together and Frisch described this particular visit as the most momentous event of his life.

In the foregoing year a whole series of products of nuclear bombardment processes had been discovered, which appeared in some instances to violate the principle that collision of a fundamental particle with a nucleus could do no more than drive out of it an alpha particle (identical to a helium nucleus) or a beta particle (an electron); the predicted and hitherto observed result was formation of an element with a nuclear charge (atomic number) two less or one more than that of the parent. Hahn and Strassmann had detected, as they thought, radium isotopes as bombard- ment products of uranium. (Isotopes are forms of an element differing only in the number of neutrons in the nucleus; because the number of posi- tively charged protons in the nucleus and therefore the number of the electronegative electrons outside it are the same in all, the isotopes are in chemical terms identical.) This appeared an inexplicable result, for radium has a smaller nucleus than uranium and Lise Meitner urged Hahn to make absolutely sure he had got it right before publishing such an unaccountable anomaly.

When Otto Frisch called on his aunt on his first day in the small Swedish town of Kungälv, where she was relaxing with friends, he found her pon- dering Hahn's latest letter. Here is how he described the encounter:

I wanted to tell her of a new experiment I was planning, but she wouldn't listen; I had to read that letter. Its content was indeed so startling that I was at first inclined to be sceptical. Hahn and Strassmann had found that those three substances were not radium, chemically speaking; indeed they had found it impossible to separate them from the barium which, routinely, they had added to facilitate the chemical separations. They had come to the con- clusion, reluctantly and with hesitation, that they were isotopes of barium [half the size of uranium].

Was it just a mistake? No, said Lise Meitner; Hahn was too good a chemist for that. But how could barium be formed from uranium? No larger frag- ments than protons or helium nuclei (alpha particles) had ever been chipped away from nuclei, and to chip off a large number not nearly enough energy was available. Nor was it possible that the uranium nucleus could have been cleaved right across. A nucleus was not like a brittle solid that can be cleaved or broken; George Gamow [81] had suggested early on, and Bohr had given good arguments that a nucleus was much more like a liquid drop. Perhaps a drop could divide itself into two smaller drops in a more gradual manner, by first becoming elongated, then constricted, and finally being torn rather than broken in two? We knew that there were strong forces that would resist such a process, just as the surface tension of an ordinary liquid drop tends to resist division into two smaller ones. But the nuclei differed from liquid drops in one important way: they were electrically charged, and that was known to counteract surface tension. At that point we both sat down on a tree trunk (all that discussion had taken place while we walked through the wood in the snow, I with my skis on, Lise Meitner making good her claim that she could walk just as fast without), and started to calculate on scraps of paper. The charge of a uranium nucleus, we found, was indeed large enough to overcome the effect of the surface tension almost completely; so the uranium nucleus might indeed resemble a very wobbly, unstable drop, ready to divide itself at the slightest provocation, such as the impact of a single neutron.

But there was another problem. After separation the two drops would be driven apart by their mutual electrostatic repulsion and would acquire high speed and hence a very large energy, about 200 MeV [mega electron volts] in all; where could that energy come from? Fortunately Lise Meitner remembered the empirical formula for computing the masses of nuclei and worked out that the two nuclei formed by division of a uranium nucleus together would be lighter than the original uranium nucleus by about one-fifth of the mass of a proton. Now whenever mass disappears energy is created, according to Einstein's formula, $E = mc^2$, and one-fifth of a proton mass was just equivalent to 200 MeV. So here was the source for the energy; it all fitted!

A couple of days later I travelled back to Copenhagen in considerable excitement. I was keen to submit our speculations—it wasn't really more at the time—to Bohr, who was just about to leave for the U.S.A. He had only a few minutes for me, but I had hardly begun to tell him when he smote his forehead with his hand and exclaimed: 'Oh what idiots we have all been! Oh but that is wonderful! This is just how it must be! Have you and Lise Meitner written the paper yet?' Not yet, I said, but we would at once and Bohr promised not to talk about it before the paper was out. Then he went off to catch his boat.

Frisch asked an American biologist in the laboratory what the process was called by which a single cell divides into two. 'Fission' was the reply, and Frisch coined the term 'nuclear fission'.

See Keith J. Laidler, *The World of Physical Chemistry* (Oxford University Press, Oxford, 1993); Leo Szilard, *The Collected Works of Leo Szilard: Scientific Papers*, ed. B. T. Feld and G. W. Szilard (MIT Press, Cambridge, Mass., 1972); and Otto Frisch, *What Little I Remember* (Cambridge University Press, Cambridge, 1979).

21 Of life and death

Isidor Rabi, born in Poland in 1898, grew up in some poverty in New York and became one of the world's great physicists. He won a Nobel Prize in

1944 for discovering a phenomenon that led ultimately to nuclear magnetic resonance spectroscopy, one of the most powerful methods for the study of molecular structure and later of generating images of living tissue. This involved observing the flip in an oscillating field of atomic nuclei, which had been discovered to possess a magnetic moment, like minuscule bar magnets.

Rabi, who spent most of his working life at Columbia University in New York, became after his Nobel Prize a statesman of science. His enthusiasm for the laboratory seemed to have deserted him; he once said of the Nobel Prize: 'Unless you are very competitive you aren't likely to function with the same vigour afterwards. It's like the lady from Boston who said, "Why should I travel when I'm already there?" The prize also attracts you away from your field because other avenues open up.'

But, it seemed, somewhere in the depths of his mind, Rabi still revolved deep problems of scientific truth. Like Einstein, he had been worried about the physical meaning of quantum theory. Rabi lived into his ninetieth year, and here is how he approached his end:

> One day, in December 1987, a colleague came into my office in the Rockefeller University, to inform me that he had just seen Rabi, who had told him that he wanted to talk with me. I knew where Rabi was: across the street in Memorial Sloan-Kettering's Hospital, and why: he was terminally ill with cancer. I went there at once, suspecting that he had some final message to convey to me. There he was in remarkably good humor. What did he want to talk about? The foundations of quantum mechanics, which, as told, had troubled him for decades and, in these last weeks, were still on his mind. We argued for maybe half an hour, then I said good-bye to him, forever. On January 11, 1988, Rabi died.

The reminiscence comes from Abraham Pais, *The Genius of Science* (Oxford University Press, Oxford, 2000).

22 Mathematical peril

George Gamow, the physicist [81] who escaped to the United States from Stalinist Russia, tells the following tale of what can befall an innocent scholar in times of political turbulence.

Here is a story told to me by one of my friends who was at that time a young professor of physics in Odessa. His name was Igor Tamm (Nobel Prize laureate in Physics, 1958). Once when he arrived in a neighbouring village, at the period when Odessa was occupied by the Reds, and was negotiating with a villager as to how many chickens he could get for half a dozen silver spoons, the village was captured by one of the Makhno bands, who were roaming the country, harassing the Reds. Seeing his city clothes (or what was left of them), the capturers brought him to the Ataman, a bearded fellow in a tall black fur hat with machine-gun cartridge ribbons crossed on his broad chest and a couple of hand grenades hanging on the belt.

'You son-of-a-bitch, you Communistic agitator, undermining our mother Ukraine! The punishment is death.'

'But no', answered Tamm. 'I am a professor at the University of Odessa and have come here only to get some food.'

'Rubbish!', retorted the leader. 'What kind of professor are you?'

'I teach mathematics.'

'Mathematics?' said the Ataman. 'All right! Then give me an estimate of the error one makes by cutting off Maclaurin's series at the nth term. Do this and you will go free. Fail, and you will be shot!'

Tamm could not believe his ears, since this problem belongs to a rather special branch of higher mathematics. With a shaking hand, and under the muzzle of the gun, he managed to work out the solution and handed it to the Ataman.

'Correct!' said the Ataman. 'Now I see that you really are a professor. Go home!'

Who was this man? No one will ever know. If he was not killed later on, he may well be lecturing now on higher mathematics in some Ukrainian university.

The dangers to scholars did not vanish with the Revolution. Mark Azbel, a theoretical physicist, who after years of persecution and imprisonment reached sanctuary in Israel, offers another example.

I heard a story about that time from Professor Povzner, who taught a course at the Military Academy for Engineers. He walked into a class one day, ready to start his lecture with a routine spiel about Russian primacy in mathematics, and then settle down to a serious session of really teaching mathematics. But to his alarm, the minute he got up in front of the class he saw that among the audience was a general, the chief of the Academy. He pulled up short and decided that he had better devote the whole lecture to the subject of early Russian genius in mathematics. Luckily, he was a very talented man, good at thinking on his feet, so on the spur of the moment he invented a wonderful lecture on Russian mathematics in the twelfth century. He engaged in flights of fancy for the entire hour, stopping only five minutes before the end to ask, as was customary, 'Are there any questions?' He saw that one of the students had raised his hand.

'Yes?'

'This is so interesting, about medieval Russian mathematics. Could you tell us, please, where we could get more information about it—what the reference books would be? I would like to learn more.' Having no time to think, the professor immediately answered: 'Well, that's impossible! All the archives were burned during the Tatar invasion!'

When the class was over, the general came up to the lecturer and said, 'So, Professor . . . All the archives were burned?' Only then did poor Povzner realize what he had said. The unspoken question hung in the air: If all the evidence of Russian primacy in this science was burned, how in the world did the professor *himself* know the history of pre-invasion mathematics? He was ready to panic when, unexpectedly, the general smiled at him sympathetically, turned around, and left. This high-ranking commander was a clever and decent person; otherwise Professor Povzner would have been in deep trouble.

See George Gamow, *My World Line* (Viking, New York, 1970) and Mark Ya. Azbel, *Refusnik* (Hamish Hamilton, London, 1982).

~

23 Fortune favours the ham fist

Fortune, Louis Pasteur tells us, favours the prepared mind. The American physicist, who observed that when his children dropped their toast on the carpet it invariably contravened the malign principle of nature by falling buttered-side up, did not dismiss it as a statistical anomaly. He investigated further and found an explanation, consistent with the laws of physics: his children buttered their toast on both sides. The history of science is replete with examples of accidental discovery, the result of fine judgement, when an anomalous and seemingly useless experimental result was examined more closely rather than discarded.

In the organic chemistry laboratory, breaking a thermometer was always considered a major infraction. Yet when Otto Beckmann (1853–1923), an assistant at the end of the nineteenth century in the laboratory of one of the mandarins of German chemistry, Wilhelm Ostwald, broke a precious thermometer, specially made by the departmental glassblower with a stem so long and uniform that the temperature could be read to one-hundredth of a degree, he turned it to good account: he fell to wondering how one might construct a less-vulnerable instrument with similar precision, and so avoid such disasters and escape the wrath of the professor. The result was the Beckmann thermometer, familiar to all chemists (at least until the age of electronic thermometers), which has a short stem with a mercury reservoir at the top so that the amount of mercury in the bulb can be adjusted to select a chosen narrow temperature range.

The most sensational outcome of a mishap with a thermometer was a commercial revolution. The first synthetic dyes were prepared from substances found in coal-tar distillates and they started a great new industry, associated with the illustrious organic chemist Adolf von Baeyer [84] and two disciples—August Wilhelm Hofmann and W. H. Perkin, both of whom established schools of chemistry in England.

Indigo was a highly prized dye from ancient times. In India, for instance, some two million acres were given over to the cultivation of the parent plant. Von Baeyer occupied himself with the dye for 20 years, eventually determined its structure and by a tour de force of his craft synthesized it from simple starting materials in 1883. But this was a complex process of many steps, quite unsuitable for industrial development. A commercial synthesis, starting from the coal-tar product, naphthalene, was achieved by chemists at the giant BASF concern in Bavaria during the following decade, but the cost was still barely competitive. Then in 1896 an insignificant worker at BASF, by the name of Sapper, was heating naphthalene with fuming sulphuric acid (a potent brew of sulphuric acid with sulphur trioxide) and apparently stirring it with a thermometer. The thermometer broke, spilling the mercury into the reaction mixture, and lo! the reaction took a different course: the naphthalene was converted to phthalic anhydride, the sought-after intermediate in the synthesis of indigo. It transpired that the mercury, or rather the sulphate into which it had been converted by the sulphuric acid, was a catalyst for this previously undiscovered reaction. Cheap BASF indigo came on the market the following year and the Indian indigo industry collapsed.

The history of dyestuffs is rich in examples of accidental discovery. Probably the earliest came when Friedlieb Ferdinand Runge (1794–1867), the German chemist, attempted to keep the dogs of the neighbourhood out of his garden in a Berlin suburb by erecting a wooden fence, which he painted with coal tar (creosote) as a preservative. Then, to discourage the dogs from lifting their legs against his fence, he scattered bleaching powder (calcium chlorohypochlorite) all around, to diffuse a noxious smell of chlorine. Inspecting the fence the next day he was surprised to observe blue streaks on the white powder, following all too obviously the trajectories of the canine urine jets. Runge investigated and found that the blue colour was the result of oxidation by the hypochlorite of some constituent of the coal tar. The dogs had merely provided the water to dissolve the reactive principle. Runge called the blue substance Kymol. Some years later Hofmann showed that the parent compound in the coal tar was aminobenzene, or aniline, and Kymol was the first synthetic prototype of a dye.

It was another accident that led to the discovery of the important intermediate in many organic syntheses—the sulphur-containing benzene-like ring compound, thiophene. Victor Meyer (1848–97), a noted German chemist, would demonstrate when lecturing to his students at the University of Zurich, a beautiful test for benzene, devised by Baeyer: the sample suspected of containing benzene was shaken with sulphuric acid and a crystal of isatin (the final precursor in Baeyer's famous synthesis of indigo). A rich blue colour was taken to betray the presence of benzene. But on this occasion in 1882 no blue colour eventuated. Meyer must have been disconcerted, but on looking into the matter he found that in place of the usual kind of benzene sample, derived from coal tar, his assistant had handed him a very pure synthetic benzene. By the next year Meyer had run to earth the impurity (no more than 0.5 per cent) in the coal-tar product responsible for the blue colour. So began another chapter in organic chemistry.

John Read worked with Baeyer during his formative years around the beginning of the twentieth century and later with Sir William Jackson Pope [1] in Cambridge, and became eventually Professor of Chemistry at the University of St Andrews. He told of the 'unsung hero'—a lazy laboratory assistant (or lab boy, as they used to be called)—of a crucial advance in stereochemistry. The chemists had been trying unsuccessfully to resolve optically active compounds by a new trick. Optical activity, the capacity to rotate the plane of polarized light to the right or the left, is a property of molecules with an intrinsic asymmetry—that is, with a structure that cannot be superimposed on its mirror image. Laboratory reactions, unlike those that occur in living organisms, deliver a mixture of both forms (antipodes) of such compounds, and to separate (resolve) the left from the right-handed components ingenuity is required. Pope's way was to react

the mixture with an already purified asymmetric reagent; this would discriminate between the two forms by generating different crystals. The
errant lab boy had omitted to clear away and wash a heap of glassware,
coated with sticky deposits, each representing a failed attempt at crystallization. Read demanded again that the sad detritus be cleared away, but the
lad took his time and, while he waited, Read's eye fell on a white sliver in a
puddle on one of the dishes. He peered at it with a hand lens and saw that it
was a crystal. Amid rejoicing, he added it to a solution of the mixed reaction
product, and behold, it 'seeded' the crystallization of its own asymmetric
form.

See Royston Roberts, *Serendipity: Accidental Discovery in Science* (Wiley, New York, 1989),
and John Read, *Humour and Humanism in Chemistry* (G. Bell, London, 1947).

24 To discern a true vocation

The doings of students in the laboratory have been a source of many treas-
ured tales of Grand Guignol. The following was recorded by John Nelson
(1876–1965), Professor of Chemistry at Columbia University in New York
for close on half a century.

Like many of his generation Nelson mistrusted women in the laboratory,
so when a young graduate of a women's college applied for a place as a
research student, he invited her to demonstrate her practical competence
by performing a simple organic synthesis, following the instructions of the
laboratory manual. She was to prepare bromobenzene from benzene and
bromine. After an hour Nelson entered the laboratory and asked how she
was progressing. Not well, it seemed, for the reaction was refusing to proceed as the manual foreshadowed. Nelson looked at the apparatus—a flask,
in which liquid was vigorously boiling under a reflux condenser (in essence
a water-cooled tube that allows the vapour to condense and the liquid to
run back into the flask and continue boiling). The flask contained benzene,
but, Nelson asked, where was the bromine (which would have imparted a
yellow tinge to the liquid)? In the flask, was the reply, floating in the liquid.
Nelson peered more closely, and observed white solid matter churning in
the boiling liquid. Looking around, he found a tin, labelled bromine, and
further questioning elicited the explanation: the student had thrown the

white packing material into the flask, having never discovered the vial of liquid bromine buried within. Nelson urged her to seek a different career. (It's only fair to add that there have always been talented and successful theoreticians whose presence in the laboratory spelled death to all apparatus and disaster to their colleagues' experiments.)

See Ralph E. Oesper, *The Human Side of Scientists* (University Publications, University of Cincinnati, Cincinnati, Ohio, 1975).

∽

25 The Pauli principle

Wolfgang Pauli (1900–58)was one of the titans who presided over theoretical physics during its golden age in the early decades of the twentieth century. He was celebrated not only for his devastating intelligence but also for his indiscriminate rudeness, or perhaps more properly unyielding candour. He became known in his profession as 'the scourge of the Lord'. According to Victor Weisskopf [95], who served a happy term as Pauli's assistant, one could ask him any question without worrying whether he might think it idiotic, for Pauli found all questions idiotic. In his most accommodating mood, and while still a student, Pauli began his remarks in the discussion that followed a lecture by Albert Einstein: 'You know, what Einstein says is not so stupid!' Weisskopf recalled what happened when he arrived in Zurich and called on Pauli in his office.

> Finally, he lifted his head and said, 'Who are you?'. I answered, 'I am Weisskopf. You asked me to be your assistant.' He replied, 'Oh, yes. I really wanted Bethe, but he works on solid state theory, which I don't like, although I started it.'

Weisskopf had fortunately been forewarned by Rudolf Peierls [42], who knew Pauli well. Pauli and Weisskopf then had a brief discussion, during which Weisskopf indicated that he would be happy to work on anything except a controversial approach, which he could not understand, to the Theory of Relativity. Pauli had been pondering this very topic but was beginning to tire of it and so assented. He then presented the new arrival with a problem and the next week asked him what progress he had made. 'I showed him my solution', Weisskopf recalled in his memoirs, 'and he said, "I should have taken Bethe after all".' Weisskopf and Pauli became and remained friends. About Peierls, Pauli remarked, 'He talks so fast that by the time you understand what he is saying he is already asserting the opposite'.

Pauli, whose name is enshrined in the Pauli Principle, also known as the Exclusion Principle (no two electrons in an atom can occupy the same quantum state), was also famous for the second Pauli Principle—that his approach spelled destruction to any scientific apparatus or mechanical device. Most famously, a devastating explosion in the physics department at the University of Berne was found to have coincided with the passage through the town of a train bearing Pauli to his home in Zurich. His driving of a motor car, enlivened by a ceaseless self-congratulatory commentary, occasioned such alarm in his passengers that many refused to travel with him more than once. To a plaintive remark by his student and associate, the Dutch physicist H. G. B. Casimir [53], he replied: 'I will make a bargain with you: don't comment on my driving and I won't comment on your physics.' In his memoirs Casimir reproduced the following graphic description of the Second Pauli Principle in action, recollected by the Belgian physicist, Léon Rosenfeld (whom Pauli liked to describe as 'the choirboy to the Pope', as Niels Bohr was sometimes called):

Heitler, by lecturing on the theory of the homopolar bond, unexpectedly excited his wrath: for, as it turned out, he had a strong dislike to this theory. Hardly had Heitler finished, than Pauli moved to the blackboard in a state of great agitation, pacing to and fro he angrily started to voice his grievance, while Heitler sat down on a chair at the edge of the podium. 'At long distance', Pauli explained, 'the theory is certainly wrong, since we have there van der Waals attraction; at short distance, obviously, it is also entirely wrong.' At this point he had reached the end of the podium opposite to where Heitler was sitting. He turned round and was now walking towards him, threateningly pointing in his direction the piece of chalk he was holding in his hand: 'Und nun', he exclaimed, 'gibt es eine an den guten Glauben der Physiker appellierende Aussage, die behauptet, dass diese Näherung, die falsch ist in grossen Abständen und falsch ist in kleinen Abständen, trotzdem sie in einem Zwischengebiet qualitativ richtig sein soll!' (And now there is a statement, invoking the credulity of the physicists, that claims that this approximation which is wrong at large distances and is wrong at short distances yet is qualitatively true in an intermediate region.) He was now quite near to Heitler. The latter leaned back suddenly, the back of the chair gave way with a great crash, and poor Heitler tumbled backwards (luckily without hurting himself too much).

Casimir, who was there, notes that George Gamow was the first to shout: 'Pauli effect'. And as an afterthought he adds: 'Sometimes I wonder whether Gamow (a noted *farceur*) [81] had not done something to the chair beforehand.'

Jeremy Bernstein, a physicist and the most vivid and illuminating of writers on physics and the ways of physicists, recalls in his fascinating memoirs a singular incident, which took place during the last year of Pauli's life.

Pauli had been engaged in a peculiar enterprise with his former collaborator Werner Heisenberg [180], another of the great architects of the quantum theory. For a while they claimed to have solved all the unsolved problems in elementary particle theory; they reduced everything to a single equation. When calmer spirits examined the matter, they concluded that the whole thing was a chimera. The dénouement, for Pauli, came at a lecture he delivered at Columbia University in the large lecture hall in Pupin Laboratory. Even though there had been an attempt to keep the talk secret, the room was filled to capacity. The audience was studded with past, present and future Nobel Prize winners, including Niels Bohr [79]. After Pauli delivered his lecture, Bohr was asked to comment. There then occurred one of the most unusual, and in its unearthly way most moving, demonstrations I have ever witnessed. Bohr's basic point was that as a fundamental theory it was crazy, but not crazy enough. The great advances, like relativity and the quantum theory, do seem—at first sight, and especially if one has been brought up in the physics that preceded them—to be crazy, to violate common sense in a fundamental way. On the other hand, Pauli's theory was just bizarre, a strange-looking equation that stared at you like a hieroglyph. Pauli objected to Bohr's assessment; he said the theory *was* crazy enough.

At this point these two monumental figures in modern physics began moving in a conjoined circular orbit around the long lecture table. Whenever Bohr faced the audience from the front of the table, he repeated that the theory was not crazy enough, and whenever Pauli faced the group, he would say it was. I recall wondering what anyone from the other world—the non-physicists' world—would make of this. [Freeman] Dyson [52] was asked to comment and refused. Afterward he remarked to me that it was like watching the 'death of a noble animal'. He was prescient. Pauli died not many months later in 1958, at the age of fifty-eight, of a previously undetected cancer. Before that he had renounced 'Heisenberg's theory', as he now referred to it, in the most acidulous manner. One could only wonder whether Pauli's brief love affair with it was a sign that he was already ill.

Pauli died not many months later. Whether he was conscious of fading intellectual and imaginative powers is unclear. But he did remark mournfully to one of his associates: 'Ich weiss viel. Ich weiss zu viel. Ich bin ein Quantengreis.' (I know much. I know too much. I am a quantum-dotard.) Pauli, as already related, passed his last days in a hospital room with the number 137, a 'magic number' of wave theory, and the coincidence disturbed him.

For the first story see Victor Weisskopf, *The Joy of Insight: Passions of a Physicist* (Basic Books, New York, 1991). Casimir's reminiscences of Pauli are told in his memoirs, H. G. B. Casimir, *Haphazard Reality: Half a Century of Science* (Harper and Row, London and New York, 1983). The last story is from Jeremy Bernstein's *The Life it Brings: One Physicist's Beginnings* (Ticknor and Fields, New York, 1987).

~

26 The first eureka

Archimedes (287–212 BC) was a remarkable polymath—a scientist, inventor, and mathematician. He was a son of Syracuse and supposedly related to King Hieron II. Among his many practical inventions were a water snail (used to raise water for irrigation), the compound pulley, and several engines of war, which included the legendary burning glasses with which he was supposed to have destroyed the Roman invasion fleet as it approached his city.

One of Archimedes's mathematical diversions was the 'Sand Reckoner', which facilitated the series of multiplications that allowed him to compute the number of grains of sand on the Sicilian coastline, and also the number that would fill the prevailing model of the universe (10^{63} was the answer). According to Plutarch, he asked that at his death his grave should be marked by a sphere circumvented by a cylinder, with an inscription giving the difference between the volumes within and outside the sphere.

Plutarch, Livy, and Valerius Maximus all agree, although the details differ, that Archimedes met his end at the hands of a Roman soldier after the capture of Syracuse. The Emperor had given instructions for Archimedes to be brought to him unharmed, but the scholar, it was said, was absorbed in a calculation and would not heed the tap on the shoulder; and so the soldier, incensed, slew him.

The story for which Archimedes is now remembered is of uncertain authenticity, but was told by a Roman historian, Vitruvius. The tale goes that King Hieron asked Archimedes to determine whether the substance of a crown that had been fashioned for him was indeed pure gold, or had been adulterated with silver:

> While Archimedes was turning the problem over, he chanced to come to a place of bathing, and there, as he was sitting down in the tub, he noticed that the amount of water that flowed over the tub was equal to the amount by which his body was immersed. This indicated to him a method of solving the problem, and he did not delay, but in his joy leapt out of the tub, and rushing naked towards his home, he cried out in a loud voice that he had found what he sought, for as he repeatedly shouted in Greek, *heureka, heureka*.

Archimedes's Principle, as it is still called, states, of course, that the upthrust of an immersed object is equal to the weight of water displaced. So when the crown was lowered into a vessel full of water the amount of water displaced, or the apparent weight of the immersed crown, would give a measure of the volume of the metal; this, with the weight of the crown in air, would deliver the density of the metal and thus its composition.

Scientists of later eras often tried to reproduce Archimedes's contrivances, most famously the burning glasses. The debate about whether the artifice could really have sunk the Roman fleet rumbled on for centuries and many celebrated savants proffered their opinions (including Descartes, who discounted the story). But then, in 1747, it was at last put to the experimental proof by the great French polymath, the comte de Buffon [118]. Buffon erected his apparatus in Paris, in what is now the Jardin des Plantes (then the Jardin du Roi, of which he was the director). About 150 concave mirrors were mounted on four wooden frames and adjusted with screws to focus on a plank of wood some 50 metres away. A large crowd watched as the sun emerged from the clouds: within a few minutes smoke was seen to issue from the plank and the point was made. Later the same year Buffon, to great acclaim, set some houses on fire in the presence of the monarch himself, and received the compliments not only of Louis XV but also the intellectual Frederick the Great of Prussia.

The source of the stories about Archimedes's discoveries and inventions is Vitruvius, in *De Architectura*, Book IX, Chapter 3. For Buffon's experiments see Jacques Roger, *Buffon: A Life in Natural History* (Cornell University Press, Ithaca, New York, 1997).

27 Lordly disdain

Lord Rutherford of Nelson [16], Cavendish Professor of Physics in Cambridge, was a majestic figure. Here is how he dismissed a theory he simply did not care for. The setting is the Clarendon Laboratory in Oxford and the writer, R. V. Jones [106].

> Rutherford came across occasionally from Cambridge, and his lectures were always memorable. In the question session after one of them E. A. Milne [64] [a noted theoretician and cosmologist] asked for his views on what was known as the Tutin atom. This concept had been started by a Dr Tutin, who argued that the Rutherford model of the atom [the planetary system, with a nucleus, made up of protons and neutrons, as the sun, and the electrons in orbits around it at defined radii] was all wrong, for everyone knew that if you spin a mixture of light and heavy particles, the heavy ones fly to the outside while the light ones stay close to the centre. So in an atom the electrons should be in the middle with the protons orbiting outside. The theory had achieved notoriety because F. W. Soddy [149], the Professor of Inorganic Chemistry, who had given isotopes their name, had communicated Tutin's paper to the Chemical Society, which had refused to publish the paper in its Journal, whereupon Soddy had resigned from the Society and had promptly advertised the sale of all his volumes of *Journal of the Chemical Society* in *Nature*. Milne asked Rutherford how he could be sure that Tutin was wrong and he was right. I can still see Rutherford, who was a big man, bending over towards Milne, who was a small one, and booming 'When you've got an elephant and a flea, you assume it's the flea that jumps'.

Rutherford's almost unerring instinct about the elementary particles is again illustrated by the following recollection during an interview, recorded some 70 years after the event, by M. L. Oliphant:

> 'We were doing experiments with all the possible projectiles in order to produce transformations in elements. It was natural to try to use heavy hydrogen [29] and, indeed, the results were very interesting. The experiments with heavy water brought about two discoveries, one was helium-3 and the other was tritium [isotopes of helium and hydrogen respectively].'
>
> On his interactions with Rutherford in connection with these discoveries: 'Rutherford was the greatest influence on me and on so many other people at that time in Cambridge. He was my scientific father in every sense of the word. Rutherford didn't like his associates keeping long hours in the lab. He thought it was silly to overdo it. But this didn't mean that Rutherford stopped

working at any time. One day we went home without having understood the results of an experiment, and our telephone rang during the night at 3 A.M. My wife told me that the Professor wanted to speak to me. Rutherford said, 'I've got it. Those short-range particles are helium-3'. I asked him about his reasons and he said, 'Reasons! Reasons! I feel it in my water!'"

Oliphant (later Sir Mark) became Professor of Physics in Birmingham, where he created a notable department, in which, among other important achievements, the source of centimetric radar, the cavity magnetron [12], was developed. Oliphant later returned to his native Australia and was the principal begetter of the Australian National University in Canberra. One of his less-successful enterprises was the construction of a particle accelerator, which never worked and was eventually abandoned. Because of this expensive fiasco the machine was dubbed 'The White Oliphant'.

The first passage is from an article by R. V. Jones in *The Making of Physicists*, ed. R. Williamson (Adam Hilger, Bristol, 1987). The interview with Oliphant is by I. and M. Hargittai, *The Chemical Intelligencer*, **6**, 50 (2000).

~

28 A martyr to science

Pliny the Elder, or Gaius Plinius Secundus, was the author of *Natural History*, the most comprehensive survey of scientific learning in the ancient world. He met his end, a martyr to scientific curiosity, during the eruption of Vesuvius that destroyed Pompeii and Herculaneum in AD 79. The disaster was vividly described by his nephew, Pliny the Younger. The family was at Misenum on the other side of the Bay of Naples and the elder Pliny had command of the fleet, then stationed in the bay. He was made aware of a large and rapidly growing cloud, ascending on a column of smoke, which was ramifying, 'like a Mediterranean pine tree'. 'My uncle,' the nephew recorded, 'who was a great scholar, just could not keep away.' He took ship and headed for the source of the smoke. As the ship approached Pompeii hot ash and pumice rained down, but Pliny would not turn back and ordered his helmsman to sail on into the gathering gloom until the ship became trapped in the shallows by an unfavourable wind. His nephew continues:

My uncle decided to go out onto the shore and see for himself if the sea would let them sail. But it was still angry and against them. His slaves put down a sheet for him to lie on and he demanded and drank one or two cups of cold water. The flames and the smell of sulphur, which always tells you the flames are coming, made the others run away. These flames made him wake up. He stood up, leaning on two young slaves, but he fell down straight away. I suppose the thick fumes had blocked his windpipe and closed his gullet, which was always weak and giving him trouble. When they found his body in the light two days later, there was not a mark on it.

Source: *Pliny: A Selection of his Letters*, trans. Clarence Greig (Cambridge University Press, London, 1978).

~

29 The marble and the mop

In 1934, physicists around the world were in a state of high excitement over the transmutation of elements. It was already known that the nuclei of some heavy atoms could capture a flying neutron and thereby form a new and heavier isotope [20]. The energy given up by the neutron on colliding with the nucleus was emitted as a gamma ray—the signature of the process. Enrico Fermi (1901–54), the great Italian physicist, had set in train a programme to examine systematically the behaviour of a range of elements under neutron bombardment. Elation at what had seemed an early success with a light element (sodium) was tempered by a baffling feature of the result: the gamma-ray emission occurred with a time delay far beyond what theory allowed. A better proof of neutron capture was required. Two of Fermi's bright young associates, Emilio Segrè and Edoardo Amaldi, thought they had settled the matter when they found that the next element they studied, aluminium, not only captured a neutron but in doing so gave rise to a radioactive isotope with a lifetime (measured by gamma-ray emission) of nearly three minutes. Fermi, delighted, reported the results at a meeting in London.

But then Segrè caught cold and stayed home for some days leaving Amaldi to continue the experiments. To general dismay he could not repeat the original observations. Fermi, greatly vexed at the prospect of a humiliating retraction, vented his displeasure on his juniors, who were now getting persistently erratic and, as it seemed, meaningless results. Then another remarkable young physicist joined the lab: Bruno Pontecorvo was to become famous 20 years later when he defected to the Soviet Union, taking with him valuable information about atomic weapons development. Pontecorvo and Amaldi set about calibrating the efficacy of neutron activations, using a standard of silver, which was known to generate on neutron capture an isotope with a conveniently long, easily measured, lifetime. To their astonishment and consternation they found that the results depended on where in the laboratory the measurements were made. Here is how Amaldi put it: 'In particular, there were certain wooden tables near a spectroscope in a dark room which had miraculous properties, since silver irradiated on those tables gained much more activity than when it was irradiated on a marble table in the same room.'

Here was a phenomenon that required investigation. The first step was to try shielding the apparatus with lead. But the experiment was suspended when Fermi's associates had to supervise student examinations. Fermi, ever impatient, decided to continue on his own. Here is how he described what happened, in a letter to his Chicago colleague of later years, the celebrated cosmologist Subrahmanyam Chandrasekhar:

> I will tell you how I came to make the discovery which I suppose is the most important one I have made. We were working very hard on the neutron-induced radioactivity and the results we were obtaining made no sense. One day, as I came into the laboratory, it occurred to me that I should examine the effect of placing a piece of lead before the incident neutrons. Instead of my usual custom, I took great pains to have the piece of lead precisely machined. I was clearly dissatisfied with something: I tried every excuse to postpone putting the piece of lead in its place. When finally, with some reluctance, I was going to put it in its place, I said to myself: 'No, I do not want this piece of lead here; what I want is a piece of paraffin.' It was just like that with no advance warning, no conscious prior reasoning. I immediately took some odd piece of paraffin and placed it where the piece of lead was to have been.

The result was a violent surge in activation of the target. Segrè and the others were summoned back to the laboratory to witness the startling effect. Segrè later wrote that he thought the radioactive counter must have gone wrong before being persuaded otherwise. Fermi ruminated over lunch, which he invariably took at home with his wife: if the paraffin caused such a huge effect, and the activation was also affected by whether the target sat on a wooden or a marble table, then perhaps the neutrons were being slowed down by collisions with hydrogen nuclei (protons, of the same mass as the neutrons), which abounded in paraffin and wood; and what if, contrary to the unquestioned prevailing assumption, it was slow rather than fast neutrons that were most easily captured?

Fermi returned to the laboratory and he and his team took their neutron source and silver target down to the pond in the garden. The hydrogen in the water and in the resident goldfish acted just like that in the paraffin wax. Other light elements were tried and also worked, though none as well as hydrogen with the lightest nucleus of all, which best accepted the momentum of the colliding neutron. A paper was at once written and despatched to the leading Italian physics journal, and opened a new chapter in the history of atomic physics (and the thinking that led to the atomic bomb). Hans Bethe [62], the famous theoretician, mused that the slow-neutron phenomenon might never have been discovered if Italy were not so rich in marble that it was used even for laboratory furniture.

But recently a new light has been thrown on the story. Two Italian physicists discovered that the caretaker, who had looked after the physics laboratory in 1934 and had witnessed the critical experiment in October of that year, was still alive in Fermi's centenary year of 2001. His recollection was that a cleaner by the name of Cesarina Marani, having mopped the marble hallway outside, had left three buckets of water under the laboratory bench. They were spotted by Fermi's young associates and the water vapour was quickly identified as a source of the all-important hydrogen.

The story is related in all its fascinating detail in Richard Rhodes's remarkable book, *The Making of the Atomic Bomb* (Simon & Schuster, New York, 1988).

~

30 Pythagoras's theme

Pythagoras of Samos (d. 510 BC), known to every schoolchild for the square on the hypotenuse, founded a great school of mathematics, dedicated to both practical and philosophical ends. The Pythagorean Brotherhood consisted of some 600 votaries, who abjured worldly ambitions and dedicated themselves to the advancement of learning. Pythagoras is now viewed by historians as a shadowy figure of uncertain identity (which appears to imply that Pythagoras's theorem may have been discovered not by Pythagoras but by another scholar of the same name). Yet Porphyry, the philosopher, wrote a biography of Pythagoras, admittedly some 800 years later, full of circumstantial detail. Pythagoras's fame rests not only on his mathematics but also on his discovery of the laws of music—the numerical relation between the intervals of the harmonic scale. According to legend and the account of one of Porphyry's followers, Iamblichus,

> he happened to pass by a brazier's shop, where he heard the hammers beating out a piece of iron on an anvil, producing sounds that harmonised, except one. But he recognised in these sounds the concord of the octave, the fifth and the fourth. He saw that the sound between the fourth and the fifth [the augmented fourth, as we would now call it] was a dissonance, and yet completed the greater sound among them.

Investigating further, he established, the tale continues, that the intervals between the notes struck by the different hammers were in proportion to their weights. He then supposedly went on to hang weights on a gut string and found that the same relation obtained between the weight (or tension in the string) and the note sounded when the string was plucked. He was also thereafter said to have used a monochord (a primitive single-stringed instrument) to demonstrate the relation between musical interval and length of the string, and so link music to the abstract world of numbers. This would have been seen as strong support for the Pythagorean doctrine that all observable phenomena are governed by the laws of mathematics.

One of the guiding principles of Pythagoras's system was the rationality of all numerical constants of nature (for instance, π, the ratio between the circumference and the diameter of a circle); that is to say, they must be expressible as a ratio of two whole numbers, and so the natural world could be described in all its aspects by integers and their ratios.

But Pythagoras was wrong. The story goes that Hippasus, a young student of Pythagoras, was seeking a rational expression for the square root of 2, when a proof came to him that there could be none—that the square root of 2, in fact, was irrational. Hippasus must have been delighted at such a fundamental discovery, but Pythagoras refused to countenance this subversion of his *Weltanschauung*, and, unable to refute Hippasus's argument, eliminated the problem by ordering Hippasus killed by drowning. 'The father of logic and the mathematical method', says Simon Singh, 'had resorted to force rather than admit he was wrong. Pythagoras' denial of irrational numbers is his most disgraceful act and perhaps the greatest tragedy of Greek mathematics. It was only after his death that irrational numbers could be safely resurrected.'

It is only fair to say that irrational numbers still exercise some mathematicians to this day. Two Russians in New York, the Chudnovsky brothers, have computed π to eight billion decimal places and are heading for a trillion, in the hope of detecting a recurring pattern.

See Simon Singh's highly readable and entertaining book, *Fermat's Last Theorem* (Fourth Estate, London, 1997).

~

31 New ways with barometers

The following is a question in a physics degree exam at the University of Copenhagen:

'Describe how to determine the height of a skyscraper with a barometer.'

One student replied: 'You tie a long piece of string to the neck of the barometer, then lower the barometer from the roof of the skyscraper to the ground. The length of the string plus the length of the barometer will equal the height of the building.'

This highly original answer so incensed the examiner that the student was failed. The student appealed on the grounds that his answer was indisputably correct, and the university appointed an independent arbiter to decide the case.

The arbiter judged that the answer was indeed correct, but did not display any noticeable knowledge of physics.

To resolve the problem it was decided to call the student in and allow him six minutes in which to provide a verbal answer which showed at least a minimal familiarity of the basic principles of physics. For five minutes the student sat in silence, forehead creased in thought. The arbiter reminded him that time was running out, to which the student replied that he had several answers, but didn't know which to use.

On being advised to hurry up the student replied as follows:

'Firstly, you could take the barometer up to the roof of the skyscraper, drop it over the edge, and measure the time it takes to reach the ground. The height of the building can then be worked out from the formula $H = 0.5gt^2$. But bad luck on the barometer.'

'Or if the sun is shining you could measure the height of the barometer, then set it on end and measure the length of the shadow. Then you could measure the length of the skyscraper's shadow, and thereafter it is a simple matter of proportional arithmetic to work out the height of the skyscraper.'

'But if you wanted to be highly scientific about it, you could tie a short piece of string to the barometer and swing it like a pendulum, first at ground level and then on the roof of the skyscraper. The height is worked out by the difference in the gravitational restoring force $T = 2\pi(l/g)^{1/2}$.'

'Or if the skyscraper has an outside emergency staircase, it would be easier to walk up it and mark off the height of the skyscraper in barometer lengths, then add them up.'

'If you merely wanted to be boring and orthodox about it, of course, you could use the barometer to measure the air pressure on the roof of the skyscraper and on the ground, and convert the difference in millibars into feet to give the height of the building.'

'But since we are constantly being exhorted to exercise independence of mind and apply scientific methods, undoubtedly the best way would be to knock on the janitor's door and say to him "If you would like a nice new barometer, I will give you this one if you tell me the height of this skyscraper".'

The student was Niels Bohr [79], the only Dane to win the Nobel Prize for Physics.

The story comes from an anonymous source on the Internet, and it is wrong on the last point, for Niels Bohr's son, Aage, won the prize in 1975. What precedes it may be true, although it is not mentioned in Abraham Pais's definitive biography. Certainly Bohr had the habit of thinking with unbreakable concentration before producing an answer to a question. Here is how the physicist James Franck (as quoted by Pais) remembers Bohr in action:

Sometimes he was sitting there almost like an idiot. His face became empty, his limbs were hanging down, and you would not know this man could even see. You would think he must be an idiot. There was absolutely no degree of life. Then suddenly one would see that a glow went up in him and a spark came, and he said: 'Now I know.' It is astonishing, this concentration . . . You have not seen Bohr in his early years. He could really get an empty face; everything, every movement was stopped. That was the important point of concentration. I am sure it was the same with Newton [15].

Bohr was, in the view of many, the deepest thinker of all in the rarified world of theoretical physics. In speech he seldom communicated his thoughts clearly to the listeners. His singular style of discourse in public lectures has been most vividly described by his protégé and friend Abraham Pais, who explains it thus:

The main reason was that he was in deep thought as he spoke. I remember how that day he had finished part of the argument, then said 'And . . . and . . . ', then was silent for at most a second, then said, 'But . . . ', and continued. Between the 'and' and the 'but' the next point had gone through his mind. However, he simply forgot to say it out loud and went on somewhere further down the road.

Here is another description of Bohr at the lectern:

It was in Edinburgh that I first heard the greatest, Niels Bohr. At the end of the session devoted to the foundations of quantum mechanics he made a brief but striking contribution.

I had shamelessly propelled myself into the front row because I did not want to miss a single word of what the great man was going to say; I had been warned that he was not easy to understand. (I learned later that at a large international conference with simultaneous translation, when Bohr spoke in 'English' there was another channel with simultaneous translation of his speech into . . . English.) He spoke for a few minutes in a low guttural voice, which was more like a deep whisper, hammering out each word with tremendous emphasis, and punctuating his speech from time to time with a gesture of the hand. Even a layman could not have missed the importance of the far-reaching conclusions which he was drawing from the day's session. I did not miss the importance but I missed the meaning; in fact, I did not understand one single sentence. When the applause subsided, I asked my neighbour, Leon Rosenfeld, a physicist of Belgian origin who spoke French, English, German, Danish and 'Bohr' [for he had worked with Bohr in Copenhagen as his principal assistant]: 'What did he say in his conclusion?' 'He said that we have had a long and interesting session, that everyone must be very tired, and that it is time for refreshment.'

Bohr, of course, was completely unaware of his deficiencies in the line of communication. Pais recalls that Bohr was utterly taken aback when a colleague dropped a hint: 'Imagine', he said incredulously to Pais, 'he thinks I am a bad lecturer.'

Bohr was universally loved and revered. He was a man of unyielding moral courage and intellectual honesty and totally devoid of vanity:

When Niels Bohr visited the Physics Institute of the Academy of Sciences of the USSR, to the question of how he had succeeded in creating a first-rate school of physicists he replied: 'Presumably because I was never embarrassed to confess to my students that I am a fool . . .'

On a later occasion when his colleague E. M. Lifshitz read out this sentence from a translation of the speech it emerged in the following form: 'Presumably because I was never embarrassed to declare to my students that they are fools . . .'

This sentence caused an animated reaction in the auditorium, then Lifshitz, looking at the text again, corrected himself and apologised for his accidental slip of the tongue. However, P. L. Kapitsa [170] who had been sitting in the hall very thoughtfully noted that this was not an accidental slip of the tongue. It accurately expressed the principal difference between the schools of Bohr and of Landau to which E. M. Lifshitz belonged.

Lev Davidovich Landau [137] was a great physicist, noted for arrogance and political recklessness. The multi-volume textbooks by Landau and Lifshitz remains a bible for physicists. Landau was arrested for his political indiscretions and would probably have lost his life but for Kapitsa's intercession with Stalin. The acid-tongued Landau met his match in Wolfgang Pauli [25]: after expounding his work to a sceptical Pauli, he angrily demanded whether Pauli thought his ideas were nonsense. 'Not at all, not at all', came the reply. 'Your ideas are so confused I cannot tell whether they are nonsense or not.'

For Bohr's biography, see Abraham Pais, *Niels Bohr's Times* (Oxford University Press, Oxford, 1991). The helpless listener of Bohr's lecture was the Russo-French physicist, Anatole Abragam, writing in *Time Reversal: An Autobiography* (Clarendon Press, Oxford, 1989). The story about Bohr and Lifshitz is quoted from a Russian publication, *Physicists Continue to Laugh* (MIR Publishing House, Moscow, 1968) in *A Random Walk in Science*, ed. R. L. Weber (The Institute of Physics, London, 1973).

~

32 The professor remembers

The American mathematician Norbert Wiener (1894–1964) was noted for his coruscating intellect, his childlike vanity and legendary unworldliness, and absence of mind. Wiener passed most of his life at the Massachusetts Institute of Technology, where tales of his strange behaviour still circulate.

Among the stories often told of his abstraction was his confusion one evening when trying to find his way home after he and his family had moved house. Accosting a small girl who was approaching in the opposite direction, he inquired whether she might not be able to direct him towards Brattle Street. The child giggled: 'Yes, daddy', she said, 'I'll take you home.'

Another oft-related instance was an encounter with a student who sought him out with a problem. Finding the professor in a deep reverie, he waited, but after an interval decided to intrude: 'Hello, Professor Wiener', he began. Wiener jumped: 'That's it!', he cried, 'Wiener!' More convincing perhaps is the image of Wiener slouching along the corridor, his nose in a book. It was his practice to keep to his course by sliding a finger along the wall. On one such occasion the finger led him through an open door, around the walls of a lecture room, in which a lecture was in progress, back to the door and out into the corridor again.

Wiener did not mention these proclivities in his memoirs. Stories abound, but a good portrait of the man can be found in the article by Steven G. Krantz, *The Mathematical Intelligencer*, 12, 32 (1990).

~

33 Dalton's daltonism

John Dalton (1766–1844) was a Mancunian, who, in the closing years of the eighteenth century, gave rational expression to the atomic theory of matter. His conclusions were based on careful experimentation on combining weights of elements, from which he inferred that the atoms of any element were identical and that they combined in fixed proportions with those of other elements.

Dalton was a Quaker, who was supposed to have scandalized his soberly attired co-religionists by appearing in the streets of Manchester in his scarlet doctor's gown. For Dalton was colour-blind and, indeed, gave his name to the French expression for the condition—*le daltonisme*—and to one who suffers from it—a *daltonien*. Here is how he eventually discovered his disability and its hereditary nature:

I was always of the opinion, though I might not often mention it, that several colours were injudiciously named. The term *pink*, in reference to the flower of that name, seemed proper enough; but when the term *red* was substituted for pink, I thought it highly improper; it should have been *blue*, in my apprehension, as pink and blue appear to me very nearly allied [the pink in question must have been closer to mauve, for Dalton would have been insensitive to the red component]; whilst pink and red have scarcely any relation.

In the course of my application to the sciences, that of optics necessarily claimed attention; and I became pretty well acquainted with the theory of light and colours before I was apprized of any peculiarity in my vision. I had not, however, attended much to the practical discrimination of colours, owing, in some degree, to what I conceived to be a perplexity in their nomenclature. Since the year 1790, the occasional study of botany obliged me to attend more to colours than before. With respect to the colours that were *white, yellow*, or *green*, I readily assented to the appropriate term. *Blue, purple, pink*, and *crimson* appeared rather less distinguishable; being according to my idea, all referable to *blue*. I have often seriously asked a person whether a flower was blue or pink, but was generally considered to be in jest. Notwithstanding this, I was never convinced of a peculiarity in my vision, till I accidentally observed the colour of the flower of the *Geranium zonale* by candle-light, in the autumn of 1792. The flower was pink, but it appeared to me almost an exact sky-blue by day; in candle-light, however, it was astonishingly changed, not having then any blue in it, but being what I call red, a colour which forms a striking contrast to blue. [It would have appeared essentially grey or black]. Not then doubting but that the change of colour would be equal to all, I requested some of my friends to observe the phenomenon; when I was surprised to find that they all agreed, that the colour was not materially different from what it was by day-light, except my brother, who saw it in the same light as myself. This observation clearly proved, that my vision was not like that of other persons.

The story of Dalton's colour-blindness had to wait a century and a half for its denouement. Dalton's theory was that he viewed the world through a blue filter—that his vitreous humour (the gelatinous substance inside the eyeball) would indeed be blue. He therefore gave instructions that on his death his assistant, Joseph Ransome, should remove his eyes and test the conjecture. Ransome did as he was bid: he cut open one eyeball and expelled its contents into a watch-glass, but the vitreous was 'perfectly pellucid'. He then made a hole in the other eye and peered through it to see whether red and green appeared identical and grey. The result again was negative and Ransome inferred that the defect must have lain in the optic nerve connecting the retina to the brain.

The mutilated eyeballs were then put in a jar of preservative and left in the care of the Manchester Literary and Philosophical Society. There they reposed undisturbed until in 1995 a group of Cambridge physiologists begged leave from the society to take a small sample of a retina in order to extract and amplify the DNA by the polymerase chain reaction, or PCR [108], and examine the genes (by then already fully characterized) of the three types of retinal cones implicated in colour vision. (The cone types contain pigments of different wavelength-sensitivity, a vindication of the three-colour theory of colour perception advanced at the end of the eighteenth century by Dr Thomas Young.) Dalton, it transpired, was indeed a 'deuterope', with a defect in the middle-wavelength optical pigment (and not, as Thomas Young thought, a 'protanope' with a defect in the short-wavelength pigment). Dalton would no doubt have been well pleased by the result so long after his death.

Dalton's cogitations can be found in *Memoirs of the Manchester Literary and Philosophical Society*, 5, 28 (1798); the report on Dalton's genetic defect is by D. M. Hunt *et al.*, *Science* 267, 984 (1995).

~

34 The trick of the tick

David Keilin (1887–1963) was a highly respected Russian-born biologist, who spent most of his working life in Cambridge. His fame rested on his early studies of parasitic insects and, more especially, on his investigations of the iron-containing haem proteins. The following reminiscence comes from an affectionate memoir of Keilin by Max Perutz [88].

> In 1931, when Keilin had succeeded Nuttall as Quick Professor of Biology, an elderly lecturer called Warburton complained that he had been appointed before the University Superannuation Scheme had been instituted, so that he had no pension and would have to die in harness. When Keilin told the University Treasurer that Warburton was in his late seventies and had no pension, the Treasurer agreed that in view of Warburton's advanced age the University could afford to be generous. He failed to foresee that 24 years later we would celebrate Warburton's 100th birthday! On that occasion he told us a wonderful story. In his prime Warburton had been the world's authority on ticks. One day in the twenties, some of his students were eating their lunch of

bread and cheese when they found a tick in their butter. They brought it to Warburton who identified it as a Siberian tick. That discovery was to provoke a diplomatic crisis. The students had bought their butter at Sainsbury's, not knowing where it came from. Impressed by entomology's detective powers that could trace the butter's origin to Russia, they told their story to a don who mentioned it to a visiting MP and he in turn related it to a journalist. The outcome was a headline in one of the London evening papers: 'Disease-Carrying Tick Imported with Russian Butter'. Questions were asked in Parliament, the horse-drawn milk carts which in those days also distributed butter in London bore placards reassuring housewives that they carried no Russian butter, the Soviet Ambassador called on the Foreign Secretary to protest against the campaign of slander against his country's agricultural exports and Pravda condemned Warburton's deliberate lies. Years later, Russian parasitologists visiting the Molteno Institute reproached Keilin for allowing it to become a tool of anti-Soviet propaganda and refused to believe that Warburton was just an unworldly scholar who had happened to come across a curiosity. Secure in his generous pension, Warburton continued to live in good health in Grantchester to the ripe old age of 103.

M. F. Perutz, 'Keilin and the Molteno' in *Selected Topics in the History of Biochemistry: Personal Recollections, V* (Comprehensive Biochemistry Vol. 40), ed. G. Semenza and R. Jaenicke (Elsevier Science, Amsterdam, 1997).

~

35 Comfort in adversity

Physical handicap has been known to ignite, rather than extinguish, the ambition to learn and discover. Stephen Hawking, cosmologist and occupant of Newton's Lucasian Chair of Mathematics in Cambridge, embodies this truth. Consider also Solomon Lefschetz (1884–1972), the much-admired American topologist, who, destined to be an engineer until both his hands were severed in a laboratory accident, turned instead to mathematics. He was fitted with prosthetic claws, always sheathed in black gloves. A student was charged with thrusting a stick of chalk into one claw at the beginning of the working day and taking out the stub at its conclusion.

Prison and the madhouse are scarcely the most congenial environments for the advancement of learning. Yet science has flowered after a fashion in internment and prisoner-of-war camps, and even captives in solitary confinement have added new chapters to human knowledge. Jean Victor

Poncelet (1788–1867), a French mathematician, was perhaps the most celebrated of these indomitable spirits. An officer in the Corps of Military Engineers in the Napoleonic campaigns, he was captured by the Russians after a skirmish during the retreat from Moscow in 1812 and incarcerated in a camp at Saratov on the Volga river. There he remained for nearly two years. To divert himself he turned his mind to his youthful interest, mathematics, and especially geometry. Having, in the absence of books, reconstructed the elements of the subject from the beginning, he set out on a programme of research on projections of conic forms. This set the course for his most important later work, which he pursued after his release, while continuing to serve as a military engineer—a specialist in fortifications. Late in his life he published his definitive work, *Applications d'analyse et de la géométrie*, the first volume of which bore the title, *Cahiers de Saratov*.

Another notable prisoner was one of the founders of geological science: Déodat de Gratet de Dolomieu, who gave his name to the Dolomites. Dolomieu was born in 1750 into a French military family and was destined for the army, but instead entered the religious Sovereign and Military Order of the Knights of Malta. He was evidently of a fiery disposition for in 1768 he killed a brother-officer in a duel. Condemned to life imprisonment, he was reprieved by the intervention of the Pope, who secured his release. But the Grand Master of the Order did not welcome his turbulent follower, and Dolomieu was despatched to the military garrison at Metz. There he had leisure enough to study, and, under the tutelage of an apothecary, he devoted himself to the sciences, and particularly to geology, to such effect that he was soon elected a corresponding member of the Académie des Sciences.

While in Metz Dolomieu had the good fortune to fall in with two powerful patrons, the Duc de la Rochefoucauld and the Prince de Rohan. The Duke encouraged his young protégé's interest in geology and Dolomieu soon began his study of rock formations, especially those of basaltic rocks. When de Rohan was appointed ambassador to Portugal he took Dolomieu as his secretary. The duties were evidently not arduous, for it was during this period that Dolomieu carried out some of his most important work. He welcomed the Revolution, but his ardour was abruptly extinguished by the brutal murder of Rochefoucauld. The Republic nevertheless offered him an appointment at the École des Mines, where (apart from an interval, in company with many other leading men of science, on Napoleon's Egyptian expedition) he remained for 15 years, inspecting mines and geologizing. But he was then called on to assist Napoleon in wresting Malta from the Order of Knights, and on the return journey his ship was driven into Taranto, where he was seized by Calabrian revolutionaries and delivered into the hands of his enemies, the Knights of Malta.

For 21 months Dolomieu was kept in solitary confinement in Messina, and in those harsh conditions he pondered and wrote. On his eventual release in 1801 and return to Paris he was greeted with public demonstrations, much as Arago had been after his imprisonment by another hostile power [166]. While in prison Dolomieu had been elected to a chair at the Muséum National d'Histoire Naturelle in Paris, but his health had suffered and he died the same year, a celibate, in accordance with his vows as a member of the Order, which he never renounced.

Perhaps the strangest and most tragic case of a working life spent in confinement is that of the mathematician André Bloch. He was born in Besançon in 1893, one of three brothers of Jewish parents. Orphaned at an early age, André and the youngest of the brothers, Georges, showed outstanding talents and both gained places by competitive examination in the École Polytechnique in Paris. Their studies were interrupted by the First World War, in which Georges was severely wounded and lost an eye, while André, serving as an artillery officer, fell from an observation post under shell-fire. After several spells in hospital, he was given indefinite leave in 1917 and resumed his studies at the École.

Then in November of that year, while dining *en famille* in Paris, he fell on his brother Georges, and his aunt and uncle with a knife and stabbed them to death. He then ran into the street, shouting, and was arrested without resistance. With the country in the grip of a desperate war, the affair, which after all involved two army officers, was kept quiet, and the perpetrator was immured in a psychiatric hospital, the Maison de Charenton, in a suburb of Paris. There he remained until his death from leukaemia in 1948.

To a doctor at the Charenton, André Bloch calmly explained that he had had no option but to eliminate a branch of his family that had been touched with mental illness. The laws of eugenics, he insisted, were ineluctable and it had been his duty to act as he did. He chided the doctor for his emotional reaction; 'You know very well', he declared, 'that my philosophy is inspired by pragmatism and absolute rationality. I have applied the example and principles of a celebrated mathematician from Alexandria,

Hypatia.' There is, of course, no evidence that Hypatia [168] entertained any such radical notions, nor was it ever established whether Bloch's derangement stemmed from his experience in the war. But in all other respects he appeared perfectly sane, and from his cell in the Charenton there proceeded a series of important mathematical communications— mainly in algebraic analysis, number theory, and geometry, although he also wrote a paper on the mathematics of tides. One paper was prepared with another mathematician, confined for a period in the Maison de Charenton.

Bloch's achievements are the more remarkable in that he was com- pletely self-taught and only later established contacts through letters and rare visits with some of the leading mathematicians of the day (who were initially unaware that they were corresponding with an inmate of an asy- lum). He also developed an interest in economic theory and wrote several letters to President Poincaré (kinsman of the celebrated mathematician and physicist, Jules Henri Poincaré) with suggestions for the management of the national economy. During the German occupation in the Second World War, Bloch was shrewd enough to conceal his Jewish name and published under two pseudonyms. In the year of his death he was awarded the Becquerel Prize of the Académie des Sciences. The story of 'the mathe- matician of Charenton', as a prominent French psychiatrist called him, irresistibly recalls the Surgeon of Crowthorne (the subject of a book with that title by Simon Winchester published by Penguin Books in 1999)—the paranoid doctor, who, having murdered an innocent passer-by in a London street, contributed with such deep learning and dedication to the first *Oxford English Dictionary* from his cell in an asylum during the latter part of the nineteenth century.

A reminiscence of Solomon Lefschetz is included in the entertaining article by Steven G. Krantz in *The Mathematical Intelligencer*, 12, 32 (1990). For the life and work of Jean Poncelet see René Taton in *Dictionary of Scientific Biography*, ed. C. C. Gillespie (Scribner, New York, 1975), and for the biography of Déodat Dolomieu see Kenneth L. Taylor, also in the *DSB* (1971). The facts of the tragic life of André Bloch are recorded in an absorbing article by two French mathematicians, Henri Cartan (whose father, the famous Élie, corresponded with Bloch) and Jacqueline Ferrand, *The Mathematical Intelligencer*, 10, 23 (1988).

36 Winter in Paris: Becquerel and the discovery of radioactivity

Henri Becquerel (1852–1908) was a member of an august scientific dynasty, the third of his line to hold the Chair of Physics at the Muséum National d'Histoire Naturelle in Paris (in which his son, Jean, later succeeded him).

In 1896, Becquerel was busy pursuing a mirage. Like all physicists he had been shaken and impressed by Röntgen's discovery of X-rays [4]. If cathode rays striking glass could elicit a secondary radiation, might not visible light do the same when it fell on a phosphorescent material? To test this wholly incorrect conjecture Becquerel chose a crystal of a uranium compound as a phosphor. He placed a cross made of copper on a photographic plate, enclosed the whole in black paper, laid the crystal on top, and allowed sunlight to fall on it. And, indeed, when the plate was developed it revealed an exposed area, on which the white outline of the cross could be plainly discerned.

Becquerel must have been delighted with a result that seemed to confirm his theory. Then, like any conscientious experimenter, he set out to repeat his triumphant observation. But it was February and the sun did not shine in Paris, so Becquerel put the assembly—photographic plate and copper cross in black paper, the crystal of uranium salt on top—in a drawer, where it reposed for several days. Now phosphorescence generally endures for some time—a luminous watch dial will continue to glow in the night after exposure to sunshine hours before—so Becquerel developed the plate to see whether weak blackening might still have occurred, or so at least one supposes. Sir William Crookes, an English physicist who happened to be visiting the laboratory when these events unfolded, wrote that after some days of dull weather, his host developed the plate because 'he was tired of waiting (or with the unconscious prevision of genius)'. What Becquerel saw, at all events, was a picture no less intense than had, as he had inferred, been generated by the effect of sunlight.

Becquerel realized that whatever was blackening his plates had nothing to do with sunshine and he went on to show that other uranium compounds exerted a like effect—all but one and this was the mineral, pitchblende, which proved vastly more potent. The inference was that this contained another substance with much higher radioactivity, as Pierre and Marie Curie [9] later termed it. Becquerel subsequently found that he could observe the radiation by placing the source next to an electroscope, a simple instrument which responds to an induced electrostatic charge. Becquerel's radiation charged up the metal conductor in the electroscope, which meant that it was generating ions (charged particles) in the air as it passed. Yet he never recognized the significance of his discovery and remained wedded to the conviction that what he had observed was a new and unusual form of phosphorescence, arising, that is to say, from the emission (after storage in the molecule) of energy in the form of visible light. It remained for the Curies to track down the origin of the emissions and for Ernest Rutherford [16] in Cambridge to identify their nature.

A curious historical footnote to Becquerel's discovery is the long-forgotten fact of a similar observation made in Paris 40 years earlier. Abel Niepce de Saint-Victor achieved fame for innovations in photography, especially his invention of the albumin print. His interests in the properties of light and in the chemistry of colour led him to the laboratory of the veteran organic chemist Michel Eugène Chevreul. Chevreul, Director of the Natural History Museum, was scientific adviser to the Gobelin tapestry works and influenced the theories of Seurat and the *pointilliste* school of painters. (Chevreul had perhaps the longest career in the history of science, for he was active up to his death at the age of 103.) With Chevreul's encouragement, Niepce undertook a study of fluorescent and phosphorescent substances, and in 1857 reported that a drawing on cardboard with a solution of uranium nitrate would produce an image of itself on sensitized (silver chloride-impregnated) paper—a kind of primitive photographic film. The trick worked with the drawing in darkness, as much as in sunlight; more, it would do so at a distance of up to 3 centimetres from the paper. Several reports were published on the phenomenon, from 1857 onwards, and excited considerable interest, not least on the part of Henri Becquerel's father, Edmond. Had Henri forgotten about Niepce de Saint-Victor by 1896 when he performed his fortuitous experiment? And was he perhaps influenced by some dim recollection when he developed his photographic plate?

The discovery of radioactivity, like that of X-rays, aroused some scepticism. The English physiologist Sir Henry Dale (1875–1968) recalled a meeting of the Cambridge Natural Science Club at which the Hon. R. J. Strutt, son of the great Lord Rayleigh, and himself later a considerable physicist, spoke about Becquerel's observations; his remarks drew the following comment from an undergraduate, a future theoretician of note: 'Why,

Strutt, if this story of Becquerel's were true it would violate the law of the conservation of energy!' This remark cut to the heart of a critical issue [16]. It was only when the nature of radioactivity—the transmutation of a radioactive into an inert element (often through other radioactive intermediates) until no radioactivity remains (even if this should take years or millennia)—was understood that the paradox vanished.

There have been many descriptions of Becquerel's discovery, as well as, recently of Niepce de Saint-Victor's earlier work. See, for instance, a paper, 'Hasard ou mémoire dans la découverte de la radioactivité', by P. and J. Fournier, *Revue de l'Histoire des Sciences* 52, 51 (1999). For the exchange with R. J. Strutt, see H. H. Dale, *British Medical Journal* ii, 451 (1948).

~

37 The unbreakable cypher

Otto Frisch [20], the German physicist, was in Birmingham, England, at the beginning of the Second World War, preoccupied with the possibilities of achieving a nuclear fission bomb and denying the German side access to the Norwegian stocks of heavy water (deuterium oxide), which would be required for building such a weapon. A committee was set up to study the prospects, and this anecdote of Frisch's illustrates the overwrought temper of the time.

The report which Peierls [another emigré physicist in Birmingham] [42] and I had sent to (Sir) Henry Tizard on Oliphant's advice had triggered the formation of a committee, with (Sir) George Thomson [son of J. J., the discoverer of the electron] as chairman, which was given the code name 'Maud Committee'. The reason for that name was a telegram which had arrived from Niels Bohr [79], ending with the mysterious words 'AND TELL MAUD RAY KENT'. We were all convinced that this was a code, possibly an anagram, warning us of something or other. We tried to arrange the letters in different ways and came out with mis-spelt solutions like 'Radium taken', presumably by the Nazis, and 'U and D may react', meant to point out that one could get a chain reaction by using uranium in combination with heavy water, a compound of oxygen and the heavy hydrogen isotope called deuterium, abbreviated D. [Frisch does not mention that a cryptographer was enlisted to study the problem and came up with 'Make Ur day nt'.] The mystery was not cleared up until after the war when we learned that Maud Ray used to be a governess in Bohr's house and lived in Kent.

A somewhat similar misunderstanding led to the detention, and nearly the death, of André Weil, the French mathematician, in Finland in 1939. He and his wife had entered the country in June of that year to visit two Finnish mathematician friends. On 30 November Weil, alone now, his wife having departed on a tourist trip to the north, was seized by the police. That day the Russo-Finnish war had broken out, the first bombs had fallen on Helsinki, and Weil had sought safety during the raid outside the city. The raid over, he made for his hotel, but stopped to turn his myopic gaze on what turned out to be a group of soldiers manning anti-aircraft machine-guns. Weil's clothes proclaimed him a foreigner and he was at once arrested on suspicion of spying for the Soviet Union.

In his cell Weil consoled himself with the recollection that Sophus Lie, the great Norwegian mathematician, had been similarly imprisoned as a spy—while visiting Paris during the Franco-Prussian war in 1870. A search of Weil's effects brought forth a letter in Russian from one mathematician, and a second, which ended: 'I hope your illustrious colleague M. Bourbaki will continue sending me the proofs of his magisterial work.' Weil was indeed a member of the famous group of French mathematicians who met in a café in Paris and published collectively under the name N. Bourbaki, with the 'Poldevian Academy of Sciences' as byline. (Poldevia was a country invented earlier by a group of student hoaxers from the École Normale in Paris.)

To the police such gibberish implied a coded message. Weil was unable to reach his hosts, but one of them, the mathematician Rolf Nevanlinna, was a colonel in the reserve and had been called to the colours. By good fortune he found himself at dinner one evening with the chief of the Helsinki police, who cheerfully revealed to him that they were planning the next morning to execute a spy—a spy, in fact, who had claimed to know Nevanlinna. With some difficulty Colonel Nevanlinna prevailed on the policeman to take Weil to the border and expel him from the country instead of shooting him. It was an option that had not occurred to the guardian of the law. Weil was despatched in a sealed railway compartment and released at a Swedish frontier post. From there he managed eventually to reach England, where he was again arrested for failure to report for military duty in France. Sent back to face the French police, he spent some time in prison (during which he was struck by a mathematical inspiration and was able to write a paper for publication) and was then tried by a military court, which contented itself with putting him into uniform for the duration.

The affair of the Maud Committee is recounted by Otto Frisch in *What Little I Remember* (Cambridge University Press, Cambridge, 1979); the wartime tribulations of André Weil are described in his memoir, *The Apprenticeship of a Mathematician* (Birkhäuser, Basel, 1992); see also O. Pekonen in *The Mathematical Intelligencer*, 21, 16 (1999).

~

38 Two hundred monks aleaping and the devil in the bottle

The discovery of electricity in the eighteenth century caused great excitement, and attracted not only scholars, but the public and a swarm of charlatans. An appetite grew for spectacular scientific entertainments. So Stephen Gray, for instance, who expounded his work on conductors and insulators before the Royal Society of London in 1720, took to performing demonstrations on human subjects, and especially the boys in the Charterhouse charitable foundation. He 'caught an urchin, hung him up with insulating cords, electrified him by contact with rubbed glass, and drew sparks from his nose'.

Such public diversions became all the rage, and led in due course to the belief that electrical discharges might have therapeutic properties and might even revive the recently dead (as in Mary Shelley's novel, *Frankenstein*). Considerable voltages were generated by rapidly revolving glass cylinders, and then Pieter van Musschenbroek (1692–1761) in Holland invented the famous Leyden jar. This was a water-filled vessel, coated with conducting material on both the inside and outside. A wire poked through a stopper in a hole at the top and when this was brought into contact with a charged object (a rubbed glass rod, for example) charge would be conducted down the wire to accumulate in the jar. Discharges from such jars could attain considerable power. Indeed, van Musschenbroek warned against the danger after he had himself experienced a traumatic shock.

A device of this kind was used by the Abbé Nollet, court electrician to Louis XV, to conduct a series of telling experiments. Nollet was a vain and truculent man, who later engaged in a prolonged polemic with Benjamin Franklin [47], the engaging American who became a favourite at the royal court, and thus inflamed Nollet's jealousy.

Commanded by Louis XV to demonstrate the wonders of electricity, Nollet set up his apparatus in the Palace of Versailles. Here, one day in

1746, 148 French Guards were paraded in the Grande Galerie, and ordered to link hands with the man on either side. The first and the last in the line then grasped a metal wire which was attached to Nollet's apparatus. When all was ready the accumulated charge was sent through the wire; all 148 guardsmen jumped simultaneously as the shock hit them. The demonstration was surpassed when a group of Carthusian monks in Paris were formed up for Nollet in a line 900 feet long, each man connected to the next by a length of iron wire. Like the guardsmen, when the current was applied all leapt in synchrony like a corps de ballet. 'The exclamations of surprise were simultaneous,' reported Nollet, 'even though they came from two hundred mouths.'

The success of this rudimentary timing mechanism led to an important conclusion, for it showed that the electricity was instantaneously transmitted over the distance of 900 feet. It took another century for James Clerk Maxwell [44] to determine that electricity travels at the same speed as light.

It was six years after Nollet's display that Benjamin Franklin 'fetched lightning out of the sky' with a key, attached to a wire trailing from a kite. Franklin had apparently been drawn to the study of electricity by an article in the London *Gentleman's Magazine*, which had reached him in Philadelphia; it was written by Albrecht von Haller, a Swedish biologist, and described a variant of Stephen Gray's experiment, using a boy as capacitor. The boy in this arrangement stood on an insulating bed of pitch and was charged up by an electrical machine. When approached by a third party a crackling discharge would pass between them and both would experience a sharp spasm of pain. Franklin pondered the implications of this phenomenon, devised other similar experiments and soon began to wonder whether such discharges resembled natural lightning.

Franklin's experiments with lightning were insanely dangerous. The most famous, aimed at demonstrating the principle of the lightning conductor, was conducted at Marly-le-Ville near Paris. There he erected a long metal spike and when the thunder clouds gathered above he instructed the local caretaker, a retired soldier, to touch the metal with a wire, the end of which rested in a glass bottle. There was a fierce spark, a savage crackling and hissing, and a sulphurous smell arose. The terrified veteran dropped the bottle and ran to find the village priest, for he had drawn the conclusion that the devil must have appeared in person. The following year a professor at the University of St Petersburg, G. W. Richmann, tried the same experiment, driven by the conviction that 'in these times even the physicist has an opportunity to display his fortitude', and was duly killed. The practical value of Franklin's discovery was quickly recognized and secured him the devotion of Louis XV and the hatred of the Abbé Nollet, who had been convinced, until Franklin's arrival in Paris, that the American was a fictitious figure, invented by Nollet's many enemies—such as Buffon [118]—just to vex him.

The passage about Nollet's experiments is taken from Ronald W. Clark's *Benjamin Franklin: A Biography* (Weidenfeld and Nicolson, London, 1983). See also the excellent article by J. L. Heilbron, 'Franklin's physics', in *History of Physics*, ed. Spencer W. Weart and Melba Phillips (American Institute of Physics, New York, 1985).

~

39 The success of the operation and the death of the patient

It was Francis Bacon (1561–1626) who was commonly credited with the first attempt to formulate a 'scientific method'. To understand nature it was first necessary to purge the mind of preconceptions. Truth was to be sought by inductive reasoning, by maintaining rigorous scepticism, and devising experiments to test all inferences. It was Bacon's dedication to the experimental approach that did for him in the end.

Francis Bacon was a wily politician, who was ennobled for his services to the Crown, but he attracted enemies, in part at least for his too-zealous investigations into the workings of nature—for 'being too prying into the then receiv'd philosophy', as Robert Hooke [63] later wrote. And so he fell from grace; accused of corruption, he was stripped of his public offices and banished from London. After the death of King James, Charles I relaxed the restrictions and permitted Bacon to visit London. On such an occasion one snowy day in March of 1626, Bacon was travelling in a coach with the King's physician; the conversation turned to the effect of cold on the preservation of food. Could meat be preserved in ice as effectively as in salt? Bacon and his companion resolved to try an experiment: at Highgate, then a village north of London, they stopped the coach and purchased a chicken from a woman, who killed and gutted it. The two men stuffed the carcass with snow and packed it in more snow.

The exercise chilled Bacon, who took sick. He was taken to the nearby home of the Earl of Arundel, who was away, serving time in the Tower of London. Bacon was put to bed, but the bed, despite the application of a warming pan, was damp. Bacon had probably caught pneumonia and he wrote a last letter to the Earl, recording that though he was now mortally ill the experiment with the chicken had 'succeeded excellently well'. Some hours later he was dead, truly a martyr to science.

For an account of Bacon's life, work, and death by Mary Hesse, see the *Dictionary of Scientific Biography*, Vol. 1, ed. C. C. Gillespie (Scribner, New York, 1970).

~

40 Pendant professor

Ernest Shackleton's expedition to the Antarctic in 1908 was accompanied by two intrepid geologists, who aimed to determine the position of the south magnetic pole. They were Edgeworth David, Professor at the University of Sydney, and Douglas Mawson, later Professor of Geology in Adelaide. In pursuit of their quarry they traversed glaciers and made observations of the local formations. Here is how Mawson described what happened one day, after they had pitched tent. David took his sketchbook and went to record the outline of a range of hills, while Mawson sat in the tent changing the photographic plates in the plate-holder of his camera. To do this he snuggled into his sleeping bag with the photographic paraphernalia and was hard at work, when

> I heard a voice from outside—a gentle voice—calling:
> 'Mawson, Mawson.'
> 'Hullo!' said I.
> 'Oh, you're in the bag changing plates, are you?'
> 'Yes, Professor.'
> There was a silence for some time. Then I heard the Professor calling in a louder tone:
> 'Mawson!'
> I answered again. Well the Professor heard by the sound I was still in the bag, so he said:
> 'Oh, still changing plates, are you?'
> 'Yes.'
> More silence for some time. After a minute, in a rather loud and anxious tone:

'Mawson!'

I thought there was something up, but could not tell what he was after. I was getting rather tired of it and called out:

'Hullo. What is it? What can I do?'

'Well, Mawson, I am in a rather dangerous position. I am really hanging on by my fingers to the edge of a crevasse, and I don't think I can hang on much longer. I shall have to trouble you to come out and assist me.'

I came out rather quicker than I can say. There was the Professor, just his head showing and hanging on to the edge of a dangerous crevasse.

The Professor was pulled out and the expedition resumed the next day, climbing over two glaciers and making 10 miles a day with their sledge. Mawson inferred from the variability of the compass bearing that they were now practically on top of the magnetic pole, but David decided they should continue to the estimated position, another 13 miles away. After a day's forced march they pitched their tent. Then the next morning

We were up at about 6 a.m. and after breakfast we pulled our sledge for two miles. We then deposited all our heavy gear and pulled on for two miles and fixed up the legs of the dip circle to guide us back on our track, the compass moving in a horizontal plane being now useless for keeping us on our course. At two miles further we fixed up the legs of the theodolite, and two miles further put up our tent, and had a light lunch. We then walked five miles in the direction of the Magnetic Pole so as to place us in the mean position calculated for it by Mawson, 72°25′ South latitude, 155°16′ East longitude. Mawson placed his camera so as to focus the whole group, and arranged a trigger which could be released by means of a string held in our hands so as to make the exposure by means of a focal plane shutter. Meanwhile, Mackay [botanist and the third member of the party] and I fixed up the flag-pole. We then bared our heads and hoisted the Union Jack at 8.30 p.m. with the words uttered by myself, in conformity with Lieutenant Shackleton's instructions. 'I hereby take possession of the area now containing the Magnetic Pole for the British Empire.' At the same time I fired the trigger of the camera by pulling the string. We then gave three cheers for His Majesty the King.

One would probably not see the like at the conclusion of such a scientific quest today.

See A Geological Miscellany, ed. G. Y. Craig and E. J. Jones (Orbital Press, Oxford, 1982; Princeton University Press, Princeton 1985).

41 Feud

Public brawls between scientists are rare today; animosities are conducted by way of a little discreet disparagement in the committee room when the rival's grant comes up for discussion. In more uninhibited times learned disputes could make newspaper headlines. So it was with the two most prominent American palaeontologists of their time, Edward Drinker Cope (1840–97) and Othniel Charles Marsh (1831–99).

The late nineteenth century was the high noon of the fossil hunters; Cope, professor at the University of Pennsylvania, and Marsh, a professor at Yale and President of the National Academy of Sciences, were of their number. The excitement stemmed partly from the debate ignited some years before by the publication of Charles Darwin's *On the Origin of Species*, which provoked a search for evidences of evolution from fossils, and partly from the discovery of dinosaur remains in increasing abundance. A violent animus developed between Marsh and Cope, who had started out close professional allies. The origins are unclear, but the sincerity of the opinions that they held of each others' characters and abilities is not in doubt. One of his associates recorded the following rumination by Cope: 'One day . . . he slyly opened the lower right-hand drawer of his study table and said to me: "Osborn, here is my accumulated store of Marshiana. In these papers I have a full record of Marsh's errors from the very beginning, which at some future time I may be tempted to publish." ' The quarrel became public; accusations of poaching and downright theft of specimens, of plagiarism and other forms of skulduggery flew, and broke surface in the press in 1890. The following account details some examples of what the two professors got up to:

> Cope had an uncanny visual memory [which, however, did not always lead him to the right conclusion, for in a reconstructive drawing of a pleiosaurus he had placed the head at the wrong end and had been humiliatingly put right in print by Marsh]. I remember Leonard Stejneger, late Head Curator of Biology in the U.S. National Museum, telling me that Cope stood looking over his shoulder at a curious little lizard which the old collector, John Xanthus, had sent from Lower California. Stejneger was studying this lizard when Cope entered the room, indeed he was preparing to write out his description, for nothing like it had ever been known before. Cope glanced at the specimen for a few moments, put on his coat, walked to the telegraph office, and wired a perfectly accurate description of the beast to the *American*

Naturalist, thus gleefully stealing the credit of the discovery for himself.

In addition to their own efforts Marsh and Cope employed other collectors who travelled far and wide, gathering fossils and ruthlessly destroying material which they did not have time to take up before the approach of winter so that no rival would chance to find it subsequently.

Samuel Garman was a protégé of Alexander Agassiz, who was also sent out into the field. On one trip he reached Fort Laramie just as Professor Marsh brought in a collection which was to be shipped east. As I remember it, neither Marsh nor Garman knew that Cope was in town. Since lodgings were scarce, Garman bunked down in the empty station.

Late one night, after he had turned in, he heard someone stealthily enter the room. The intruder made a careful examination of the slatted crates containing Marsh's material. This went on for some time, then at last the figure departed empty-handed. In the morning Marsh arrived. Garman described what had happened and Marsh said, 'Oh, I foresaw that possibility. That was Cope. He likes to describe from skulls, and all the good skulls I got this season are in the stove.' Marsh then went to the stove, opened the door, extracted a bushel or so of treasures, wrapped them, boarded the train, and went east with the cream of his catch. Marsh didn't dare keep them in his lodging, but put them for safe keeping in a place where he felt sure they would be undisturbed—as they were.

From Thomas Barbour's *Naturalist at Large* (Little, Brown, New York, 1943); see also *The Bone Hunters' Revenge* by D. R. Wallace (Houghton Mifflin, Boston and New York, 1999) and Hal Hellman's *Great Feuds in Science* (Wiley, New York, 1998).

~

42 The man of few words

Paul Adrien Maurice Dirac (1902–84) was one of the giants of twentieth-century physics. 'There is no God and Dirac is his prophet', Wolfgang Pauli used to say. He was venerated as a theoretician of incomparable insight. Elegance was said to lurk in his equations. When Emilio Segrè and Enrico Fermi [29] were teasing each other about their accomplishments, Segrè challenged his illustrious colleague like this: 'I bet you would exchange all your life's work for one paper of Dirac's'. Fermi thought for a moment, and then answered, 'yes'.

Dirac occupied the Lucasian Chair of Mathematics in Cambridge, once graced by Isaac Newton, and many legends sprang up around him. He was a kind man, noted for his extreme economy with words. His vocabulary in conversation was in general limited to 'yes', 'no', and 'I don't know'. On one famous occasion, in the discussion after one of Dirac's seminars, a questioner began: 'Professor Dirac, I did not quite understand your derivation of . . .' When he had finished his question a long silence ensued. Would the lecturer answer the question, the chairman finally asked. 'It wasn't a question', was Dirac's rejoinder, 'it was a statement.' This was no deliberate discourtesy, but rather a reflection of an unconventional directness of mind. The astrophysicist Dennis Sciama has told how, as a research student in 1950, he entered Dirac's office, flushed with the exhilaration of a deep insight. 'Professor Dirac', he burst out, 'I've just thought of a way of relating the formation of stars to cosmological questions. Shall I tell you about it?' 'No', said Dirac.

Leopold Infeld, a Polish theoretical physicist, won a scholarship to work in Cambridge, and his first encounter with Dirac, as recounted in his wonderful memoirs, is typical of the experience of many:

> When I visited Dirac for the first time I did not know how difficult it was to talk to him as I did not then know anyone who could have warned me.
>
> I went along the narrow wooden stairs in St. John's College and knocked at the door of Dirac's room. He opened it silently and with a friendly gesture indicated an armchair. I sat down and waited for Dirac to start the conversation. Complete silence. I began by warning my host that I spoke very little English. A friendly smile but again no answer. I had to go further:
>
> 'I talked with Professor Fowler. He told me that I am supposed to work with you. He suggested that I work on the internal conversion effect of positrons.'
>
> No answer. I waited for some time and tried a direct question:
>
> 'Do you have any objection to my working on this subject?'
>
> 'No.'
>
> At last I had got a word out of Dirac.
>
> Then I spoke of the problem, and took out my pen in order to write a formula. Without saying a word Dirac got up and brought paper. But my pen refused to write. Silently Dirac took out his pencil and handed it to me. Again I asked him a direct question to which I received an answer in five words which took me two days to digest. The conversation was finished. I made an attempt to prolong it.
>
> 'Do you mind if I bother you sometimes when I come across difficulties?'
>
> 'No.'
>
> I left Dirac's room, surprised and depressed. He was not forbidding, and I

should have had no disagreeable feeling had I known what everyone in Cambridge knew. If he seemed peculiar to Englishmen, how much more so he seemed to a Pole who had polished his smooth tongue in Lwow cafés!

In 1931, Dirac was on sabbatical leave at the University of Wisconsin. Here is the account of an interview he granted to 'Roundy' for a local newspaper.

An enjoyable time is had by all

I been hearing about a fellow they have up at the U. this spring—a mathematical physicist, or something, they call him—who is pushing Sir Isaac Newton, Einstein and all the others off the front page. So I thought I better go up and interview him for the benefit of State Journal readers, same as I do all other top notchers. His name is Dirac and he is an Englishman. He has been giving lectures for the intelligentsia of the math and physics departments— and a few other guys who got in by mistake.

So the other afternoon I knocks at the door of Dr Dirac's office in Sterling Hall and a pleasant voice says 'Come in'. And I want to say here and now that this sentence 'come in' was about the longest one emitted by the doctor during our interview. He sure is all for efficiency in conversation. It suits me. I hate a talkative guy.

I found the doctor a tall youngish-looking man, and the moment I seen the twinkle in his eye I knew I was going to like him. His friends at the U. say he is a real fellow too and good company on a hike—if you can keep him in sight, that is.

The thing that hit me in the eye about him was that he did not seem to be at all busy. Why if I went to interview an American scientist of his class— supposing I could find one—I would have to stick around for an hour first. Then he would blow in carrying a big briefcase, and while he talked he would be pulling out lecture notes, proofs, reprints, books, manuscript, or what have you out of his bag. But Dirac is different. He seems to have all the time there is in the world and his heaviest work is looking out the window. If he was a typical Englishman it's me for England on my next vacation.

Then we sat down and the interview began.

'Professor,' says I, 'I notice you have quite a few letters in front of your last name. Do they stand for anything in particular?'

'No', says he.

'You mean I can write my own ticket?'

'Yes', says he.

'Will it be all right if I say P.A.M. stands for Poincaré Aloysius Mussolini?'

'Yes', says he.

'Fine', says I, 'We are getting along great! Now doctor will you give me in a few words the low down on all your investigations?'

'No', says he.

'Good', says I, 'Will it be all right if I just put it this way—"Professor Dirac solves all problems of mathematical physics, but he is unable to find a better way of figuring out Babe Ruth's batting average"?'

'Yes', says he.

'What do you like best in America?', says I.

'Potatoes', says he.

That knocked me cold! It was a new one on me.

Then I went on: 'Do you go to the movies?'

'Yes', says he.

'When?', says I.

'In 1920—perhaps also in 1930', says he.

'Do you like to read the Sunday comics?'

'Yes', says he, warming up a bit more than usual.

'This is the most important thing yet, doctor', says I. 'It shows me and you are more alike than I thought. And now I want to ask you something more: They tell me that you and Einstein are the only two real sure-enough high-brows and the only ones who can really understand each other. I wont ask you if this is straight stuff for I know you are too modest to admit it. But I want to know this: Do you ever run across a fellow that even you can't understand?'

'Yes', says he.

'This will make great reading for the boys down at the office', says I. 'Do you mind releasing to me who he is?'

'Weyl', says he.

The interview came to a sudden end just then, for the doctor pulled out his watch and I dodged and jumped for the door. But he let loose a smile as we parted and I know that all the time he had been talking to me he was solving some problem that no one could touch. If that fellow Professor Weyl ever lectures in this town again I sure am going to take a try at understanding him! A fellow ought to test his intelligence once in a while.

The Weyl alluded to by Dirac is the German mathematician Hermann Weyl (1885–1955), who left Germany after Hitler's accession to power in 1933 for Princeton, where he became a close colleague of Einstein's. Roundy omitted to ask Dirac how he arrived at those ideas that changed the course of physics. Had he done so, he would probably have been rewarded by the master's standard reply: he lay on his study floor with his feet in the air so that the blood ran to his head.

Shortly after Dirac's election to his chair in Cambridge, Niels Bohr [79] asked the doyen of British physicists, J. J. Thomson [73], whether he was pleased with the appointment. Thomson replied with the following parable. A man enters a pet shop to buy a parrot. Price was immaterial, but the bird must talk. Some days later, the parrot not having uttered a word, the man returns to the shop to complain. 'Ah', says the shopkeeper, 'I must have made a mistake. I thought he was a talker, but now I see he was a thinker.'

A well-known episode, which characterizes Dirac, concerns his meeting with E. M. Forster. Forster was then an ageing bachelor, still living in King's College, Cambridge. A friend of the Diracs had been surprised to find Dirac reading *A Passage to India* and thought it might be interesting to bring the two taciturn old men together. A tea was arranged, and an introduction effected. There was a long silence and then Dirac spoke: 'What happened in the cave?' Forster replied: 'I don't know.' Whereupon both lapsed into silence and in due course parted. It makes a pleasant story, but according to the physicist Rudolf Peierls, who knew Dirac well and asked him about the occasion, it is inaccurate. As Dirac remembered it, he had asked Forster whether there was a third person in the cave. 'No', came the answer, and to the question, 'What happened?', Forster had replied 'Nothing'. But memories, of course, are fallible.

Peierls also recalls in his memoirs a social occasion at the Diracs. Margit Dirac (sister of the Hungarian physicist Eugene Wigner, and whom indeed Dirac once introduced to a visitor with the words, 'Have you met Wigner's sister?') attempted to put a student at his ease in the discouraging presence of her silent husband. 'Paul, do you have any students?', she asked. 'I had one once,' came the lugubrious reply, 'but he died.'

The first passage comes from Leopold Infeld's wonderful memoirs, *Quest: The Evolution of a Scientist* (Gollancz, London, 1941). Roundy's interview is from the *Wisconsin State Journal*, dated 31 April [sic] 1931, and is reproduced in (among other places) S. S. Schweber, *QED and the Men Who Made It* (Princeton University Press, Princeton, 1994). Rudolf Peierls's reminiscence is from his *Bird of Passage* (Princeton University Press, Princeton, 1985).

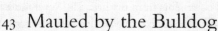

43 Mauled by the Bulldog

The most famous scientific drama of the Victorian era was the public confrontation between the forces of liberalism, led by 'Darwin's Bulldog',

Thomas Henry Huxley, and those of reaction in the shape of the high Tory Bishop of Oxford, Samuel Wilberforce, more familiarly known as Soapy Sam. *On the Origin of Species* had appeared seven months earlier, but Charles Darwin was now a sickly recluse and made no public appearances. He followed the course of the debate about his work eagerly and was kept informed by letters from his friends, especially Huxley, Joseph Hooker, and Sir John Lubbock. Huxley (1825–95) was the most pugnacious of Darwin's disciples; indeed he owned that controversy was to him as gin to a reformed drunkard. His adversary, the Bishop, was no fool: he was learned in mathematics and a keen ornithologist, but his views on religious and political matters were unyielding and out of tune with the liberalizing spirit that was then permeating the Church of England.

The clash took place in 1860 at the annual meeting of the British Association for the Advancement of Science, held that year in Oxford. On Wednesday 27 June, Darwin's most formidable opponent, Richard Owen, regarded as the foremost comparative anatomist of the day, had spoken. He had denounced the evidence for evolution of man from ape, citing his own new anatomical data on simian brains. Huxley had replied briefly from the floor. The debate featuring Wilberforce was scheduled for Saturday afternoon and in the expectation of high drama so many students and outsiders converged on the new Natural History Museum that the venue had to be changed from a lecture room to the large west gallery. Some 700 (some said a thousand) people crowded into the room, packing the aisles and window ledges. On one side of the platform were Hooker, Lubbock, and Sir Benjamin Brodie, the Queen's Physician and President of the Royal Society; in the centre sat the austere President of the meeting, the botanist, J. S. Henslow (Hooker's father-in-law), on whose other side were the Bishop of Oxford and the first speaker, Professor J. W. Draper of New York.

Draper's lecture was long and tedious, and then Henslow threw open the discussion. There is no written record of what followed, except in letters from some of the protagonists, and they by no means all agree. Huxley had complained of weariness and might have kept silent had he not been provoked by Wilberforce's flippant arrogance. The discussion began with inconsequential comments from the hall. The Reverend Richard Cresswell of Worcester College asserted that all theories of human evolution were undermined by the circumstance, noted by Alexander Pope, that 'Great Homer died three-thousand years ago'. Admiral Fitzroy, who had commanded HMS *Beagle*, when Darwin joined her many years earlier, and was now three-parts mad, denounced Darwin's works. After more such interjections a clamour arose for Wilberforce to speak. The Bishop rose and launched into a fluent but nugatory diatribe. In Hooker's words, 'Sam Oxon got up & spouted for half an hour with inimitable spirit, ugliness & emptyness & unfairness'. The audience appeared delighted. Then came his frivolous peroration, when he asked 'if anyone were to be willing to trace his descent through an ape as his *grandfather*, would he be willing to trace his descent through an ape on the side of his *grandmother*?' At this Huxley was said to have slapped his knee and muttered, 'The Lord has delivered him into mine hands!'

Huxley stood and began his reply. Here is how he himself recalled his performance in a letter to a friend, written some months later:

[H]e [Wilberforce] performed the operation vulgarly and I determined to punish him—partly on that account and partly because he talked pretentious nonsense, and when I got up I spoke pretty much to the speech—that I had listened with great attention to the Lord Bishop's speech but had been unable to discover either a new fact or a new argument in it—except, indeed, the question raised as to my personal predilections in the matter of ancestry—That it would not have occurred to me to bring forward such as that for discussion myself, but that I was quite ready to meet the Rt Rev. prelate even on that ground—If, then, said I, the question is put to me, 'would I rather have a miserable ape for a grandfather, or a man highly endowed by nature and possessed of great means and influence, and yet who employs these faculties and that influence for the mere purpose of introducing ridicule into a grave scientific discussion'—I unhesitatingly affirm my preference for the ape.

This was powerful language to use in addressing a Prince of the Church and, if these were indeed Huxley's words, it is no wonder that they caused a sensation and that the pious Lady Brewster (wife of the Scottish physicist, Sir David Brewster) swooned. Huxley was certainly pleased with the effect he thought he had created:

Whereupon there was unextinguishable laughter among the people—and they listened to the rest of my argument with the greatest of attention. Lubbock and Hooker spoke after me with great force and among us we shut up the bishop and his party. I happened to be in very good condition and said my say with perfect good temper and politeness—I assure you of this because all sorts of reports were spread about, e.g. that I had said I would rather be an ape than a bishop.

All the Oxford Dons were there & several hundred people in the room—so I think Samuel will think twice before he tries a fall with men of science again.

Others who were present remembered the scene a little differently. Huxley had not appeared as even-tempered as he made out but was 'white with rage' and too agitated to articulate effectively. According to Hooker, in the letter that he wrote soon after the event to Darwin, Huxley

> turned the tables, but could not throw his voice over so large an assembly, nor command the audience; & he did not allude to Sam's weak points nor put the matter in a form or way that carried the audience. The battle waxed hot. Lady Brewster fainted, the excitement increased as others spoke.

Hooker clearly saw himself, more than Huxley, as the hero of the day:

> my blood boiled, I felt myself a dastard; now I saw my advantage—I swore to myself I would smite that Amalakite Sam hip & thigh if my heart jumped out of my mouth & I handed my name up to the President as ready to throw down the gauntlet . . . there I was cocked up with Sam at my right elbow, & there and then I smacked him amid rounds of applause—I hit him in the wind at the first shot in 10 words taken from his own ugly mouth . . . Sam was shut up—he had not one word to say in reply & the meeting *was dissolved forthwith* leaving you master of the field after 4 hours battle. Huxley who had borne all the previous brunt of the battle & who had never before (thank God) praised me to my face, told me it was splendid, & that he did not know before what stuff I was made of. I have been congratulated and thanked by the blackest coats and whitest stocks in Oxford.

It was agreed, at all events, that 'Saponaceous Samuel', as Hooker liked to call him, had been vanquished, and Darwin, though present only in spirit, had prevailed. But as the audience left the hall, the Bishop of Worcester's wife was heard to voice her opinion of the Theory of Evolution to her companion: 'Let us hope it is not true. But if it is true, let us hope it does not become generally known.'

For excellent accounts of the debate, its historical setting and its aftermath, see, for instance, *Darwin* by Adrian Desmond and James Moore (Michael Joseph, London, 1991), *Huxley: The Devil's Disciple* by Adrian Desmond (Michael Joseph, London, 1994), and Ronald W. Clark's *The Huxleys* (McGraw-Hill, New York, 1968).

44 Damping the canine rotor

James Clerk Maxwell was the presiding genius of nineteenth-century physics. He made revolutionary contributions in many areas of the subject, most famously in clarifying the nature of electromagnetic radiation. Maxwell, who died in 1879 at the age of 48, was also a man of humanity and wit. His name still has a secure place in anthologies of comic verse. From his obituary in *Nature* comes this example of his sense of fun: Maxwell liked to construct ingenious models, theoretical and practical, to illustrate physical principles and one such was an adjustable top; this had screws to alter its moment of inertia about its geometrical axes so that the angle of its rotation could be varied and the mathematics governing its stability or instability made manifest.

> When Professor Maxwell came to Cambridge in 1857 to take his M.A. degree, he brought this top with him from Aberdeen. In the evening he showed it to a party of friends in college, who left the top spinning in his room. Next morning he espied one of these friends coming across the court, so jumping out of bed, he started the top anew and retired between the sheets. The reader can well supply the rest of the story for himself. It is only necessary to add that the plot was completely successful.

Here now is an instance of how Maxwell put the laws of dynamics to compassionate use:

> It was during the visit of the comet in 1874, when unfortunately the comet's tail was a subject of general conversation, that Maxwell's terrier developed a great fondness for running after his own tail, and though anyone could start him, no one but Maxwell could stop him until he was weary. Maxwell's method of dealing with the case was, by a movement of the hand, to induce the dog to revolve in the opposite direction and after a few turns to reverse him again, and to continue these reversals, reducing the number of revolutions for each, until like a balance wheel on a hair spring with the remaining power withdrawn, by slow decaying oscillations the body came to rest.

The first Maxwell story is from Wm. Garnett, *Nature*, **21**, 45 (1879); the second was told in *Nature*, **128**, 605 (1931).

45 Nemesis in Nancy

R. W. Wood (1868–1955), Professor of Physics at Johns Hopkins University, was a leader in his field of spectroscopy, and also a noted wit and *farceur*. His escapades became legendary. He would alarm the citizens of Baltimore by spitting into puddles on wet days, while surreptitiously dropping in a lump of metallic sodium, which would explode in a jet of yellow flame. He wrote clever verses, a collection of which was published as a slim volume with the title, *How to Tell the Birds from the Flowers*, and is still intermittently in print.

It was related that when, as a young man, Wood was staying in a rooming house in Paris, he astonished his fellow guests by sprinkling copious amounts of a white powder over the bones left on the plates after a dinner of chicken. The next night, when the soup was served, Wood brought out a small spirit burner and dropped a little of the liquid on the flame. A red flash drew a smile of satisfaction: the white powder, he explained to the other diners, was lithium chloride and the red flash indicated that it was now in the soup. His suspicion that the landlady recycled the bones was thus confirmed. It must be said, however, that what was in essence the same story has also been imputed to Georg von Hevesy [112], the pioneer of radioactive tracers: he was supposed to have laced the leftovers with a radioactive salt and detected the radioactivity in the soup with a Geiger counter. This is a much more sensitive test than the lithium flame, which would probably have required an unacceptable amount of the salty-tasting lithium chloride; but everyone to his own speciality. (Victor Moritz Goldschmidt, a distinguished geochemist, was reputed to have carried a capsule of potassium cyanide when he was planning his escape from Nazi Germany. When a friend in the engineering department of the university expressed interest, Goldschmidt was supposed to have replied that cyanide was for professors of chemistry; his friend, a professor of mechanics, would have to carry a rope.)

Another of Wood's practical jokes was also played out in Paris, when he discovered that the landlady, or concierge, who lived in the flat below his room, kept a tortoise in a pen on the balcony. Wood procured a collection of tortoises of varying sizes, retrieved the landlady's pet with a hook on a broom handle and substituted one of slightly larger size. Every few days he would change the tortoise for one the next size up. The astonished landlady told Wood about this prodigy of nature, and he urged her to consult a celebrated professor at the university and also to inform the press. The press supposedly turned up to inspect the expanding tortoise, and then Wood proceeded to reverse the process and the animal shrank as mysteriously as it had grown. It is not related whether anyone in Paris ever got to the bottom of the affair.

Wood made many important contributions to spectroscopy (including the construction of a spectrograph with a long optical path, through which he trained his cat to walk to clear it of cobwebs and dust). Yet he is especially remembered today for his part in one of the strangest episodes in the history of physics. A French physicist of high standing, René Prosper Blondlot, discovered what he believed to be a new form of electromagnetic radiation. He called this N-rays after his native city of Nancy. The visible effects of N-rays were apparent to Blondlot, his colleagues in Nancy, and several other French scientists, but to hardly any in other lands. The N-rays proved to be a figment, perceptible only to those predisposed to believe in them. The delusion was finally exposed by Wood on a visit to Blondlot's laboratory in the University of Nancy in 1903. Here is Wood's own rather smug description of how he ensnared the unfortunate Blondlot.

> Reading of his [Blondlot's] remarkable experiments, I attempted to repeat his observations, but failed to confirm them after wasting an entire morning. According to Blondlot, the rays were given off spontaneously by many metals. A piece of paper, very feebly illuminated, could be used as a detector, for, wonder of wonders, when the N rays fell upon the eye they increased its ability to see objects in a nearly dark room.
>
> Fuel was added by a score of other investigators. Twelve papers had appeared in the *Comptes rendus* [the published proceedings of the sessions of the French Academy of Sciences] before the year was out. A. Charpentier, famous for his fantastic experiments on hypnotism, claimed that N rays were given off by muscle, nerves, and the brain, and his incredible claims were published in the *Comptes*, sponsored by the great d'Arsonval, France's foremost authority on electricity and magnetism.
>
> Blondlot next announced that he had constructed a spectroscope with aluminum lenses and a prism of the same material, and found a spectrum of lines separated by dark intervals, showing that there were N rays of different refrangibility [that is, dispersed to different degrees by the prism, as the red is from yellow, green, blue, and violet when white light passes through a prism of glass] and wavelength [again, by analogy with other forms of electromagnetic radiation, such as visible light, with violet at its short- and red at its

long-wavelength limit]. He measured the wavelengths. Jean Becquerel [son of Henri, the discoverer of radioactivity] [36] claimed that N rays could be transmitted over a wire. By early summer Blondlot had published twenty papers, Charpentier twenty, and J. Becquerel ten, all describing new properties and sources of the rays.

Scientists in all other countries were frankly skeptical, but the French academy stamped Blondlot's work with its approval by awarding him the Lalande prize of 20,000 francs and its gold medal 'for the discovery of N rays'.

In September (1904) I went to Cambridge for the meeting of the British Association for the Advancement of Science. After the meeting some of us got together for a discussion of what was to be done about the N rays. Professor Rubens, of Berlin, was most outspoken in his denunciation. He felt particularly aggrieved because the Kaiser had commanded him to come to Potsdam and demonstrate the rays. After wasting two weeks in vain attempts to duplicate the Frenchman's experiments, he was greatly embarrassed by having to confess to the Kaiser his failure. Turning to me he said, 'Professor Wood, will you not go to Nancy immediately and test the experiments that are going on there?' 'Yes, yes,' said all the Englishmen, 'that's a good idea, go ahead.' I suggested that Rubens go, as he was the chief victim, but he said that Blondlot had been most polite in answering his many letters asking for more detailed information, and it would not look well if he undertook to expose him. 'Besides,' he added, 'you are an American, and you Americans can do *anything . . .*'

So I visited Nancy, meeting Blondlot by appointment at his laboratory in the early evening. He spoke no English, and I elected German as our means of communication, as I wanted him to feel free to speak confidentially to his assistant.

He first showed me a card on which some circles had been painted in luminous paint. He turned down the gas light and called my attention to their increased luminosity, when the N ray was turned on. I said I saw no change. He said that was because my eyes were not sensitive enough, so that proved nothing. I asked him if I could move an opaque lead screen in and out of the path of the rays while he called out the fluctuations of the screen. He was almost 100 per cent wrong and called out fluctuations when I made no movement at all, and that proved a lot, but I held my tongue. He then showed me the dimly lighted clock, and tried to convince me that he could see the hands when he held a large flat file just above his eyes. I asked if I could hold the file, for I had noticed a flat wooden ruler on his desk, and remembered that wood was one of the few substances that *never* emitted N rays. He agreed to this, and I felt around in the dark for the ruler and held it in front of his face. Oh, yes, he could see the hands perfectly. This also proved something.

But the crucial and most exciting test was now to come. Accompanied by the assistant, who was by this time casting rather hostile glances at me, we went into the room where the spectroscope with the aluminum lenses and prism was installed. In place of an eyepiece, this instrument had a vertical thread, painted with luminous paint, which could be moved along in the region where the N ray spectrum was supposed to be by turning a wheel having graduations and numerals on its rim. Blondlot took a seat in front of the instrument and slowly turned the wheel. The thread was supposed to brighten as it crossed the invisible lines of the N-ray spectrum. He read off the numbers on the graduated scale for a number of the lines, by the light of a small, darkroom red lantern. This experiment had convinced a number of skeptical visitors, as he could repeat his measurements in their presence, always getting the same numbers. I asked him to repeat his measurements, and reached over in the dark and lifted the aluminum prism from the spectroscope. He turned the wheel again, reading off the same numbers as before. I put the prism back before the lights were turned up, and Blondlot told his assistant that his eyes were tired. The assistant had evidently become suspicious, and asked Blondlot to let him repeat the reading for me. Before he turned down the light I noticed that he placed the prism very exactly on its little round support, with two of its corners exactly on the rim of the metal disc. As soon as the light was lowered, I moved over towards the prism, with audible footsteps, but *I did not touch the prism*. The assistant commenced to turn the wheel, and suddenly said hurriedly to Blondlot in French, 'I see nothing; there is no spectrum. I think the American has made some *dérange-ment*'. Whereupon he immediately turned up the gas and went over and examined the prism carefully. He glanced at me, but I gave no indication of my reaction. This ended the seance.

Next morning I sent off a letter to *Nature* giving a full account of my findings, not, however, mentioning the double-crossing incident at the end of the evening, and merely locating the laboratory as 'one in which most of the N-ray experiments had been carried out'. *La Revue scientifique*, France's weekly semipopular scientific journal started an inquiry, asking French scientists to express their opinions as to the reality of N rays. About forty letters were published, only a half dozen backing Blondlot. The most scathing one by Le Bel [a chemist and one of the founders of stereochemistry] said, 'What a spectacle for French science when one of its distinguished savants measures the position of the spectrum lines, while the prism reposes in the pocket of his American colleague!'

The Academy at its annual meeting in December, when the prize and medal were presented, announced the award as given to Blondlot 'for his life's work, taken as a whole'.

Wood's intervention in the N-ray affair was devastating, to be sure, and his letter to *Nature* brutal enough. But his account, given to his biographer many years after the event, is undoubtedly embroidered. The story about Rubens's failed demonstration to the Kaiser was a canard, apparently started by a professor in Paris, and denied by Rubens himself, so Wood's memory must have played him false when he asserted that he had it from Rubens in Cambridge. No more was heard of N-rays, but Blondlot never conceded that they were an illusion. He retired early from his university position and apparently continued to pursue the elusive radiation in solitude in his private laboratory.

From the biography of R. W. Wood by William Seabrook, *Dr Wood, Modern Wizard of the Laboratory* (Harcourt Brace, New York, 1941). For more on N-rays see, for instance, Irving Klotz, *Diamond Dealers and Feather Merchants: Tales from the Sciences* (Birkhäuser, Basel, 1986).

~

46 Mathematician's melodrama

Évariste Galois, one of the great mathematical geniuses of his time, had a brief and tragic life, for he was killed in a duel in 1832, at the age of 20. Passionate and irascible, his short temper and uncompromising nature kept him out of the leading French academies and brought him to the verge of despair. His paranoia had been fed by an unfortunate episode, when at the age of seventeen he submitted his work on the solution of quintic equations [equations in which the unknown quantity appears as its fifth power and for which there was at that time no known solution] to the Académie des Sciences. The reviewer was Baron Cauchy, one of the mandarins of the French mathematical establishment, who discerned in Galois's work an exceptional talent, and suggested that it should be submitted for the Academy's mathematics prize. Galois reworked his paper and despatched it to the secretary of the Academy, Joseph Fourier. But Galois was not awarded the prize, nor even a mention, for Fourier died before he could forward the manuscript, which was never afterwards found. A later paper, submitted to the Academy, was rejected by one of the pioneers of statistical analysis, Siméon-Denis Poisson, on the grounds that it was insufficiently clear or fully developed, although 'we have made every effort to understand [his] proof'.

Galois's fervent republican convictions, culminating in a threat against Louis-Philippe's life, landed him in prison. And then a romantic attachment, or so at least the surmise runs, led to his untimely death. The woman in question is supposed to have been Stéphanie-Félicie Poterine du Motel, and, according to the evidence of no less an authority than Alexandre Dumas *père*, witness to a quarrel in a restaurant, Galois's nemesis was her fiancé, Pécheux d'Herbinville.

The offended swain challenged his rival to a duel, and throughout the preceding night the unhappy Galois tried to get the results of his researches onto paper, so that they would not be lost, should his presentiment prove correct and he be killed. Here is how he expressed his feelings in a letter to a friend:

> I beg my patriotic friends not to rebuke me for dying otherwise than for my country. I die the victim of an infamous coquette and her two dupes. My life is extinguished amidst trivial gossip. Oh, why die for so little, for something so contemptible? I call on Heaven to witness that only under compulsion and force have I yielded to a provocation which I have tried by all possible means to avert. I regret having told such dangerous truth to those incapable of hearing it with composure. I shall take with me to the grave a spotless conscience, untouched by lies, untainted with patriotic blood.

But then he adds: 'Adieu! Life was dear to me for the common good. Pardon those who have killed me, they are of good faith', which has been taken by some to imply that the quarrel was in actuality with a republican colleague (for which there is indeed some written evidence).

That night Galois scribbled down his equations frantically, with many crossings-out, neurotic doodles, and asides—'*une femme*' and 'Stéphanie', and then the despairing words, '*je n'ai pas le temps*'. Early the next morning Galois and his opponent, without seconds and no doctor in attendance, faced each other with pistols at 25 paces. Galois was hit in the abdomen and died in hospital the following day of peritonitis. His last words to his brother at the bedside were: 'Don't cry—I need all my courage to die at twenty.'

His interment, in the common ditch at the cemetery of Montparnasse, attended by 3000 republicans, was the occasion for a riot, for the police were in attendance and clashed with the mourners, many of whom evidently believed that Galois had been the victim of a conspiracy—that d'Herbinville and Stéphanie had been engaged by the Government to bring down their turbulent opponent.

Galois had entrusted his papers to a friend, Auguste Chevalier, with the following injunction:

My dear friend.

I have made some new discoveries in analysis. The first concerns the theory of quintic equations and other integral functions.

In the theory of equations I have researched the conditions for the solvability of equations by radicals; this has given me the occasion to deepen this theory and describe all the transformations possible in an equation even though it is not solvable by radicals. All this will be found here in three memoirs . . .

In my life I have often dared to advance propositions about which I was not sure. But all I have written down here has been clear in my head for over a year, and it would not be in my interest to leave myself open to the suspicion that I announce theorems of which I do not have a complete proof.

Make a public request of Jacobi and Gauss [the leading German mathematicians] to give their opinions, not as to the truth, but as to the importance of these theorems. After that, I hope some may find it profitable to sort out this mess.

I embrace you with effusion—E. Galois.

The mess, of course, was the state of Galois's hastily scribbled notes. Chevalier, together with Galois's brother, did their best to edit his chaotic writings and forwarded them as they had been enjoined. It was to be a decade, however, before there was any response, and this came not from Jacobi or Gauss, but from Galois's illustrious compatriot, Joseph Liouville (1809–82). When the notes came into Liouville's hands he recognized the mark of genius and after much laborious interpretation he despatched an edited version to the leading French mathematical journal. He prefaced the papers by an introduction, in which he explained that Galois's meaning was often obscured by an exaggerated and ill-advised striving after conciseness. He concludes:

My zeal was well rewarded, and I experienced an intense pleasure at the moment when, having filled in some slight gaps, I saw the complete correctness of the method by which Galois proves, in particular, this beautiful theorem.

Galois's achievements were immediately and at long last recognized.

There are many accounts of Galois's brief and turbulent life and of his work. One of the most accessible to those with minimal grasp of mathematics is in Simon Singh's excellent book, *Fermat's Last Theorem* (Fourth Estate, London, 1997); and see also Ian Stewart's *Galois Theory* (Chapman and Hall, London, 1972).

~

47 Ben Franklin stills the waves

The voracious curiosity of Benjamin Franklin (1706–90) extended over all branches of science and much more besides. He had a special interest in what became known as surface forces and devised a favourite trick of waving his walking stick over a turbulent stream. The surface at once became smooth, for the stick was hollow and when shaken released a few drops of oil. Here is how his interest in the phenomenon was captured while he was en route for England as the diplomatic representative of the Assembly of the State of Pennsylvania.

> In 1757, being at sea in a fleet of 96 sail bound against Louisbourg [in Nova Scotia], I observed the wake of two of the ships to be remarkably smooth, while all the others were ruffled by the wind, which blew fresh. Being puzzled with the differing appearance, I at last pointed it out to the captain, and asked him the meaning of it? 'The cooks, says he, have, I suppose, been just emptying their greasy water through the scuppers, which has greased the sides of those ships a little;' and this answer he gave me with an air of some little contempt, as to a person ignorant of what every body else knew. In my own opinion I at first slighted his solution, tho' I was not able to think of another. But recollecting what I had formerly read in PLINY [28], I resolved to make some experiment on the effect of oil on water, when I should have the opportunity.

Franklin some time later also observed a striking effect of oil on the surface of water in the bottom of a lamp that he hung in his cabin at sea. The promised experiment was carried out in London on the Round Pond on Clapham Common:

> At length being at CLAPHAM where there is, on the common, a large pond, which I observed to be one day very rough with the wind, I fetched out a cruet of oil, and dropt a little of it on the water. I saw it spread itself with surprising swiftness upon the surface; but the effect of smoothing the waves was not

produced; for I had applied it first on the leeward side of the pond, where the waves were largest, and the wind drove my oil back upon the shore. I then went to the windward side, where they (the waves) began to form; and the oil, though not more than a teaspoonful, produced an instant calm over a space of several yards square, which spread amazingly, and extended itself gradually till it reached the lee side, making all that quarter of the pond, perhaps half an acre, as smooth as a looking glass.

After this, I contrived to take with me, whenever I went into the country, a little oil in the upper hollow joint of my bamboo cane, with which I might repeat the experiment as opportunity should offer; and I found it constantly to succeed.

Franklin made many other observations on this and related phenomena, and inferred something close to what many years later proved to be the truth—that the oil formed a film on the water only one molecule thick.

For an account of Franklin's life, science, and personality, which puts his observations into a modern context, see Charles Tanford's fascinating book, *Ben Franklin Stilled the Waves* (Duke University Press, Durham, NC and London, 1989).

~

48 Fraternal fire

Here is a reminiscence by the biochemist Martin Kamen, from his student years at the University of Chicago. Jean Picard, the Swiss deep-sea explorer and inventor of the bathyscape, was visiting the university in 1933:

As a former professor of organic chemistry at the university, Picard was invited to lecture to his old colleagues and the students in the Kent Chemical Laboratory theater one afternoon. He arrived with his brother, Dr Auguste Picard, a physicist who pursued cosmic ray studies in the stratosphere. One could call Jean the 'down-going Picard' and Auguste the 'up-rising Picard'. Professor Arthur Compton [who shared the 1927 Nobel Prize for his studies on cosmic rays] sat with Auguste in the back of the room as Jean delivered a speech on one of his favorite topics—explosions and explosives. To ensure that the audience knew what was meant by 'explosion waves', he began with great energy mimicking the operation of digging a hole in which to plant a stick of dynamite. After a while he paused, breathing heavily from his labors and appeared satisfied with the results. Then he went on to dig several more imaginary holes at appropriate intervals until he had traversed the whole width of the lecture platform. Finally he said, 'We put sticks of dynamite in each hole, set off the first one and then, va-voom! Explosion wave!' He accompanied this statement with a fast shuffle across the stage, flapping his arms to give the effect of a series of rapid detonations. Collecting himself, he returned to the blackboard and began a learned discourse on the mechanism of the explosion wave, based unfortunately on an irrelevant classical theory of sound wave propagation. Almost immediately from the back of the room there came Auguste's shout 'But no!' Instantly Jean wheeled about, pointed a finger in the direction of Auguste and shouted, 'But yes!' There ensued a violent and incomprehensible argument in excited French between the two brothers, over the heads of the stunned and embarrassed audience. Professor Compton tried to intervene and lower the heat, but to no avail. The argument raged on for a few minutes, then was broken off suddenly by Jean who turned his back on the audience, and with folded arms stared darkly at his scratch-ings on the blackboard. Then, slowly letting his arms drop to his sides and turning to face the audience, he said with a gesture to indicate how one must put up with all manner of idiots, 'Let us proceed'.

He talked in a most fascinating way about the manufacture of explosives, and in particular about how mercury fulminate was used as a filler in detona-tion caps. This highly hazardous material had formerly been stacked in piles on a table, he said, at which an operator sat and selected an empty cap which he or she filled by pushing a bit of the fulminate in with vigorous effort. 'Thus', Jean remarked with infinite regret, 'many were killed, but not all!' . . . He finished by assuring us that much more caution was now observed in these procedures, with operators moving about quietly and with great care. He illustrated this by tiptoeing off the stage with his finger to his lips, mur-muring 'Shhh'.

M. Kamen, *Radiant Science, Dark Politics* (University of California Press, Berkeley, 1985).

~

49 The wages of sin

To leave bottles of laboratory preparations unlabelled is an offence against the deities of research. The English physiologist A. S. Parkes, noted for his work on human and animal fertility, related how such a transgression led to an advance that changed the face of the field.

In the autumn of 1948 my colleagues, Dr Audrey Smith and Mr C. Polge [at the National Institute for Medical Research in London], were attempting to repeat the results which [others] had obtained in the use of laevulose [fruit sugar, now called fructose] to protect fowl spermatozoa against the effects of freezing and thawing. Small success attended the efforts, and pending inspiration a number of the solutions were put away in the cold-store. Some months later work was resumed with the same material and negative results were again obtained with all the solutions except one which almost completely preserved motility [the wriggling of the sperm] in fowl spermatozoa frozen at −79°C. This very curious result suggested that chemical changes in the laevulose, possibly caused or assisted by the flourishing growth of mould which had taken place during storage, had produced a substance with surprising powers of protecting living cells against the effects of freezing and thawing. Tests, however, showed that the mysterious solution not only contained no unusual sugars, but in fact contained no sugar at all. Meanwhile, further biological tests had shown that not only was motility preserved after freezing-thawing but, also, to some extent, fertilizing power. At this point, with some trepidation, a small amount (10–15 ml.) of the miraculous solution remaining was handed over to our colleague Dr D. Elliott for chemical analysis. He reported that the solution contained glycerol, water and a fair amount of protein! It was then realised that Mayer's albumen—the glycerol and albumen of the histologist—had been used in the course of morphological work on the spermatozoa at the same time as the laevulose solutions were being tested, and with them had been put away in the cold-store. Obviously there had been some confusion with the various bottles, although we never found out exactly what had happened. Tests with new material very soon showed that the albumen played no part in the protective effect, and our low temperature work became concentrated on the effects of glycerol in protecting living cells against the effects of low temperatures.

In retrospect the sloppiness of the laboratory technique is quite breathtaking: had this work been carried out by biochemists, not only would the solutions have been properly labelled, but no one would have dreamed of experimenting with stored sugar solutions containing a visible growth of mould. The glycerol (nothing other than glycerine), as biochemists would also have known, does not freeze at −79 °C and so the cells are not damaged by ice crystals when they are brought back to room temperature. The use of glycerol as what is now called a cryoprotectant initiated a new era in artificial insemination and the study of fertility, thanks to this sublime case of carelessness.

This should not, of course, be seen as justification for such acts of professional delinquency; which seldom have such a happy outcome. An embarrassing instance was brought to light in 2001. In the aftermath of the 'mad cow' disaster in Britain, fears surfaced of new human diseases. Sheep are prone to scrapie—harmless to humans but the surmised cause of BSE (bovine spongiform encephalopathy) by way of cattle feed derived from sheep carcasses. Could a scrapie mutant, infectious like BSE to humans, not make an appearance? A public laboratory was charged with determining whether sheep already harboured an agent resembling BSE. Testing of pooled brain paste from sheep began in 1987, with cow brain paste, containing the BSE agent, as control. After three years it transpired that the sheep brain material was liberally contaminated with cow brains. Was it a simple case of mislabelling of samples? Accusations, denials, and counter-accusations have not established the truth.

A. S. Parkes's account is in *Proceedings of the 3rd International Conference on Animal Reproduction*, Cambridge, 25–30 June 1956.

~

50 Loving an enzyme

Arthur Kornberg is one of the great biochemists of our time. His Nobel Prize in 1959 came for his work on the synthesis of DNA. The hallmark of his approach to his science has ever been the isolation of highly pure materials from biological tissues, and the analysis of their functions in minute and fastidious detail. He called his memoirs *For the Love of Enzymes*. Here is an episode that he has related from his apprenticeship in the New York laboratory of his mentor, the Spanish biochemist Severo Ochoa.

The purification of an enzyme was (and often still is) an arduous undertaking, involving generally a long succession of treatments that, for instance, rendered some components (generally the unwanted proteins extracted from the tissue) insoluble, while other material remained in solution. The presence of the enzyme was recognized by its activity in the reaction that it catalysed in the cell, so as more and more of the contaminating components were eliminated the activity relative to the total protein in the preparation increased.

Now [in December of 1946] we were completing a very large scale preparation starting with several hundred pigeon livers. Four of us . . . had worked for several weeks to reach the last step in which successive additions of alcohol finally yielded the precipitate which we believed, from small scale trials, would have the enzyme in an adequate state of purity. We had only to fill in some details in a paper we had prepared for publication.

Late one night, Ochoa and I were dissolving the final enzyme fraction, which had been collected in many glass centrifuge bottles. I had just poured the contents of the last bottle into a measuring cylinder that contained the entire enzyme fraction. Then I brushed against and overturned one of the empty, wobbly bottles on the crowded bench. That bottle knocked over another and the domino effect reached the cylinder with the enzyme. It fell over and all the precious material spilled on the floor. It was gone forever. Ochoa tried to be reassuring, but I remained terribly upset. By the time I got home by subway train an hour later, Ochoa had called several times because he was so worried about my safety.

The next morning back in the laboratory I glanced at the supernatant fluid beyond the last fraction. I might have discarded it because in our trial procedures it had been inactive. However, I had saved and stored it in the freezer at −15°C and now noticed that the previously clear fluid had become turbid. I collected the solid material, dissolved it, and assayed it for activity. 'Holy Toledo', I shrieked. This fraction had the bulk of the enzyme activity and was severalfold purer than the best of our previous preparations. Severo came running over to share my relief and pleasure, greatly amused by the 'Holy Toledo'.

Why did I save and assay the fraction we assumed was inactive? Because Ochoa's enthusiasm and optimism was infectious. Rather than suffusing a blinding intelligence, Ochoa taught me that with an ethic of unremitting experimental work, good things eventually happen. I believed they would for me as they had for him.

He might have added that it was also a matter of prudence and caution.

A. Kornberg, *Journal of Biological Chemistry*, **276**, 10 (2001).

~

51 The poltergeist next door

The Institute for Advanced Study at Princeton has at various times harboured many famous scholars—Albert Einstein, of course, included. It has the reputation of being remote and rarified, with no students and little communication with the academic world outside. Here is a vignette of life in the institute; the protagonists are a young physicist, Andrew Lenard, and C. N. (Frank) Yang, a famous theoretician, who had shared the 1957 Nobel Prize for Physics with his compatriot, Tsung-Dao Lee, a professor at Columbia University in New York. (When the award was announced the small Chinese restaurant nearby, where they were accustomed to take lunch together every week, placed a notice in the window, proclaiming: 'Eat here, get Nobel Prize'.) There was in general, it seems, little contact between the junior members of the institute and the grandees who should have been their inspiration.

> Lucky for Andrew Lenard, then, that one day [in 1966] Yang should happen to knock on his door to come in for a chat. Yang wanted to know what Lenard was working on, and so he tells him about the stability of matter problem [the deep question of why matter, which is composed of atoms that are almost entirely empty space between widely separated fundamental particles, is solid and stable]. Yang thinks this is curious: 'Very interesting', he says. 'That's either a trivial problem or a very difficult problem.'
>
> So Yang goes back to his office—which is right next door—and Lenard starts to hear this *tapping* on the wall. He realizes it's Frank Yang writing on his blackboard. This goes on for a while—*tap, tap, tap*—chalk banging against slate, and Lenard thinks no more about it.
>
> Then all of a sudden the tapping stops, as if the poor man had a heart attack. Dead silence.
>
> A few minutes later Yang pops his head back into Lenard's office: 'It's difficult', he says, then vanishes.

Yang's reaction recalls that of the applied mathematician Sir Harold Jeffreys, when he was a consultant to ICI (Imperial Chemical Industries). On one of his visits the company's physicists outlined in exhaustive detail a problem with which they hoped he would help them. Jeffreys listened patiently in total silence. When the presentation was over there was more silence and then he spoke: 'Well, I'm glad it's your problem, not mine', and quickly took his leave.

The Yang/Lenard exchange is from Ed Regis, *Who Got Einstein's Office* (Simon and Schuster, London, 1988).

52 The problem solver

Freeman Dyson, one of the most admired theoretical physicists and applied mathematicians of our time, has described himself as 'a problem solver', implying probably, with much modesty, that his strength lies in resolving, rather than conceiving, the deep questions of physics. (He has also written widely and lucidly about the progress of science and the future of our species; he paints with a broad brush, believing, for instance, that, faced with eventual extinction by 'heat death', we should think about moving our planet into a more hospitable orbit and perhaps even breaking through into a 'parallel universe'—assuming such exists; for confinement within the universe we know, says Dyson, gives him claustrophobia.)

In his memoirs Dyson recaptures the idyll of his early years among the patricians of the theoretical physics community in the United States in the period immediately after the Second World War. In 1948, at the end of his time working with Hans Bethe [62] at Cornell University, the young Dyson was bound for the Institute for Advanced Study in Princeton. Meanwhile Bethe had arranged for him to attend the annual physics summer school at the University of Michigan in Ann Arbor, a five-week event at which young physicists could listen to lectures by the luminaries in their field, and could interrogate and dispute with them. Then two weeks before the start of the school Dyson met Richard Feynman [89], who announced that he was driving to Albuquerque in New Mexico and invited him to come along.

For four days Dyson and Feynman talked and argued. Their philosophies of physics were antipodal: Dyson believed in equations, Feynman in a picture that the mind could grasp; he had an almost mystical view of the unity of nature and physical law—the kind of unity that Einstein unsuccessfully pursued for the last four decades of his life—while Dyson merely wanted a theory that would work within its set limits. Feynman distrusted Dyson's mathematics, and Dyson suspected Feynman's intuition. Feynman had formulated an intuitive picture of what became known as quantum electrodynamics—the rules governing the interactions of particles, for which he developed the famous Feynman diagrams, now a stock-in-trade of all particle physicists. Julian Schwinger [12], on the other hand, was known to have constructed an elaborate and, to most of the interested parties, impenetrable mathematical theory of such processes, and would be expounding his results at the Ann Arbor summer school. Dyson completed his journey by Greyhound bus and settled down to listen to Schwinger. Then he tackled him in private. Schwinger was accommodating:

I could talk to him at length, and from these conversations more than from the lectures I learned how his theory was put together. In the lectures his theory was a cut diamond, brilliant and dazzling. When I talked with him in private, I saw it in the rough, the way he saw it himself before he started the cutting and polishing. In this way I was able to grasp much better his way of thinking.

I filled hundreds of pages with calculations, working through various simple problems with Schwinger's methods. At the end of the summer school, I felt that I understood Schwinger's theory as well as anybody could understand it, with the possible exception of Schwinger. That was what I had come to Ann Arbor to do.

Departing, Dyson continued his journey west by Greyhound bus, with sojourns in Utah and California. And then, on his way back east, came the *coup de foudre*:

I got onto a Greyhound bus and traveled nonstop for three days and nights as far as Chicago. This time I had nobody to talk to. The roads were too bumpy for me to read, and so I sat and looked out of the window and gradually fell into a comfortable stupor. As we were droning across Nebraska on the third day, something suddenly happened. For two weeks I had not thought about physics, and now it came bursting into my consciousness like an explosion. Feynman's pictures and Schwinger's equations began sorting themselves out in my head with a clarity they had never had before. For the first time I was able to put them all together. For an hour or two I arranged and rearranged the pieces. Then I knew how they all fitted. I had no pencil or paper, but everything was so clear I did not need to write it down. Feynman and Schwinger were just looking at the same set of ideas from two different sides. Putting their methods together, you would have a theory of quantum electrodynamics that combined the mathematical precision of Schwinger with the practical flexibility of Feynman. Finally there would be a straight-forward theory of the middle ground [as Dyson called the state of matter between the large-scale, that of objects like heavenly bodies the behaviour of which is governed by gravitation, and the smallest—the most elusive, short-lived subatomic particles, occurring in high-energy collisions and in the atomic nucleus, and dominated by the so-called strong nuclear force]. It was my tremendous luck that I was the only person who had had the chance to talk at length to both Schwinger and Feynman and really understand what both of them were doing. In the hour of illumination I gave thanks to my teacher Hans Bethe, who had made it possible. During the rest of the day as we watched the sun go down over the prairie, I was mapping out in my head the shape of the paper I would write when I got to Princeton.

The following account of Dyson, the mathematical virtuoso, in action comes from the memoirs of Jeremy Bernstein. Bernstein came as a young theoretical physicist to the Institute of Advanced Study in Princeton in 1957 and was working with Marvin Goldberger (known as Murph), sometime President of the California Institute of Technology, but then still hewing at the scientific coalface. They were grappling with a problem in electromagnetic interactions between fundamental particles.

It was early in the morning—by Institute standards. Most people worked at night and were not seen until after noon. Murph had come up with a nasty-looking integral equation. It doesn't matter much what that is except that it was very nasty. He had divided the terms into two groups; one was labelled $G(x)$, for 'good of x', and the other was labelled $H(x)$, for 'horrible of x'. We were standing at the blackboard, staring morosely at horrible of x, when Dyson came in with his morning cup of coffee. He studied our equation. Murph asked, 'Freeman, have you ever seen one like this?' Dyson said no, but that he felt particularly strong that morning. He copied down our equation and disappeared. In about twenty minutes he was back with the solution. It was rediscovered by other people later and bears their names, but I saw what seemed to me, and still seems to me, like an incomprehensible conjuring trick. Over the years I have watched Dyson solve many different kinds of mathematical problems, and I cannot imagine what it must be like to be able to think with that rapidity and clarity in mathematics. Does everyone else appear to be going in slow motion? It is something that surely cannot be taught, at least to me. But I have learned enough mathematics to get pleasure and delight every time I see it happen.

From *The Life It Brings* by Jeremy Bernstein (Ticknor and Fields, New York, 1987). Freeman Dyson's memoirs have the title, *Disturbing the Universe* (Harper and Row, New York, 1979).

~

53 The Resonance Bridge

Hendrik Casimir, a distinguished Dutch physicist and for many years director of research at the Philips company in Eindhoven, did the circuit in his formative years of the great European centres of theoretical physics. Like all who worked with him, Casimir became devoted to Niels Bohr [79]. Here he recalls an example of the great man's pawky sense of humour.

Close to Bohr's Institute there is a body of water—I hesitate to call it a lake or pond—about three kilometers long and between 150 and 200 meters wide, the Sortedamsø. It is crossed by several bridges. One day Bohr took me on a stroll along that lake and across one of the bridges, 'Look,' he said, 'I'll show you a curious resonance phenomenon.' The parapet of that bridge was built in the following way. Stone pillars, about four feet high and ten feet apart, were linked near their tops by stout iron bars (or rather more likely, tubes) let into the stone. Halfway between each two pillars an iron ring was anchored to the stonework of the bridge, and two heavy chains, one on each side, were suspended between shackles welded to the top bar close to the stone pillars and that ring. Bohr grasped one chain near the top bar and set it swinging, and to my surprise the chain at the other end of the top bar began to swing too. 'A remarkable example of resonance', Bohr said. I was much impressed, but suddenly Bohr began to laugh. Of course, resonance was quite out of the question; the coupling forces were extremely small and the oscillations were strongly damped. What happened was that Bohr, when moving the chain, was rotating the top bar, which was let into, but not fastened to, the stone pillars, and in that way he had moved the two chains simultaneously. I was crestfallen that I had shown so little practical sense, but Bohr consoled me, saying that Heisenberg had also been taken in; he had even given a whole lecture on resonance.

The bridge became known in Bohr's institute as the Resonance Bridge. Casimir uses the tale as an illustration not only of Bohr's sense of humour but also of his strong practical sense. 'As a young man', says Casimir, 'he had done beautiful experiments on surface tension and had built most of the apparatus with his own hands, and his grasp of orders of magnitude went all the way from the atomic nucleus to engineering problems of daily life.'

From H. G. B. Casimir, *Haphazard Reality: Half a Century of Science* (Harper and Row, London and New York, 1983).

54 A laboratory libation

A. R. Todd, later Sir Alexander, and still later Lord Todd of Trumpington, was often known in Cambridge, where he held the Chair of Organic Chemistry, as Lord Todd Almighty. He was famous for many tours de force of structure determination, most notably that of the nucleotides—the building blocks that are connected together to form the long chains of DNA (and of RNA) [88]. Todd was an earthy Scotsman (born in 1907 in Glasgow) and known as something of an authoritarian. A jingle by one of his colleagues ran:

> Do you not think it is odd
> That a commonplace fellow like Todd
> Should spell, if you please,
> His name with two d's,
> When one is sufficient for God?

And an (approximate) Cambridge clerihew ran:

> Alexander Todd
> Thinks he's God;
> But Nevill Mott
> Knows he's not.

Like all ambitious young organic chemists of his generation he spent a period in Germany. In 1945, he was sent there again, this time in uniform, to report on the state of chemistry in the demoralized and prostrated country. Finding himself in Frankfurt, where he had worked 20 years before, he went to the chemistry institute to look around. He inspected his old laboratory and then took a stroll through the building.

In the course of this tour I went down to the basement, where the store which issued chemicals and equipment used to be. It was still there and still functioning as in the old days, with a queue of students filing past to purchase or borrow things. What really surprised me was that I could see the white-coated figure behind the store counter was the same Herr Müller who had been storekeeper in my day. So I joined the queue of customers, and in due course, came up to the counter. Müller looked up, gazed at me in silence for nearly a minute, and then in his ripe Frankfurt dialect said, 'Good God, who would have believed it—it's Herr Todd'. And with that he pulled down the shutter over the counter, emerged from the side-door, seized me by both hands and said 'Come in, come in—this calls for a drink!' So I went in and sat on one of the two stools in the store. Müller, meanwhile, took two beakers from a cupboard, put a generous amount of laboratory alcohol in each, diluted them with a roughly equal volume of distilled water [a liquid which has a curiously bitter, metallic taste], handed one to me, then sat down on the other stool. We toasted one another and the old days several times in this ghastly potion and then Müller began to chat about the laboratory and its inhabitants. 'Herr Todd,' he said, 'in our days we had chemists in the place, eh? You remember them—von Braun, Borsche and the others! Ah! Things have changed! Do you know, some of the people the Nazis sent here were so small that you could hardly see them!'

From Alexander Todd's memoirs, *A Time to Remember: The Autobiography of a Chemist* (Cambridge University Press, Cambridge, 1983).

~

55 Slot machine yields jackpot

A remarkable example of how inspiration can come unbidden from an extraneous source is related by the geneticist Salvador Luria (1912–91). An Italian, driven out of his country by Mussolini's antisemitic laws, he found refuge at the University of Indiana, where he launched himself into the study of bacteriophage genetics. This proved an immensely fruitful move and Luria, together with a small group of other pioneers, including Jim Watson [**88**], his student, and the spiritual leader of the movement, Max Delbrück in California, laid the foundations of molecular genetics, as it has evolved today.

Luria had been studying the fate of bacteria infected by a bacteriophage (a virus that attacks a bacterium and multiplies within, until its numerous progeny burst open the cell and spill out, ready to attack more bacteria); he had noticed that a few colonies of bacteria in his dishes of nutrient agar gel survived the onslaught. They were clearly mutants and the question now was whether these bacteria had been transformed by the action of the bacteriophage or whether they arose by occasional spontaneous mutations that rendered them resistant to attack.

I struggled with the problem for several months, mostly in my own thoughts, and also tried a variety of experiments, none of which worked. The answer finally came to me in February 1943 in the improbable setting of a faculty dance at Indiana University, a few weeks after I had moved there as an instructor.

During a pause in the music I found myself standing near a slot machine, watching a colleague putting dimes into it. Though losing most of the time he occasionally got a return. Not a gambler myself, I was teasing him about his inevitable losses, when he suddenly hit a jackpot, about three dollars in dimes, gave me a dirty look, and walked away. Right then I began giving some thought to the actual numerology of slot machines; in so doing it dawned on me that slot machines and bacterial mutations have something to teach each other.

What Luria had suddenly grasped is that a jackpot return could not be predicted, even if one knew that it would occur on average, say, once every 50 throws. Similarly, if bacterial mutations were random events a phage-resistant colony would spring up at unpredictable intervals. The progeny of the survivors, being all resistant, would give rise to clusters of flourishing colonies on the culture plate. If, by contrast, it was the action of the bacteriophage that generated a few resistant bacteria, while killing all the rest, then their colonies would be disposed about the culture plates in a random manner, in accordance with laws of random statistical distributions. Luria resumes:

Realizing the analogy between slot-machine returns and clusters of mutants was an exciting moment. I left the dance as soon as I could (I had no car of my own). Next morning I went early to my laboratory, a room I shared with two students and eighteen rabbits. I set up the experimental test of my idea— several series of identical cultures of bacteria, each started with a very few bacteria. It was a hard Sunday to live through, waiting for my cultures to grow. I still knew almost no one in Bloomington, so I spent most of the day in the library, unable to settle down with any book. Next day, Monday morning, each culture contained exactly one billion bacteria. The next step was to count the phage-resistant bacteria in each culture. I proceeded to mix each culture with phage on a single test plate. Then I had again a day of waiting— but at least I was busy teaching. Tuesday was the day of triumph. I found an average of ten resistant colonies per culture, with lots of zeros and, as I hoped to find, several jackpots. I had also set up my control. I had taken many individual cultures and pooled them all together, then divided the mixture again into small proportions and counted the resistant colonies in each portion. Complete success: this time the number of resistant colonies was again about the same, but the individual numbers were distributed at random and there were no jackpots.

Thus was the principle of spontaneous mutation established. The technique became famous as the 'fluctuation test'. It allowed the frequency of spontaneous mutation to be determined, and brought an understanding of how such attributes as antibiotic resistance can emerge.

From Salvador Luria's scientific autobiography, *A Slot Machine, a Broken Test Tube* (Harper and Row, New York, 1985).

~

56 Shooting down Venus

The following passage occurs in a letter from J. Robert Oppenheimer, director of the atomic bomb project at Los Alamos, to Eleanor Roosevelt:

> Very shortly before the test of the first atomic bomb, people at Los Alamos were naturally in a state of some tension. I remember one morning when almost the whole project was out of doors staring at a bright object in the sky through glasses, binoculars and whatever else they could find; and nearby Kirtland Field reported to us that they had no interceptors which had enabled them to come within range of the object. Our director of personnel was an astronomer and a man of some human wisdom; and he finally came to my office and asked whether we would stop trying to shoot down Venus. I tell this story only to indicate that even a group of scientists is not proof against the errors of suggestion and hysteria.

Oppenheimer (1904–67) was a remarkable theoretical physicist, possessor of a scientific intellect of formidable scope. Most of the stories that gathered around him made him out to be an intimidating personage. He evidently also had an engaging side. Here is a vignette from Martin Kamen, who, as a young postdoctoral scientist at the University of California in Berkeley, became friendly with him.

> 'Oppie', as he was affectionately known, once took me to a New Year's party in the city given by Estelle Caen, a pianist and the sister of the popular newspaper columnist Herb Caen. On our way there, Oppie remarked that he was not sure of her address, but he knew her apartment was on Clay Street and that the number was made up of double digits divisible by seven—1428, 2128, 2821, and so forth. So we tooled up Clay Street peering at houses on the way until we found Estelle's apartment at 3528.

The first passage is quoted by Richard Rhodes in *The Making of the Atomic Bomb* (Simon & Schuster, New York, 1986); the second is from Martin D. Kamen, *Radiant Science, Dark Politics* (University of California Press, Berkeley, 1985).

~

57 None so blind

F. A. Lindemann, born in Germany into a family of Alsatian origins but brought up in England, was a protégé of Walther Nernst [14], the great German physical chemist. In 1919, aged 33, he was appointed to the Chair of Experimental Philosophy (otherwise known as physics) in Oxford. He had acquired much credit for his work on military aircraft during the First World War; most spectacularly he conceived a theory for how to recover control during a tailspin—a lethal hazard for the early aviators. To test his reasoning he taught himself to fly and deliberately threw his plane into a tailspin. Perhaps unfortunately for Oxford, he survived. Lindemann had an austere, saturnine personality. Independently rich, he was more at home in the country seats of the aristocracy and the committee rooms of Whitehall than in the laboratory. He was a bachelor and lived in a suite of rooms in his college, Christ Church.

His colleague and biographer, the economist Roy Harrod, remembered calling on Lindemann one morning to find him in a characteristic pose: he was sitting in a tall chair, tying his bow-tie, with one valet kneeling before him polishing his shoes while another took dictation.

Though a man of penetrating intelligence, Lindemann made little impact on physics in the university. It was conjectured that he essentially withdrew from active research because he feared competing with the triumphal Rutherford, who was changing the face of science in Cambridge. His main achievement was to recruit a remarkable group of low-temperature physicists, displaced from Germany by the Nazis. When war came again Lindemann became Churchill's science adviser, a capacity in which he made mistakes of a grand order, especially in promoting the policy of bombing of the German centres of population. He was ennobled as Lord Cherwell for his services. His colleagues in Christ Church and elsewhere in the Oxford gave their reactions in slyly malicious, donnish verse. Thus the waspish historian Hugh Trevor-Roper (later himself elevated to the ermine as Lord Dacre):

> Lord Cherwell, when the war began,
> Was plain Professor Lindemann,
> But now, midst ministerial cheers,
> He takes his place among the peers.
> The House of Christ with one accord
> Now greets its newly-risen Lord.

And when the Prof (as he was generally known) was advanced from Baron to Viscount, and honoured by the Vatican for work he had done on meteors some 30 years earlier, an Oxford chemistry don, D. Ll. Hammick, added the following:

> But now a greater honour yet:
> He gets a leg-up in Debrett,
> And so becomes a nobler Lord
> Than Ernest, Baron Rutherford.
> At last his lordship's cup is full,
> Crammed to the fill with papal bull.

When the young Lindemann arrived in Oxford, Einstein and his Theory of Relativity pervaded physics and spilled over, albeit in diluted form, into philosophy and, indeed, into public consciousness. Lindemann was a champion of Einstein [161] and was happy to spread the gospel when he was invited to confront the Oxford philosophers in a lecture and debate before the Jowett Society (named after Benjamin Jowett [71] and dedicated to the discussion of philosophical questions). Roy Harrod, then an undergraduate, was present, and one of the philosophers was his tutor, a well-known Oxford character, H. W. B. Joseph. Harrod's account is a perfect illustration of the chasm that separated C. P. Snow's two cultures, and yawned most widely in Oxford—not without reason celebrated as 'the Home of Lost Causes' (for which see also [71]).

Lindemann gave, according to Harrod, a model exposition, crisp and fluent, taking 'the attitude that we were presumably highly intellectual people, if he gave us an essential point lucidly, we would grasp it and appreciate its bearing'. Then J. A. Smith, Waynflete Professor of Metaphysical Philosophy, took the floor. With his flowing white locks and drooping moustache he cut a majestic figure, standing at his ease before the fireplace. Smith, to Harrod's astonishment, set out to prove that the Theory of Relativity was false. He asserted that it embodied a demonstrably incorrect assumption (not identified by Harrod). But scientists in the audience knew that, whatever this assumption was, it formed no part of the theory and they loudly contradicted Smith, who 'looked extremely vexed. He knocked out his pipe in the fireplace and sat down'.

Harrod was dismayed. He knew that Lindemann had discussed relativity with Einstein himself and with a galaxy of famous physicists. 'Had there been a technical error, would all these men of genius have missed it? Was it left to an unmathematical, unscientific don at Magdalen College . . . to detect a technical error? This idea seemed to argue an attitude of mind totally remote from the world of reality, utterly provincial and parochial and incredibly complacent.' Smith, it seemed, was immured in 'some remote fastness of Greats'. Was he then the man to direct the future of the young, destined to inherit the running of the country and of the Empire? But worse was to follow, for now Joseph was preparing to take on Lindemann:

He [Joseph] had come into the room with a schoolboy satchel slung across his shoulders; to judge from its dilapidated appearance, it might well have been his own schoolboy satchel . . . To the dismay of his audience he proceeded to extract from the satchel a thick wad of manuscript. Balancing a pair of *pince-nez* on the lower end of his nose, he made ready to read from his papers. We steeled ourselves for a long session.

In one respect Joseph was at one with J. A. Smith; it appeared from an earlier stage that he was setting out to prove that the Theory of Relativity was 'wrong'. But he did not, like Smith, proceed at the technical level . . .

Joseph had the idea that certain words that are commonly used express some genuine apprehension by the mind. On this particular evening he was much concerned with such words as 'greater than', 'less than', 'before', 'after', 'simultaneously', 'moving relatively to'. What was meant by these words was based on a definite intellectual apprehension. One must not import different meanings, which violated these original apprehensions, into them. There they were, tokens of the mind's power of grasping certain things. And so he proceeded forward, in a lengthy and elaborate demonstration, to show that among the mind's original powers to grasp certain things, powers indicated by the use of words, powers which you could only challenge by using words in senses that were manifestly improper, was the mind's knowledge that space was Euclidean in character. Therefore the Theory of Relativity must be wrong.

Undergraduates, hand-picked although they are for the great seats of learning, do not always get things right. There was current in New College a limerick about Joseph:

There was an old person called Joseph
Whom nobody knows if he knows if
He knows what he knows, which accounts I suppose
For the mental condition of Joseph.

They were right in thinking that the question of what he knew or did not know was quite essential to the inner personality of Joseph. But the whole point about him was quite the opposite to that suggested by the limerick. What was peculiar was his very great degree of assurance that he *did* know certain things. For instance, he knew, absolutely and unshakably, that the space in which we live is in fact Euclidean.

At long last his discourse ended. One felt, as he proceeded, that the smooth polished surface of his phrases ought really to be interminable; and yet they did in fact terminate, no one quite knew why. All eyes were turned upon Professor Lindemann. What on earth was he to say? He had had lengthy discussions with Einstein, Max Planck, Broglie and other great men of thought. But I do not suppose that he had ever heard anything like this before. It was a hot-house Oxford product. He had been fairly and squarely challenged. He had been told that the Theory of Relativity was quite wrong, and this great chain of argument had been furnished.

There is a further point that must be made about Joseph. His paper against Relativity must have caused him much arduous work and taken much time to compose. But he did not show the slightest sign of his ever having seriously tried to understand either what the theoretical considerations were, or what the experimental results had been, that had led these distinguished physicists to feel the need to expound these tiresome theories of the relativity of space and time. It was evident that, in the ordinary sense of language, he knew nothing whatever about the Theory of Relativity. Since it was so evident to him that the conclusion reached was untrue, on quite different and sufficient philosophical grounds, there was no need for him to bother with the reasons why certain persons had been induced to frame such a theory. Indeed, I would go further. I have doubts myself as to whether Joseph, who had very limited intellectual capacity, despite his quite extraordinary linguistic acrobatics, would ever have been capable of understanding the Theory of Relativity.

So what was Professor Lindemann to do about it? He resumed, in his previous style of brief staccato sentences. He reiterated certain points. He gave some further illustrations. Then with the corners of his lips turned down and an ironic expression on his face, he said, 'Well, if you really suppose that you have private inspiration enabling you to know that . . .' But that was precisely what Joseph claimed. When there was a pause in Professor Lindemann's utterances, Joseph began unwinding again. And so each time. The Prof never really got to grips with his argument; none of the real points of interest in relation to relativity had been touched on; the whole game must have seemed to him to be perfectly futile. None the less there was this distinguished audience listening, and he was by no means winning the debate.

I suppose that some others among the learned persons present must have contributed something to the discussion. If so, that is entirely effaced from my memory. The spotlight was upon the Lindemann-Joseph interaction.

I mingled among the audience as they finally left the room. The Wykehamist Greats men were jubilant; a scientific professor had been torn to pieces; the Theory of Relativity had been shown to be untrue. But I was reluctant to join in their jubilation. I had already had an unhappy experience with Joseph. Unlike those Wykehamists, I had read a great deal of philosophy at my school (Westminster), and had come up to Oxford full of theories and earnestness to learn more. My arguments with Joseph had led to nothing but frustration. He had successfully shown that I was unable to express my thoughts in clear English and that sometimes what I had written for him meant nothing at all. But he seemed totally indifferent to what I had *tried* to mean, of the thought behind my words, just as he had been totally indifferent to the question of what were the theoretical considerations and the empirical facts that had led to the Theory of Relativity. Thus I had a certain fellow feeling with the unfortunate Professor Lindemann. I remember that I turned to an old friend, N. A. Beechman, a Balliol Greats man, afterwards President of the Union, and still later Minister of the Crown. He had a certain worldly shrewdness such as is necessary to those who interest themselves in politics. I asked him, 'Which was right?' He answered at once, 'Of course Professor Lindemann was right'.

The affair led Roy Harrod to question what he had been doing, reading Greats, 'the crown of all human studies', as it was regarded in Oxford. 'Professor Lindemann', he concludes, 'remained in my mind after that evening as a sort of symbol of the free advance of the human spirit.'

The passage is from Roy F. Harrod's biography of Lindemann, *The Prof: A Personal Memoir of Lord Cherwell* (Macmillan, London, 1959).

58 Raising the dead

The discovery of electricity implanted in the minds of some biologists the notion that it was somehow involved in life processes. This was why Luigi Galvani (1737–98), professor in the medical school in Bologna, was trying to induce a physiological preparation of frogs' legs, with their attached sciatic nerves and part of the spinal cord, to react to an electrical stimulus.

As he was setting up such an experiment an assistant accidentally touched the nerve with a scalpel blade and was startled to see the muscle respond with a convulsive contraction. The effect, it soon transpired, occurred only when the assistant's fingers (presumably wet) touched the iron nails that secured the blade to the bone handle, so as to close a circuit between nerve and earth. Galvani then decided to look into the effects of atmospheric electricity, which Benjamin Franklin and others had used during thunderstorms to charge a Leyden jar [38]. Galvani strung a row of frogs' legs along an iron railing in his garden with brass hooks, and was astonished to see them twitch with no external stimulus. (The story took root that the experiment was a by-product of Galvani's intent to prepare a nourishing frogs'-leg broth for his invalid wife; she, indeed, was the daughter of an eminent physiologist and probably participated in her husband's researches.) Galvani leapt to the conclusion that he had discovered an 'electric fluid', analogous perhaps to the source of the 'animal magnetism' that Franz-Anton Messmer (who gave his name to mesmerism) and other quacks were purporting to demonstrate in France.

Galvani published his observations and the interpretation he had put on them, under the title *De Viribus Electricitas*, in 1791 and his fame quickly spread, to the irritation of more critical spirits. Chief among these was a professor at the University of Pavia, a sceptical physicist, named Alessandro Volta (1745–1827). Volta repeated Galvani's observation, but recognized that his explanation was nonsense: the electricity was generated by the conjunction of iron and brass, separated by a conducting solution of physiological fluid in the muscles. Volta further noted that the bimetallic couple could go on producing a low level of electric current without any external charge, and he went on to join a succession of such couples in series, separated by wads of paper, soaked in salt solution—the voltaic pile, soon taken up by Humphry Davy [123] in London, who used it to bring about electrolysis of water (the chemical breakdown of H_2O into gaseous hydrogen and oxygen, liberated at the electrodes).

Galvani, evidently a man of limited imagination, never relinquished his belief in animal electricity. He became embittered by the rejection of his theory, the death of his wife, and the political persecution that he was made to endure; for he stoutly opposed the subjugation by Napoleon of northern Italy, which, as the Cisalpine Republic, became a French satrapy. But at least his ideas about electricity were vigorously, if misguidedly, promoted by his pupil and nephew, Giovanni Aldini. Aldini went so far as to collect freshly severed human heads from the foot of the guillotine and insert electrodes into the brains. This, he reported, provoked grimaces, twitching of the lips, and opening of the eyes. Volta, for his part, eschewed such theatricality and achieved wider recognition. He exhibited his battery at the Academy of Sciences in Paris in the presence of the Emperor himself, who grasped its potential and awarded its inventor a gold medal. Later he became a respected politician and statesman. Volta's name is commemorated in the unit of electrical force, the volt, and Galvani's in the galvanometer and most of all in the expressive verb, to galvanize.

The story of Galvani and Volta has been often told. A vivid brief account is to be found in *The Scientific Traveler: A Guide to the People, Places and Institutions of Europe* by Charles Tanford and Jacqueline Reynolds (Wiley, New York, 1992); for further details see J. F. Fulton and H. Cushing, *Annals of Science*, 1, 593 (1936).

~

59 Vibrios in Vienna

The agglutination test was for many decades one of the mainstays of laboratory and clinical immunology. It consisted in adding to a suspension of unknown bacteria, say, an antiserum against a known species of bacterium. If a solid mass formed and sank to the bottom of the tube then the identity of the bacterium was established. Cultures of unknown bacteria would be tested with a variety of antisera, prepared by immunizing animals, such as rabbits or sometimes goats or horses. A preparation from one animal could serve as a standard antiserum for years. The test was discovered in the laboratory of Max Gruber in Vienna by his English student, H. E. Durham. Durham remembered:

It was a memorable morning in November 1894, when we had all made ready with culture and serum provided by Pfeiffer to test his diagnostic reaction *in vivo*. Professor Gruber called out to me 'Durham! Kommen Sie her, schauen Sie an!' Before making our first injections of serum and vibrios [cholera bacilli] he had put a specimen under the microscope and there agglutination was displayed. A few days later, we had been making our mixtures in small sterilized glass pots; it happened that none were ready sterilized, so I had to make use of sterile test-tubes; these containing a mixture of culture and serum were left standing for a short time and then I called, 'Herr Professor! Kommen Sie her, schauen Sie an!' The phenomenon of sedimentation was before his eyes! Thus there were two techniques available, the microscopic and the macroscopic.

Agglutination in the test tube was thus seen only because Durham had had none of the standard sterile vessels available. Credit for the discovery was contested by the German bacteriologist Richard Pfeiffer, who provided the materials to be used for inoculation.

The story is recorded in W. I. B. Beveridge's book, *The Art of Scientific Investigation*, 3rd edn (Heinemann, London, 1960).

~

60 Drowning the telephone

R. V. Jones [106] was a physicist, latterly Professor of Physics at the University of Aberdeen, who made notable contributions to operational research during the Second World War, and was celebrated for his ingenious hoaxes on his colleagues. The following jape was conceived in Oxford in the years before the war with the connivance of two friends. Here is Jones's own description of what occurred.

I telephoned Gerald Touch's digs. Before anyone could answer I rang off again, and repeated the procedure several times to create the impression that someone was trying to ring the number but that something must be wrong. After this spell of induction, I dialled the number again, and heard a voice which I recognized as belonging to a very able research student in chemistry—in fact he had won the Senior Scholarship in Chemistry in the whole University that year. Reverting to the tongue that was my second language, the Cockney that came from my early schooling, I explained that I was the telephone engineer and had just received a complaint from a subscriber who was trying to dial the number and who had failed to get through. From the symptoms that he described I would say that either his dial was running a bit too fast or there was a leak to earth somewhere at the receiving end. I added that we would send a man round in the morning to check the insulation, but it was just possible the fault could be cleared from the telephone exchange if only we could be quite sure what it was. A few simple tests would check whether this were so, and if the victim would be good enough to help us with these tests, whoever it was who wanted to get through might be able to do so the same evening. Would the victim therefore help with the tests? Immediately, of course, he expressed a readiness to do so, and I explained that I would have to keep him waiting while I got out the appropriate manual so that we could go through the correct test sequence.

I realized that he was so firmly 'hooked' that I could even afford to clown, and I persuaded him to sing loudly into the telephone on the pretext that its carbon granules had seized up. By this time, of course, all the residents of the household had now been alerted, and watched with some amazement the rest of the performance. I told him that his last effort had cleared the microphone and that we were now in a position to trace the leak to earth.

I explained that I would put on a testing signal, and that every time he heard the signal that particular test had proved okay. The appropriate signal was very simply generated by applying my own receiver to its mouthpiece, which resulted in a tremendous squawk. As I had also asked him to listen very carefully for it, he was nearly deafened the first time I did it. I then asked him to place the receiver on the table beside him and touch it. I could, of course, hear the noise of his finger making contact, and immediately I repeated the squawk. When he picked up the receiver I told him that the test had been satisfactory and that we must now try some others, and I led him through a series of antics which involved him holding the receiver by the flex, and as far away from his body as possible, at the same time standing first on one leg and then on the other. When I had given him time to reach each position I duly transmitted the squawk, and thus got him engrossed in listening to it. After this series of tests I told him that we were now getting fairly near the source of the trouble, and that all we needed now was a good 'earth'.

When he asked what that would be I said, 'Well, sir, have you got such a thing as a bucket of water?' He said that he would try to find one, and within a minute or two he came back with the bucket. When he said, 'Well, what do we do now?' I told him to place the bucket on the table beside the telephone and to put his hand in the water to make sure that he was well earthed and then to touch the telephone again. When he did this he duly heard the appropriate squawk; and when he picked up the receiver again I told him that there was now only one final test and we would have it clinched. When he asked what this was I told him to pick up the receiver gently by the flex, and hold it over the bucket and then gently lower it into the water. He was quite ready to do so when Gerald Touch, who had been rolling on the floor with agonized laughter, thought the joke had gone far enough, and struggled to his feet. While not wishing to give the game away, he thought that he ought to stop our victim from doing any further damage, and he started to remonstrate, saying that putting the telephone into the water would irretrievably damage it. Our victim then said to me, 'I'm very sorry about this but I'm having difficulty. There is a chap here who is a physicist who says that if I put the telephone into the water it will ruin it!' I could not resist saying, 'Oh, a physicist, is he, sir. We know his kind—they think they know everything about electricity. They're always trying to put telephones right by themselves and wrecking them. Don't you worry about him, sir, it's all in my book here.' There was a great guffaw at the other end of the telephone while the victim said to Gerald Touch, 'Ha, ha, you hear that—the engineer said you physicists are always ruining telephones because you think you know all about them.' 'I'm going to do what he tells me.' As he tried to put the telephone in the water Gerald Touch seized his two wrists so as to try to stop him. They stood swaying in a trial of strength over the bucket and the victim being the stronger man was on the point of succeeding. I heard Touch's voice saying 'It's Jones, you fool!', and our victim, a manifest sportsman, collapsed in laughter.

Jones had a particular predilection for telephone jokes, and he recounts one perpetrated by a German friend, the physicist, Carl Bosch,

who about 1934 was working as a research student in a laboratory which overlooked a block of flats. His studies revealed that one of the flats was occupied by a newspaper correspondent, and so he telephoned this victim, pretending to be his own professor. The 'professor' announced that he had just perfected a television device which could enable the user to see the speaker at the other end. The newspaper man was incredulous, but the 'professor' offered to give a demonstration; all the pressman had to do was to strike some attitude, and the voice on the telephone would tell him what he was doing. The telephone was, of course, in direct view of the laboratory, and so the antics of the pressman were faithfully described. The result was an effusive article in next day's paper and, subsequently, a bewildered conversation between the true professor and the pressman.

The first story is from R. V. Jones's memoir of his wartime career, *Most Secret War* (Hamish Hamilton, London, 1978); the second is taken from a lecture by Jones reprinted in *Bulletin of the Institute of Physics*, June 1957, p. 193 (details of it are also in his memoir).

61 Trouble at t'lab

When judged by its impact on human life and happiness, the discovery of insulin was perhaps the most momentous event in the history of modern science. Up to the third decade of the twentieth century a diagnosis of diabetes—which an alert doctor could often make from the white spots of dried sugar that bespattered a male patient's shoes or trouser bottoms—presaged an early and miserable death. This could be delayed only by a starvation diet, no less agonizing for most patients than the disease.

The road to insulin was beset by misadventure, rancour, and deceit. The award of a Nobel Prize in 1923 to two of the principal actors, Frederick Banting (1891–1941) and John Macleod (1876–1935), inflamed several others, who felt (with some justice) that their efforts had been disparaged or forgotten. One of these was Nicolas Paulesco, a Romanian physiologist, who made the critical observation that linked diabetes to the dearth of an active component in the pancreas: he discovered that high levels of sugar in the blood and urine of dogs, rendered diabetic by extirpation of the pancreas, fell when the animals were injected with pancreatic extracts. Paulesco's work was interrupted for four years by the Austro-Hungarian invasion of his country before the end of the First World War, and by the time he returned to the problem Banting, Macleod, Best, and Collip in Toronto were closing in on their quarry.

A young German doctor, Georg Zuelzer, achieved what seemed to be a dramatic result when he injected pancreas extract into a dying patient, but his work, too, carried out under hopelessly unfavourable conditions, was terminated by the war. His claims to recognition were derided by a more famous German physiologist, Oskar Minkowski, who had been the first to make the connection between sugar and the pancreas. He was supposed to have been alerted to the presence of sugar in the urine of an incontinent, pancreatotomized dog when, during summer, flies gathered on the puddles on the laboratory floor. The veracity of this tale has been often asserted, by no less an authority even than the celebrated American physiologist W. B. Cannon; but Minkowski denied having made his discovery by such a stroke of chance. At all events, Minkowski, who was given the task by his superior of studying the function of the pancreas in the breakdown of fats, recognized the symptoms displayed by the pancreatotomized dog as those of diabetes mellitus. Minkowski's response to Zuelzer's plea for recognition was that he, too, regretted his failure to discover insulin.

The ultimate victory came to the research group in Macleod's Department of Physiology at the University of Toronto. Banting was the moving spirit and received initially grudging, but later enthusiastic support from Macleod. Charles Best, a student in the department, joined as Banting's assistant, and James Collip, a biochemist, was recruited and charged with isolating the elusive active factor in the pancreatic juice. Both Best and Collip felt passionately that they should have shared the Nobel Prize, while Banting, a man of intransigent and paranoid disposition, thought that the achievement was very much his own and lost no opportunity of traducing and defaming Macleod. Much of the mud stuck and it was often, and wrongly, alleged that Macleod had contributed little to the discovery and had stolen credit from the others. The division of the spoils, which left so much bitterness, was probably reasonably just, although many held that Best had been unfairly excluded (and he was soon rewarded with many prizes and honours), while Macleod, to the fury of Banting, stressed Collip's merits. He announced, indeed, that he would share his half of the prize money with Collip; to a friend he wrote: 'I think I have succeeded in getting people here to realise that his contribution to the work as a whole was not incommensurate with that of Banting.' Banting, meanwhile, had announced his intention to share his prize money with Best.

The most dramatic incident during the hunt for insulin occurred in January of 1922. Michael Bliss, author of the definitive study of the insulin story, describes it as 'one of the most remarkable personal confrontations in the history of science'. After many disheartening failures Collip had finally managed to prepare a highly active extract, which was probably largely pure insulin. (Soon after, he found himself unable to repeat his preparation and more time elapsed before he again succeeded.) Here is how Banting recalled the celebrated quarrel some 20 years later:

The worst blow fell one evening toward the end of January. Collip had become less and less communicative and finally after a week's absence he came into our little room about five thirty one evening. He stopped inside the door and said 'Well fellows I've got it'.

I turned and said, 'Fine, congratulations. How did you do it?'

Collip replied, 'I have decided not to tell you'.

His face was white as a sheet. He made as if to go. I grabbed him with one hand by the overcoat where it met in front and almost lifting him I sat him down hard on the chair. I do not remember all that was said but I remember telling him that it was a good job he was so much smaller—otherwise I would 'knock hell out of him'. He told us that he had talked it over with Macleod and that Macleod agreed with him that he should not tell us by what means he had purified the extract.

Charles Best remembered it differently:

One evening in January or February 1922, while I was working alone in the Medical Building, Dr J. B. Collip came into the small room where Banting and I had a dog cage and some chemical apparatus. He announced to me that he was leaving our group and that he intended to take out a patent in his own name on the improvement of our pancreatic extract. This seemed an extra-ordinary move to me, so I requested him to wait until Fred Banting appeared, and to make quite sure that he did I closed the door and sat in a chair which I placed against it. Before very long Banting returned to the Medical Building and came along the corridor to this little room. I explained to him what Collip had told me and Banting appeared to take it very quietly. I could, how-ever, feel his temper rising and I will pass over the subsequent events. Banting was thoroughly angry and Collip was fortunate not to be seriously hurt. I was disturbed for fear Banting would do something which we would both tremendously regret later and I can remember restraining Banting with all the force at my command.

Michael Bliss conjectures that Collip and Macleod were thoroughly affronted by Banting's antics in the preceding weeks, when he had appar-ently tried to conduct premature clinical trials with impure and possibly dangerous material, prepared by himself and Best. I presume, writes Bliss

that Collip and Macleod had little use for Banting's conduct in the past several weeks, particularly Banting's breaking the spirit of the collaboration by himself and Best making the extract for the first clinical trial. And, it appeared, Banting had appropriated some of Collip's improvements in mak-ing that extract. Banting had shown his distrust of them; now they had no reason to trust him. It was Collip's job to purify the extract, not Banting and Best's. Collip and Macleod may have decided that Banting was trying to take credit away from Collip—that if he knew the process for making the extract he would claim it as his own. They may have believed, after the misadventure of January 11 [when they had learned of Banting's surreptitious activities], that Banting could not be trusted not to try to forestall the rest of the team by applying for a patent. Paranoia begat paranoia. So Collip and Macleod decided not to tell Banting and Best the secret of making an effective anti-diabetic extract.

In later years Banting and Collip were reconciled and each paid tribute to the other's contribution to the great discovery. Indeed, in 1941, Banting, who was engaged in war work for the Canadian government, spent the last night of his life with Collip in Montreal before the bomber that was to convey him to England crashed, killing all on board.

For an enthralling account of the history of insulin, see Michael Bliss, *The Discovery of Insulin* (Macmillan, London, 1987).

62 The child is father to the man

Jeremy Bernstein, in his essay 'A child's garden of science', has collected the childhood reminiscences of a number of theoretical physicists, which reveal how their curiosity about numbers was first ignited.

The great theoretician Hans Bethe (busy into his nineties with theoretical physics, which he insists is the most interesting of all human occupations), when asked whether he had any early memories of mathematics, replied:

> Oh, yes—many. I was interested in numbers from a very early age. When I was five, I said to my mother on a walk one day, 'Isn't it strange that if a zero comes at the end of a number it means a lot but if it is at the beginning it doesn't mean anything?' And one day when I was about four, Richard Ewald, a professor of physiology, who was my father's boss, asked me on the street, 'What is .5 divided by 2?' I answered, 'Dear Uncle Ewald, that I don't know', but the next time I saw him I ran to him and said, 'Uncle Ewald, it's .25'. I knew about decimals then. When I was seven I learned about powers, and filled a whole book with the powers of two and three.

Stanislaw Ulam was a Polish mathematician (1909–84), who spent most of his working life in the United States and whose mathematical insights were critical to the construction of the hydrogen bomb. The following comes from his captivating autobiography, *Adventures of a Mathematician*:

> I had mathematical curiosity very early. My father had in his library a wonderful series of German paperback books—*Reklam*, they were called. One was Euler's *Algebra*. I looked at it when I was perhaps ten or eleven, and it gave me a mysterious feeling. The symbols looked like magic signs. I

wondered whether one day I would understand them. This probably contributed to the development of my mathematical curiosity. I discovered by myself how to solve quadratic equations. I remember that I did this by an incredible concentration and almost painful and not-quite conscious effort. What I did amounted to completing the square in my head without paper or pencil.

The following passage comes from the biography of Enrico Fermi [29] by his friend, the physicist, Emilio Segrè:

> Fermi told me that one of his great intellectual efforts was his attempt to understand—at the age of ten—what was meant by the statement that the equation $x^2 + y^2 = r^2$ represents a circle. Someone must have stated the fact to him, but he had to discover its meaning by himself.

This discovery, of the polar coordinate system, by a 10-year old must surely count as a prodigious achievement.

And here is how Freeman Dyson [52] brought to mind one of his earliest memories for Bernstein:

> He told me that, among them, was a time when he was still being put down for naps in the afternoon—he was not exactly sure of the age, but less than ten—and he began adding up numbers like $1 + \frac{1}{2} + \frac{1}{4} + \frac{1}{8} + \ldots$ and realized that this series was adding up to 2. In other words, he had discovered for himself the notion of the convergent infinite series.

Bernstein also notes that Einstein, who was always dissatisfied with his own mathematical grasp, devised a proof of Pythagoras's theorem (the square on the hypotenuse . . .) for himself when he was 12. This achievement, though, is eclipsed by the precocious prowess of Paul Erdös, the wildly eccentric Hungarian, to whom every minute not spent doing mathematics was a minute wasted; he could multiply three-digit numbers in his head at the age of three, manipulate squares and cubes at four, and by the time he was in his teens he had devised 37 proofs of Pythagoras's theorem.

The above stories are all to be found in Jeremy Bernstein's collection, *Cranks, Quarks, and the Cosmos* (Basic Books, New York, 1993), except the last, for which see Paul Hoffman's biography of Erdös, *The Man Who Loved Only Numbers: The Story of Paul Erdös and the Search for Mathematical Truth* (Fourth Estate, London, 1998).

63 Hooke's tease

Scientists have been known to complain of a want of candour in their con-frères, in pursuit of patents or merely priority. Matters were far worse in earlier days. The natural philosophers of the Enlightenment frequently sought to safeguard their claims to a discovery, while minimizing the risk of public error, by depositing their dated observations in an archive or by concealing them in a cipher. Isaac Newton's contemporary, Robert Hooke (1635–1703), was a formidable polymath, a friend of (among others) Christopher Wren, and it was he who designed the monument in Pudding Lane in the City of London to mark the spot at which the Great Fire started in 1666.

Hooke was habitually jealous in guarding what would now be called his intellectual property and mistrustful of his contemporaries. His name is enshrined in Hooke's Law of Elasticity, which states that extension of an elastic material is directly proportional to the applied stretching force. Hooke's interest in elasticity stemmed in part from his invention of the balance-spring watch. In 1665, Hooke formulated an encrypted description of this invention in the form: 'the True theory of Elasticity or Springiness, and a particular Explication thereof in several Subjects in which it is to be found: And the way of computing the velocities of bodies moved by them. *ceiiinosssttuu.*' This last was an anagram, as Hooke disclosed two years later, when he was sure of his results and satisfied that they could be applied to his balance-spring:

> About two years since I printed this Theory in an Anagram at the end of my Book of the Descriptions of Helioscopes, viz. ceiiinosssttuu, that is *Ut tensio sic vis*; That is, The Power of any Spring is in the same proportion with the Tension thereof: That is, if one power stretch or bend it one space, two will bend it two, and three will bend it three, and so forward. Now as the Theory is very short, so the way of trying it is very easie.

For an account of Hooke's work and that of his intellectual contemporaries see Lisa Jardine's *Ingenious Pursuits* (Little, Brown, London, 1999).

64 Know your adversary

J. G. Crowther (1899–1983) was a science journalist and popularizer, who seemed during his life to have met every scientist of any note. In his memoir of his life among scientists he tells stories of encounters during the two World Wars. The following, he thought, could be apocryphal but was at all events fully in character. E. A. Milne (1896–1960), the Oxford mathematician and cosmologist,

> offered his services to the War Office, and received some cyclostyled document informing him that his services would be called upon, if necessary. Milne, in view of his services in the First World War, and his eminence since, was infuriated by what he regarded as a discourteous reply. He used his connections to have his disapproval brought to the attention of the higher ranks of the War Office. He thereupon received an invitation, signed by a brigadier-general, to call at the Office. Milne arrived fulminating with criticism. He told the brigadier, who listened quietly to his harangue, that the War Office ought to know that the war would be a scientific one. In that case, was the way they had treated him the way to make the best use of science and eminent scientists?
>
> The brigadier waited until Milne had run out of his first breath, and then he asked: 'Did you win the Adams [mathematics] Prize in your year?' 'No,' replied Milne angrily, 'but what has that got to do with it?' 'I did', said the brigadier.

And here is his second tale of an interlocuter underestimated (in 1918):

> One day two tall American officers appeared at H.M.S. *Excellent* [the Admiralty gunnery establishment in Portsmouth harbour]. The authorities had not sent advance information on their names. They were placed in Richmond's charge [H. W. Richmond, mathematician of note], who brought them to our office, and began to explain to them in the simplest language what we were doing. Experience had taught him not to assume that visiting officers knew any mathematics, so he began by avoiding the use of calculus. The two Americans listened politely, and occasionally made quite sensible comments. Presently Richmond said in his gentle voice: 'Perhaps you know the calculus?' The taller of them, a fair, middle-aged man of agreeable manner, smiled slightly and said, 'Yes, we know the calculus'.
>
> Richmond sighed with relief and said that they would now be able to get on. He went into things a little more deeply, and the Americans' comments became still more intelligent. After a while Richmond looked up and said:

'Perhaps you are mathematicians?' The taller, fair American smiled quizzically and replied, 'My name is Veblen'. I was seated about four yards away, and I can still see Richmond jump in his chair, murmuring a series of inarticulate 'Oh's'. He and Oswald Veblen worked in the same field of mathematics, and he had spent half an hour trying to explain the solution of differential equations to him without mentioning the calculus.

Oswald Veblen (1880–1960), distinguished for his studies in branches of geometry, later became Director of the Institute of Advanced Study at Princeton, which gave refuge to Einstein.

There are other examples of such comic misapprehensions. Glenn Seaborg, Nobel Laureate in chemistry in 1951 for his work on the transuranic elements (radioactive elements, heavier than uranium, formed by nuclear bombardment), was scientific adviser to a succession of American presidents. On one occasion an aggressive interrogation by a congressional committee culminated in a rhetorical question from a bad-tempered Senator: 'What do you know about plutonium?' Seaborg was able to answer that it was he who had discovered the element.

The two stories told by J. G. Crowther are from his memoirs, *Fifty Years with Science* (Barrie and Jenkins, London, 1970). For the last story, see Glenn Seaborg, *A Chemist in the White House: From the Manhattan Project to the End of the Cold War* (American Chemical Society, Washington, DC, 1998).

~

65 The divine spark comes by night

Many scientists have experienced the transcendental flash of revelation while relaxing or between sleep and wakefulness. One captivating case was that of the Austrian physiologist Otto Loewi (1873–1961), Professor of Pharmacology at the University of Graz and chiefly remembered for his discovery of chemical transmission of the nerve impulse. In 1936, this brought him the Nobel Prize, which he shared with his English friend, Henry Dale [36]. A central question in neurobiology at the time was whether nerve impulses are imparted to the muscles that they control by way of a chemical mediator, for Dale had already shown that a substance found in the body, acetylcholine, could stimulate the action of a nerve; it could, for instance, slow down the beating of a heart, exactly like stimulation of the vagus nerve, which controls the heart muscle.

Loewi had fallen asleep one night over a novel, when he woke with a start, conscious that a dazzling revelation had erupted in his mind. He reached for a pencil and jotted down its essence. But when he awoke next morning he could neither recapture his great idea, nor, to his chagrin, interpret his note. All that day he sat in his laboratory, vainly hoping that the sight of all the familiar apparatus would jolt his memory, and trying without success to understand what he had scribbled. Loewi went to bed that night a disappointed man but during the small hours he woke, the idea once more effulgent in his mind. This time he was more careful in putting it on paper.

The next day he went to his laboratory and in one of the neatest, simplest and most defining experiments in the history of biology brought proof of the chemical mediation of impulses. He prepared two frogs' hearts which were kept beating by means of salt solution. He stimulated the vagus nerve on one of the hearts, thus causing it to stop beating. He then removed the salt solution from the heart and applied it to the other one. To his great satisfaction the solution had the same effect on the second heart as the vagus stimulating had on the first one: the pulsating muscle was brought to a standstill. This was the beginning of a host of investigations in many countries throughout the world on chemical intermediation, not only between nerves and muscles and the glands they affect but also between the nervous elements themselves.

The substance released by the stimulated nerve into the solution surrounding the heart was what is now known as a neurotransmitter, and was indeed acetylcholine.

The discovery that the heart could be kept beating for many hours was a happy accident: Sidney Ringer (1835–1910), a doctor at University College Hospital in London and spare-time pharmacologist, had worked for many years with excised frog hearts, which, suspended in physiological salt solution, would beat for at most a half-hour. Then one day a heart continued to beat, apparently indefinitely. Ringer was nonplussed; he thought at first that there might be a seasonal effect on the physiology of the amphibian heart, but on further investigation he discovered that his young assistant had that time been left to prepare the salt solution, and had used tap water instead of distilled water. As Henry Dale later wrote:

as Fielder himself [the errant 'lab boy'], whom I knew as an ageing man, explained to me, he didn't see the point of spending all that time distilling water for Dr. Ringer, who wouldn't notice any difference if the salt solution was made up with water straight out of the tap.

Ringer inquired of the New River Head Company, which in those days supplied the water to north London, what ions their tap water contained, and thus the essential nature of calcium ions in physiology was discovered. (Ions are the charged components, positive and negative, into which salts, such as common salt, sodium chloride, or, in this case, calcium chloride, dissociate when dissolved in water.) The medium used in physiological experiments is still referred to as Ringer solution.

As to Otto Loewi, because he was a Jew, he was driven out of his country after the *Anschluss* and found shelter in New York, not before being arrested by Austrian storm troopers and flung into gaol. Expecting the worst, and concerned about the safety of his wife and children, Loewi was nevertheless most of all tormented by the thought that his most recent laboratory results had not been prepared for publication and would be lost for ever if he were shot. He managed to scribble a brief account of his work and bribe a gaoler to post it to a scientific journal. This accomplished, he experienced an 'indescribable relief'. He was not shot: his influential friend, Sir Henry Dale, got him out of prison by threatening a boycott of Austrian scientists and Loewi and his family were eventually reunited in America.

Loewi was by no means the only Jewish scientist rescued *in extremis* by Henry Dale. In 1932, a year before Hitler came to power, Dale attended a conference in Germany and was impressed by a talk on the manipulation of acetylcholine release from nerves by a plant product (discovered by Loewi) called physostigmine. The speaker was a young physiologist by the name of Wilhelm Feldberg. The following year Feldberg was ejected from his position at Berlin University under the racial laws and was desperate to find a position in Britain or America. Hearing that an emissary of the Rockefeller Foundation was in Berlin to help those in need, he hastened to secure an interview.

He [the Rockefeller representative] was most sympathetic, but said something like this: 'You must understand, Feldberg, so many famous scientists have been dismissed whom we must help that it would not be fair to raise any hope of finding a position for a young person like you.' Then, more to comfort me, 'But at least let me take down your name. One never knows'. And when I spelt out my name for him, he hesitated, and said, 'I must have heard about you. Let me see'. Turning back the pages of his diary, he suddenly said, delighted [with] himself: 'Here it is. I have a message for you from Sir Henry Dale whom I met in London about a fortnight ago. Sir Henry told me, if by chance I should meet Feldberg in Berlin, and if he has been dismissed, tell him I want him to come to London to work with me. So you are all right', he said warmly. 'There is at least one person I needn't worry about any more.'

Feldberg went on to a distinguished career with the Medical Research Council in London, terminated only at the age of 89 by a grotesque and unfortunate episode. Feldberg had made an accidental discovery: less adroit in the laboratory than of yore, he knocked over a reading lamp, which fell on the abdomen of an anaesthetized rabbit. The heating that this occasioned resulted in a sharp rise in blood sugar. Feldberg received a grant to study this phenomenon, which was held to be of some physiological interest. At this stage a group of animal rights activists infiltrated his laboratory. Masquerading as a television crew, they asked, and were granted permission to film a programme about his research. Feldberg, assisted by an ageing technician, was plainly not in total control, and was filmed failing to anaesthetize a rabbit, and at one point apparently himself nodding off while injecting the animal. When the pictures were displayed in a national newspaper a scandal erupted, and amid much acrimony and embarrassment, Feldberg finally retired. He died a year later.

The story of Otto Loewi's great discovery comes from W. B. Cannon, *The Way of an Investigator* (Norton, New York, 1945); the recollection by Wilhelm Feldberg is to be found in *The Pursuit of Nature: Informal Essays on the History of Physiology*, ed. A. L. Hodgkin *et al.* (Cambridge University Press, Cambridge, 1977).

~

66 Following by example

Benjamin Silliman was a noted scholar, a professor at Yale University during the first half of the nineteenth century. He created the university's chemistry laboratories and became one of the country's foremost chemists of his time. Here he describes an exasperating incident, which occurred when he ordered from the local instrument manufacturer a dozen retorts. Having only a broken specimen, he sent both pieces to indicate exactly what was needed:

In due course my dozen green glass retorts, of East Haverford manufacture, arrived, carefully boxed and all sound, except that all were cracked off at the neck exactly where the pattern was fractured: and the broken neck and ball lay in state like decapitated kings in their coffins. The more than Chinese imitation afforded a curious illustration of the state of manufacture of chemical glass in this country, or rather in Connecticut; the same blunder would probably not have been made in Philadelphia or Boston.

See J. F. Fulton and E. H. Thomson, *Benjamin Silliman, Pathfinder in American Science* (Schuman, New York, 1947).

~

67 Science for survival

Science, some rational, some a ludicrous travesty, went on in the extermination camps of Nazi Germany, and saved some lives. In Auschwitz, chemists among the prisoners were recruited to work in the laboratories of the Buna-rubber factory, the Polymerization Department. They included a man who was to become a great writer and chronicler of endurance in the camp, Primo Levi. When the Kapo, the prisoner in charge of Levi's barracks, announced that a call was out for chemists to volunteer for laboratory work, Levi, already half-dead from starvation and toil, was one who responded. He was marched into the presence of Herr Doktor Ingenieur Pannwitz:

'*Wo sind Sie geboren?*'. He addresses me as *Sie*, the polite form of address. Doktor Ingenieur Pannwitz has no sense of humour. Curse him, he is not making the slightest effort to speak a slightly more comprehensible German.

I took my degree at Turin in 1941, *summa cum laude*—and while I say it I have the definite sensation of not being believed, of not even believing it myself; it is enough to look at my dirty hands covered with sores, my convict's trousers encrusted with mud. Yet I am he, the B.Sc. of Turin, in fact at this particular moment it is impossible to doubt my identity with him, as my reservoir of knowledge of organic chemistry, even after so long an inertia, responds at request with unexpected docility. And even more, this sense of lucid elation, this excitement which I feel warm in my veins, I recognize it, it is the fever of examinations, *my* fever of *my* examinations, that spontaneous mobilization of all my logical faculties and all my knowledge, which my friends at university so envied me.

The examination is going well. As I gradually realize it, I seem to grow in stature. He is asking me now on what subject I wrote my degree thesis. I have to make a violent effort to recall that sequence of memories, so deeply buried away: it is as if I was trying to remember the events of a previous incarnation.

Something protects me. My poor old 'Measurements of dielectrical constants' are of particular interest to this blond Aryan who lives so safely: he asks me if I know English, he shows me Gatterman's book, and even this is absurd and impossible, that down here on the other side of the barbed wire, a Gatterman should exist, exactly similar to the one I studied in Italy in my fourth year, at home.

Now it is over: the excitement which sustained me for the whole of the test suddenly gives way and, dull and flat, I stare at the fair skin of his hand writing down my fate on the white page in incomprehensible symbols.

Here is another example of a life saved through science:

[Paul] Langevin [a distinguished French physicist] told me how his daughter survived in Auschwitz. This was due to an S.S. officer who was a biologist, and wished to evade being sent to the Eastern front. He persuaded the German authorities that it might be worth trying to acclimatize Russian rubber plants to Poland. They allowed him to make a laboratory and garden at Auschwitz for this purpose. He picked out some biologists from the prisoners, whose usual length of life before entering the gas chambers was two weeks, to help him in the work.

One of these was a woman Jewish biologist of some standing. When the list of prisoners was scanned, the name of Langevin was noticed, and she said that Hélène Langevin was a biologist, so she was picked out. Langevin's daughter was in the camp for more than two years but survived, the rubber plant acclimatizers having slightly better conditions.

The first story is from Primo Levi's *If This is a Man* (Penguin Books, London, 1979); J. G. Crowther, in *Fifty Years with Science* (Barrie and Jenkins, London, 1970), recounts the second one.

~

68 The hounding of J. J. Sylvester

James Joseph Sylvester was a polymath, with a place in the history of his discipline as a mathematician of exceptional brilliance and versatility. He was also a qualified lawyer, a linguist, and man of letters, who wrote poetry and published a treatise on *The Laws of Verse*. He was born in humble circumstances in the Jewish East End of London and spoke with a marked cockney accent. It was probably anti-Semitism that denied him an academic position in England, although he was elected Fellow of the Royal Society while still in his twenties; and so in 1841, at the age of 27, he sailed for America to take up a professorial appointment at the University of Virginia.

The lucidity and sparkle of Sylvester's lectures on pure and applied mathematics earned him instant popularity with the more discerning students, but soon anti-Semitism reared its head: the local church newspaper deplored the influence that a Jew and, to boot, an Englishman, who could be assumed to deprecate slavery, might exert over Christian youth. Sylvester was exposed to insults from a few loutish students, especially two brothers, whom he had taken to task for their ignorance. The faculty cravenly refused to censure the brothers, for fear that student riots might result. Matters came to a head after Sylvester had received threats of violence:

> Sylvester bought a sword-cane, which he was carrying when way-laid by the brothers, the younger armed with a heavy bludgeon.
>
> An intimate friend of Dr Dabney's [the clergyman who related the story] happened to be approaching at the moment of the encounter. The younger brother stepped up in front of Professor Sylvester and demanded an instant and humble apology.
>
> Almost immediately he struck at Sylvester, knocking off his hat, and then delivered with his bludgeon a crushing blow upon Sylvester's bare head.
>
> Sylvester drew his sword-cane and lunged straight at him, striking him just over the heart. With a despairing howl, the student fell back into his brother's arms screaming out, 'I am killed!!' 'He has killed me!' Sylvester was urged away from the spot by Dr Dabney's friend, and without even waiting to collect his books, he left for New York, and took ship for England.
>
> Meanwhile, a surgeon was summoned to the student, who was lividly pale, bathed in cold sweat, in complete collapse, seemingly dying, whispering his last prayers. The surgeon tore open his vest, cut open his shirt, and at once declared him not in the least injured. The fine point of the sword-cane had struck a rib fair, and caught against it, not penetrating.
>
> When assured that the wound was not much more than a mosquito-bite, the dying man arose, adjusted his shirt, buttoned his vest, and walked off, though still trembling from the nervous shock.

Sylvester had not left for England, but remained in New York, where he had the prospect of a job at Columbia College (now Columbia University). But the board of visitors of the University of Virginia, overruling the faculty, refused Sylvester a letter of exculpation. Unemployed for a full year in New York, he then returned to London to earn his living practising at the bar. George Halsted, his later American student and admirer—for Sylvester ultimately returned to the country as a professor at Johns Hopkins University in Baltimore—wrote as follows:

> The five papers produced in the year 1841, before Sylvester's departure for Virginia, adumbrate some of his greatest discoveries. Then suddenly occurs a complete stoppage of this wonderful productivity. Not one paper, not one word, is dated from the University of Virginia. Not until 1844 does the wounded bird begin again feebly to chirp, and indeed it is a whole decade before the song pours forth again with mellow vigor that wins a waiting world.

If Sylvester lost what are most often a mathematician's most productive years, C. S. Peirce, the American philosopher, was nevertheless able to comment that his logical powers 'had never been equalled by more than two or three of all the sons of men'. In his later years Sylvester became an admired figure on the British scientific scene, a prolific writer, and much in demand as a public speaker.

This story is taken from an absorbing article by Lewis S. Feuer in *The Mathematical Intelligencer*, **9**, 13 (1987).

~

69 The quiet American

John Bardeen (1908–91) won two Nobel Prizes in physics—in 1956 and again in 1972. He was a large, serene man with a diffident manner and a soft voice. To the students who attended his lectures at the University of Illinois he was Whispering John. He shared his first Nobel Prize with two colleagues at the Bell Telephone Laboratories, one of them his lifelong friend, Walter Brattain, the other the head of the section, William Shockley. Bardeen, who had absorbed quantum mechanics during his days as a research student at Princeton, realized, when contemplating the behaviour of a semiconductor, that an electric current would undergo a disturbance at the interface between two microscopic crystalline regions. (A semiconductor has a conductivity for electricity between those of a conductor, such as a metal, and an insulator.) Such interfaces occur in crystals that contain impurities, and Bardeen's theoretical analysis showed how the current density (the concentration of electrons) would change in this region. The result of this theory, and of the experiments guided by it, was the transistor.

Bardeen's wife recalled how her husband returned from work one evening in 1948. He parked the car behind the house and came into the kitchen, where she was preparing dinner. 'As you know, his voice was always very quiet. He said, "We discovered something today".' In 1956, as he was scrambling eggs for breakfast one morning, he heard the announcement on the radio that he and his colleagues had been awarded the Nobel Prize.

After the Nobel Prize the companionship of the group deteriorated, for Shockley was evidently jealous of Bardeen's originality and theoretical acumen and chose to deny him the freedom to follow his own inclinations. (Shockley later became notorious for his vociferous advocacy of eugenic determinism and assertion of the superiority of the Caucasian race. In accordance with these views, he established a sperm bank for Nobel Laureates, so that the American gene pool might be rejuvenated with an intellectually and morally superior inheritance.) Bardeen left Bell to spend the remainder of his life at the University of Illinois. It was there that he formulated, in association with two of his students, an explanation of the phenomenon of superconductivity, which had perplexed theoreticians since its discovery some 50 years earlier [177]. This brought Bardeen his second Nobel Prize, which he shared with his two young associates. Bardeen told his faculty colleague, Charles Slichter, about the discovery:

> Bardeen stopped him in the hallway one day in the physics building at the University of Illinois, the morning after Bardeen, Cooper and Schrieffer had decided they had the BCS theory. Slichter reports: 'It was clear he had something he wanted to say, but he just stood there. I waited. Finally he spoke up. "Well, I think we've explained superconductivity!"' Although Bardeen was shy in many ways, Slichter says, 'If there was something really big that he had done, he wanted to tell someone'.

Bardeen's scientific productivity ended only with his death. The one other passion in his life was golf. Slichter tells of

Bardeen's long-time golf partner at the club remarking, 'Say, John, I've been meaning to ask you. Just what is it you do for a living?' Slichter asks, 'Can you imagine that? I think if I had won two Nobel Prizes like John had done, I would manage to work it into the conversation somewhere'.

From Gloria B. Lubetkin in *Physics Today*, **45**, April, p. 23 (1992).

~

70 Solving the insoluble

Robert Bunsen and Gustav Kirchhoff stand out as giants in the history of chemistry. Bunsen's most important monument is the science of spectroscopy—the analysis of lines or bands at characteristic wavelengths making up the light emitted or absorbed by chemical elements and compounds (for one of its later practical applications see [**45**]). He devised the famous burner that bears his name to generate a pale, almost colourless flame in which the colours of spectra could be viewed. Bunsen (born in Göttingen in 1811) was a much-loved, affable bachelor of slovenly habits; the wife of one of his colleagues at Heidelberg University declared that she wanted to kiss him, but first she would need to wash him. Kirchhoff, Bunsen's friend and collaborator, shared in much of the great work on spectral analysis and went on to contribute to many branches of physical chemistry. Bunsen and Kirchhoff occupied neighbouring laboratories in the physics building, the Friedrichsbau.

Its small beginnings in the middle of the last [nineteenth] century are marked by the name of Kirchhoff scratched on the window of what is now the private room of the senior assistant. From this window one may look out over the Rhine plain towards busy Mannheim, as Bunsen and Kirchhoff did one night when a fire was raging there, and they were able by spectroscopic examination of the flames to ascertain that barium and strontium were present in the burning mass. But the same window also looks across the Neckar to the Heiligenberg, along the slopes of which runs the 'Philosophers' Walk', the chief of the many paths among the wooded hills around the town, which the two friends were wont to traverse in their daily 'constitutionals'. Bunsen is known to have said that it was during such walks that his best ideas came to him. One day the thought occurred, 'If we could determine the nature of substances burning at Mannheim, why should we not do the same with regard to the sun? But people would say we must have gone mad to dream of such a thing.' All the world knows now what the result was, but it must have been a great moment when Kirchhoff could say, 'Bunsen, I *have* gone mad', and Bunsen, grasping what it all meant, replied, 'So have I, Kirchhoff!'

The light from the Sun, when examined in a spectroscope (a simple instrument in which a prism spreads the light into its rainbow components), had been found to be interrupted by a large number of thin black lines. In 1802, the English chemist William Hyde Wollaston (chiefly commemorated now in the biconical magnifying glass wielded by Sherlock Holmes in contemporary illustrations) was surprised to find seven of these interruptions in the Sun's spectrum; 10 years later, with better optics, Joseph Fraunhofer in Germany detected no less than 300 of these Fraunhofer lines, as they became known. Bunsen and Kirchhoff found that the wavelengths of a pair of the most prominent Fraunhofer lines coincided precisely with the lines in the yellow part of the spectrum emitted in a flame by sodium. They went on to identify the fingerprints of many other elements in the Sun's spectrum and their technique later led to the discovery of a previously unknown element, abundant in the Sun—the noble gas, helium.

To understand the significance of the story, and the reason for the two friends' elation, one has to recall that an influential philosopher and mathematician, Auguste Comte, had a few years earlier designated the composition of the Sun as one of the questions that would for ever remain beyond the reach of science. That observation—that the Sun, and as was later shown by the same method, remoter stars, were composed of the same elements as Earth—was a landmark in the history of science.

The passage about Bunsen and Kirchhoff is from an anonymous article in *Nature*, 65, 587 (1902).

~

71 A sceptic confounded

Logan Pearsall Smith was a man of letters, not a scientist, but he became a brother-in-law of Bertrand Russell. During his undergraduate years in Oxford he was befriended by Benjamin Jowett, the formidable Master of Balliol College and Professor of Greek. Jowett had strong views about the purpose of a university and was the prime advocate of the tutorial system. He saw research as a threat to the prevailing order and could discern no virtues in it—as this exchange with Logan Pearsall Smith reveals. The knee-jerk reflex was discovered by two German neurobiologists; reflexes, defined as impulses transmitted to the central nervous system and 'reflected' back to a muscle so as to induce an involuntary movement, were at the time a major preoccupation of physiologists.

> I remember once, when staying with him at Malvern [probably in 1885], inadvertently pronouncing the ill-omened word. 'Research!' the Master exclaimed. 'Research!' he said. 'A mere excuse for idleness; it has never achieved, and will never achieve any results of the slightest value.' At this sweeping statement I protested; whereupon I was peremptorily told, if I knew of any such results of value, to name them without delay. My ideas on the subject were by no means profound, and anyhow it is difficult to give instances of a general proposition at a moment's notice. The only thing that came into my head was the recent discovery, of which I had read somewhere, that on striking a patient's kneecap sharply he would give an involuntary kick, and by the vigour or lack of vigour of this 'knee jerk', as it is called, a judgement could be formed of his general state of health.
>
> 'I don't believe a word of it', Jowett replied. 'Just give my knee a tap.'
>
> I was extremely reluctant to perform this irreverent act upon his person, but the Master angrily insisted, and the undergraduate could do nothing but obey. The little leg reacted with a vigour which almost alarmed me, and must, I think, have considerably disconcerted that elderly and eminent opponent of research.

Logan Pearsall Smith, *Unforgotten Years: Reminiscences* (Constable, London, 1938; Little Brown, Boston, 1939).

~

72 Wrong experiment, right conclusion

During the apogee of atomic physics in the 1930s one of the movers of mountains was Ernest Orlando Lawrence (1901–58), who built at Berkeley in California, the first cyclotron—an instrument for accelerating charged particles, in particular protons, along a spiral path. The protons attained unprecedentedly high speeds, sufficient often to split nuclei in a target on which they impinged. The early cyclotron was the precursor of the giant atom smashers of today, located in underground caverns miles across. Lawrence was a man of demonic energy and uncontrollable impatience. There were reasons to believe (correct, as it transpired) that deuterons—the nuclei of the recently discovered heavy hydrogen or deuterium, which contain a neutron in addition to the proton—would be far more effective agents of destruction; Lawrence was accordingly consumed with the urgent desire to get his hands on some of this substance, which, as it happened, his colleague in the Chemistry Department, G. N. Lewis, was producing in the form of heavy water.

> Lawrence kept asking Lewis how much heavy water he had until about the first of March Lewis was able to show him a whole cubic centimeter. It was enough to accelerate, but at this point Lewis proved to be no physicist. Worried about whether he had manufactured a poison, he fed the whole sample to a mouse. It brought no good or harm to the mouse, but to Lawrence it almost brought apoplexy. 'This was the most expensive cocktail that I think mouse or man ever had!', he complained.

In actuality Lewis thought the mouse had shown signs of intoxication. Heavy water is, indeed, harmless. Much later, in the years after the Second World War, radioactivity entered the world of the biologists. Biological compounds, incorporating radioactive isotopes [20], became indispensable for the study of physiological reactions. (Their use depends on the principle that because isotopes of any element have the same number of electrons outside the nucleus, they are in chemical terms identical; so a small proportion of a radioactive isotope in the compound under study will act as a label by which the progress of the substance in a biological organism can be observed.) Radioactive biochemical compounds are now a commonplace but in the early days they were available to few researchers.

> It can be said that the new field of atomic medicine actually began at the University of California, where artificial radioactivity first became available for biological and medical research. Watching all the young men working around the cyclotron bombarding new targets and measuring the radiations with Geiger counters and Wilson cloud chambers, I was soon infected with the excitement of the early experiments. Very little was known of the biological effects of the neutron rays produced by the cyclotron, and this seemed an important place to start work.
> For the neutron ray exposures in Berkeley we made a small metal cylinder to house a rat so that it could be placed close to the cyclotron. After placing

the rat in position, we asked the crew to start the cyclotron and then turn it off again after the first two minutes. This two-minute exposure was arbitrary, since we had no basis for calculating how great a dose would produce an observable radiation effect on the animal. After the two minutes had passed we crawled into the small space between the Dees [the semicircular electrodes through which the spiral path followed by the accelerating particles ran] of the 37-in cyclotron, opened the cylinder, and found the rat was dead. Everyone crowded around to look at the rat, and a healthy respect for nuclear radiation was born. Now, of course, radiation protection measures are an integral part of all atomic energy research programs, but I think this incident of our first rat played a large part in the excellent safety record at the University. In fact we have had no radiation cataracts among the early cyclotron workers. We discovered later that the rat's death had resulted from asphyxiation rather than radiation. But since our failure to aerate the rat chamber adequately had brought about such a salutary effect on the crew, the post-mortem report was not widely circulated.

John H. Lawrence, the writer, records that the physicists, in hotfoot pursuit of results, were nevertheless reluctant to make time available on their instrument for animal experiments, and regarded the visiting biologists and doctors as something of a nuisance. He thinks this evaluation may have been intensified when he approached the cyclotron too closely with a pair of pliers, absentmindedly stuffed into his pocket. The magnetic field tore the pliers out and sucked them into the Dees, where they lodged for three weeks.

The first passage is from Nuel Pharr Davis, *Lawrence and Oppenheimer* (Jonathan Cape, London, 1969). Lawrence's reminiscence is in *California Monthly*, December (1957), reproduced in Robert L. Weber's *Science with a Smile* (Institute of Physics, Bristol and Philadelphia, 1992).

~

73 Old soldiers never die

It was Ernest Rutherford [16] who declared that scientists (by which he meant physicists) never grow old, for, unlike the less fortunate majority of the population who have no laboratories to play in, they retain throughout their lives the child-like pleasure of exploration. Here is a vignette of Rutherford's patron, J. J. Thomson, famous for many discoveries, most notably that of the electron. On his death at the age of 84 in 1940 his obituarist, the German theoretician Max Born (1882–1970), later Professor of Physics in Edinburgh, recalled

> It was Prof. J. J. Thomson's name which took me to Cambridge in 1906 . . .
>
> More than fifteen years later, on a visit to Cambridge, I met Thomson's son [later Sir George Paget Thomson and himself a physics Nobel Laureate], who took me to the Cavendish and to the basement room where 'J. J.' was working, surrounded by the usual complicated structures of apparatus, glass tubes and wires. I was introduced: 'Father, here is an old pupil of yours who studied with you years ago . . . ' The grey head, bent over a glowing vacuum tube, was lifted for a minute: 'How do you do. Now, look here, this is the spectrum of . . . ', and we were in the midst of the realm of research, forgetting the chasm of years, war and after-war, which lay between this rencontre and the days of our first acquaintance. This was Thomson in the Cavendish: science personified.

Thomson apparently retained his competitive disposition throughout his life. When F. W. Aston, who developed the mass spectrograph for the measurement of atomic weights, complained that Thomson would not believe the evidence for a new isotope, Rutherford told him he should be grateful. Had Thomson believed it, he said, 'the blighter would have swiped it from you'.

The obituary of J. J. Thomson by Max Born is in *Nature*, **146**, 356 (1940).

74 A case of night starvation

At 2 o'clock one morning in 1940 Andrew Nalbandov, a physiologist at the University of Wisconsin, drove home from a party. His route led him past his laboratory and, looking up, he was surprised to observe that the lights were on in the animal house. Nalbandov had been grappling for some time with an intractable problem: he was attempting to discover the function of the pituitary gland (which we now know is the source of the group of essential hormones that control a range of bodily activities). The pituitary, or hypophysis as it used to be called, is located just beneath the brain and is hard to get at surgically. All attempts to extirpate the pituitary from animals, in particular chickens, had led to death in a matter of days, so there was no chance to see whether and in what manner the animals might malfunction when deprived of the pituitary. Nalbandov describes his predicament:

Neither replacement therapy nor any other precautions helped and I was about ready to agree with A. S. Parkes [49] and R. T. Hill who had done similar operations in England, that hypophysectomized chickens simply cannot live. I resigned myself to doing a few short-term experiments and dropping the whole project when suddenly 98% of a group of hypophysectomized birds survived for 3 weeks and a great many lived for as long as 6 months. The only explanation I could find was that my surgical technique had improved with practice. At about this time, and as I was ready to start a long-term experiment, the birds again started dying and within a week both recently operated birds and those which had lived for several months were dead. This, of course, argued against surgical proficiency. I continued with the project since I now knew they could live under some circumstances which, however, eluded me. At about this time I had a second successful period during which mortality was very low. But, despite careful analysis of records (the possibility of disease and many other factors were considered and eliminated) no explanation was apparent. You can imagine how frustrating it was to be unable to take advantage of something that was obviously having a profound effect on the ability of these animals to withstand the operation. Late one night I was driving home from a party via a road which passed the laboratory. Even though it was 2 A.M. lights were burning in the animal room. I thought that a careless student had left them on so I stopped to turn them off. A few nights later I noted again that lights had been left on all night. Upon enquiry it turned out that a substitute janitor, whose job it was to make sure at midnight that all the windows were closed and doors locked, preferred to leave on the lights in the animal room in order to be able to find the exit door (the light switches not being near the door). Further checking showed that the two survival periods coincided with the times when the substitute janitor was on the job. Controlled experiments soon showed that hypophysectomized chickens kept in darkness all died while chickens lighted for 2 one-hour periods nightly lived indefinitely. The explanation was that birds in the dark do not eat and develop hypoglycaemia [low blood-sugar levels] from which they cannot recover, while birds which are lighted eat enough to prevent hypoglycaemia. Since that time we no longer experience any trouble in maintaining hypophysectomized birds for as long as we wish.

Thus was initiated a new chapter in the history of hormone research.

The story is told in W. I. B. Beveridge's book, *The Art of Scientific Investigation*, 3rd edn (Heinemann, London, 1960).

75 Fortunate furtive encounter

Max Born [**73**], one of the founders of quantum theory, was ejected from his professorial chair at Göttingen after the Nazi government promulgated its racial laws in 1933, and eventually found shelter in Edinburgh. It was a chance encounter with Rutherford [**16**] that opened this avenue of escape for him:

> In 1927 he had joined an international congress in Como. During one lecture which he did not find interesting and while they were showing some slides he had made good use of the darkness to sneak out of the auditorium. Checking the outside corridor just to make sure no one had seen him he saw another figure carefully slide out of another door also checking there was nobody around. It was Rutherford, who laughed and said to Born, 'you couldn't stand it either, let's have a trip on the lake'. In this way they spent the rest of the day and this was the beginning of their friendship which led Rutherford to invite Born to Cambridge in 1933. Later Born moved from Cambridge up to Edinburgh. This story is a simple example of how chance decided the fate of many in these troubled times.

According to George Gamow [**81**], one of the first things that met the already traumatized Born's eye, as he stepped off the train with his baggage in Cambridge, was a poster proclaiming *Born to be Hanged*. It had to be explained to him that this was merely an advertisement for a play at the local theatre.

From *Niels Bohr: Memoirs of a Working Relationship* by Stefan Rozental (Christian Ejlers, Copenhagen, 1998).

76 Eddington's disobedient conscience

At the outset, Einstein's Theory of Relativity (the General and the Special) was by no means universally accepted by physicists and astronomers. Its opponents were mainly those who could not bring themselves to discard the ether, the medium through which light waves were supposedly propagated; others could not swallow the tenet that time itself was relative, or that the velocity of light could not be exceeded. One of Einstein's most fervent proselytisers in this turbulent debate was the foremost British astronomer of his time, Sir Arthur Eddington (1882–1944).

Eddington was painfully shy but far from modest. His illustrious pupil Subramanyam Chandrasekhar recalled overhearing a conversation between Eddington and another astronomer, Ludwig Silberstein: Silberstein believed that he himself had a firm grasp of Einstein's theory and complimented Eddington for being one of the three people in the world to understand it. When Eddington hesitated Silberstein asked why he was flaunting his false modesty. 'Not at all,' came the reply, 'I am trying to think who the third one might be.' Eddington was, moreover, a Quaker and a pacifist and had a strong personal sympathy for Einstein, whose condemnation of German militarism at the start of the First World War had exposed him to widespread opprobrium. It is against this background that Eddington's determination to prove Einstein right must be seen.

Einstein, too, was eager that the predictions of his theory should be put to experimental test (more to convince the sceptics than for his own reassurance, for he was troubled by no doubts about its correctness). A testable prediction was that light would be bent by gravity; the most direct way to show this would be by observing the deflection of the light from a star when it passed close to the Sun. Stars almost in line with the Sun become visible during a total solar eclipse and such an eclipse was scheduled for the twenty-ninth of May 1919. It was partly on Eddington's initiative that two British expeditions set out to make the observations—one at Sobral in Brazil, the other, led by Eddington himself, on the island of Principe, off the west coast of Africa.

But the issue was not as simple as at first appeared, for the mighty Laplace [145] in the early nineteenth century and a German astronomer called Georg von Soldner, a little later, had both predicted, on the basis of Newtonian mechanics, that light (regarded as corpuscular in nature) would be deflected by a gravitational field. (Soldner's work lay forgotten until Einstein's adversary, the increasingly antisemitic and unhinged Philipp Lenard, rediscovered it in 1920 and used it in his diatribes against his *bête noire*.) The Newtonian model predicted a deflection of 0.875"—the units here are arc seconds, of which there are 3600 in a degree—whereas Einstein was predicting a deflection of 1.75″ (correcting an earlier calculation that had led to a value more or less indistinguishable from the Newtonian answer). But such deflections were very close to the limits of precision of the measuring techniques at the time. Could the telescopes on Sobral and Principe reliably discriminate, then, between about 0.9 and about 1.8″? Eddington obviously thought they could.

Conditions were more favourable in Brazil. The better of the two telescopes there gave an average deflection of 1.98″—rather too high for Einstein—and the lesser telescope gave 0.86″, as close as makes no difference to the Newtonian value. On Principe there had been cloud and only two of the sixteen photographic plates that could be exposed within the time of the eclipse showed measurable, though not very clear, images of stars. The average deflection inferred was 1.61″, with a generous error margin (standard deviation) of 0.3″. The results were presented at an extraordinary meeting, summoned for the purpose on 6 November 1919, of the Royal Society and the Astronomical Society, with Sir J. J. (Joseph) Thomson [73], President of the Royal Society, in the chair. The Astronomer Royal, Sir Frank Dyson, spoke first and reported as follows:

> The astrographic plates [photographic plates obtained by a telescope of one type] gave 0.97″ for the displacement at the limb when the scale-value was determined from the plates themselves, and 1.40″ when the scale-value was assumed from the check plates [photographs taken previously of the same stars by night]. But the much better plates gave for the displacement at the limb 1.98″, Einstein's predicted value being 1.75″. Further, for these plates the agreement between individual stars was all that could be expected.
>
> After a careful study of the plates I am prepared to say that there can be no doubt that they confirm Einstein's prediction. A very definite result has been obtained that light is deflected in accordance with Einstein's law of gravitation.

Dyson had made no mention of the data garnered on Principe. But Eddington, who spoke next, did not write off the Principe results, and, indeed, if one discards the answer from the less advanced Sobral telescope, the average of the remaining values—the uncomfortably high 1.98″ from Sobral and the sketchy 1.61″ from Principe—is exactly what Einstein's theory demanded. At this point Professor Silberstein spoke up: another attempt to verify the Theory of Relativity, based on a prediction about the magnitude of the red shift of light from remote stars, had failed; why should one then put one's faith in the dubious light-bending data, relying on measurements at the very limits of precision? Eddington had no convincing answer. (The red-shift anomaly was later resolved; a red shift arises from the change in oscillation frequency of radiation emanating from any receding object, and is an exact analogy to the fall in pitch of a whistle from a receding train.)

Here is a reminiscence from one of the members of the Principe expedition:

> As the problem then presented itself to us, there were three possibilities. There might be no deflection at all; that is to say light might not be subject to gravitation. There might be a 'half-deflection', signifying that light was subject to gravitation, as Newton had suggested, and obeyed the simple Newtonian law. Or there might be a 'full deflection', confirming Einstein's instead of Newton's law. I remember Dyson explaining all this to my companion Cottingham, who gathered the main idea that the bigger the result the more exciting it would be. 'What will it mean if we get double the deflection?' 'Then', said Dyson, 'Eddington will go mad, and you will have to come home alone.'

There can be little doubt that a major concern of Eddington's was to effect a reconciliation between German and Western scientists, soured by the patriotic excesses of the First World War. A host of German luminaries (but not, of course, Einstein) had signed the so-called Fulda declaration in 1914, exculpating their country from all responsibility for the war and affirming their solidarity with its army. This and later events had provoked an explosion of chauvinistic wrath in the scientific journals (such as *Nature*) in Britain, France, and the United States. The reported outcome of the expeditions caused a gratifying flurry of interest in the press— 'Revolution in Science, Newtonian Ideas Overthrown' was the headline in *The Times* and Einstein became an instant hero. There were, to be sure, other, later eclipse measurements, which gave conflicting and ambiguous results, but by then it was too late to quibble. The best professional minds were made up, and the public perception firmly established: the Theory of Relativity must be true. Eddington's conscience evidently troubled him a little. He owned much later to having been a mite biased, but in his obituary notice of his ally, Dyson, he wrote:

> The announcement of the results aroused immense public interest and the theory of relativity which had been for some years the preserve of a few spe-

cialists suddenly leapt into fame. Moreover, it was not without international significance, for it opportunely put an end to wild talk of boycotting German science. By standing foremost in testing, and ultimately verifying, the 'enemy' theory, our national Observatory kept alive the finest traditions of science; and the lesson is perhaps still needed in the world today.

A technically wrong deed perhaps, and to the purists more than reprehensible, but done for the most virtuous of reasons.

The story and the quotations are taken from an article on 'Relativity and eclipses' by J. Earman and C. Glymour, *Historical Studies in the Physical Sciences* 11, 49 (1980).

~

77 Smoking for the Führer

Fritz Houtermans's career was the stuff of fiction. He was German by birth but grew up and studied in Vienna. He was a physicist with, according to his friend Otto Frisch [20], a profound understanding of quantum theory. He pursued his theoretical work in Viennese cafés, where his prodigious capacity for coffee became legendary. His growing reputation took him to Germany, to one of the great centres of theoretical physics in Göttingen. Houtermans was one-quarter Jewish, so that, although he was proud of his ancestry—'when your ancestors were still living in the trees', he would tell his Aryan colleagues, 'mine were already forging cheques'—he was not under threat of racial persecution by the Nazis. He was, however, a committed Communist and for many years a party member, and this would have put his life in danger. He therefore decamped to England, where he worked at the EMI laboratories, and all but discovered the laser (the means, first achieved in 1960, of generating light of a single wavelength with a very high intensity). But life in England was not to his taste and he complained especially about the smell of boiled mutton. He moved again, this time to fulfil his old ambition of working in the Soviet Union. He found employment in the Physico-Technical Institute in Kharkov, which then housed a brilliant cluster of physicists, among them the great Lev Landau [137].

But Stalin's Great Terror was soon upon them; like many of his Soviet brethren Houtermans was arrested and endured the appalling privations of an NKVD prison. The appeals on his behalf of physicists in the West were disregarded. Houtermans was charged with spying for Germany and was tortured. Eventually it was a choice between death and confession, so he confessed, naming his German contacts as Messrs Scharnhorst and Gneisenau, long-dead generals who gave their names to two German battleships. His interrogators did not spot the deception, but his friends on the outside could guess under what circumstances the confession had been exacted.

Houtermans would probably have died in prison of starvation had he not been saved by the opportune signing of the Ribbentrop–Molotov pact between Germany and the Soviet Union in 1939. Asked where he wanted to be sent on his release, Houtermans opted for England, but his Soviet hosts despatched him instead to Germany and into the arms of the Gestapo. He was rescued through the intervention of the courageous Max von Laue [112], whose resolute and public opposition to the Nazi regime set him apart from the other leaders of German science. Houtermans was released from prison and put to work in the private laboratory in a Berlin suburb of a well-known physicist, inventor, and millionaire, Manfred von Ardenne. During this period Houtermans was despatched on several brief visits to his old haunts in the German-occupied Ukraine, with a commission set up by the German Navy to find out what the Soviet laboratories had been doing. On his return he would send food parcels to his friends in Kharkov, and he engaged in a dangerous game of sheltering Jews and other fugitives.

Von Ardenne's laboratory formed part of the German atomic bomb project, and Houtermans, on a visit in Switzerland, sent a telegram to England, warning that the German physicists were embarked on a programme of development. It was in Berlin that Houtermans nearly met his Waterloo once more. He was a chain-smoker (a habit that did for him in the end) and by 1945 tobacco was becoming very hard to find in Germany. So Houtermans approached Abraham Esau, the administrative head of the atomic bomb project, and persuaded him that Macedonian tobacco was rich in the heavy water [72] required for the manufacture of an atomic bomb. A sack of tobacco was accordingly procured and sent, as priority war material, to Houtermans. But when all had been consumed Houtermans overreached himself and asked for another consignment. This time suspicions were aroused and questions were evidently asked; the Gestapo instructed von Ardenne to sack Houtermans, and his arrest swiftly followed. Once more Laue, with some help from other leading physicists, managed to extract his reprobate friend, who was allowed to move to the physics institute in Göttingen. A few months later the war ended and Houtermans was finally safe.

Houtermans continued to work in Göttingen for seven years, his interests having now shifted to natural (geological) radioactivity; but he chafed under the restrictions imposed on scientists by the occupying Allies. Thus, for example, an upper limit was set on resistors permitted for laboratory use. This limit was 10^9 ohms, and an incensed Houtermans protested that even a pencil had a higher resistance than that. In 1962, he received a call to the Chair of Physics at the University of Berne. In this backwater he developed a vigorous programme of research, but four years later he died of lung cancer at the age of 63.

Fragmentary accounts of Houtermans, the man, and his tempestuous career, can be found in Otto Frisch's memoirs, *What Little I Remember* (Cambridge University Press, Cambridge, 1979); in those of George Gamow, *My World Line* (Viking Press, New York, 1970); and in Hendrik B. G. Casimir's autobiography, *Haphazard Reality* (Harper and Row, New York, 1983). An interesting brief biography of Houtermans, written by I. B. Khriplovich, appeared in *Physics Today* 45, 29 (1992).

~

78 Polish and perish

In the early years of the twentieth century it became apparent to nutritionists that most foods contain trace substances essential for life. Casimir Funk [163], a Polish biochemist, gave them the name *vitamines*, from 'vital amine'. It was a misnomer, for when the structures of several vitamins were determined they turned out not to be amines at all. The first of these substances to be discovered was what is now known as vitamin B_1, or thiamine. It came to light through a happy accident.

Beri-beri is a degenerative condition, leading to early death. It has ravaged populations throughout history, and late in the nineteenth century an epidemic of the disease swept through the Dutch East Indies. In 1886, the Government in Holland despatched a small team of experts to investigate. Two were scientists, Clemens Winkler and Cornelis Pekelharing by name, and they were accompanied by a young army doctor, Christiaan Eijkman (1858–1930). At this time the germ theory of disease, promulgated by the fathers of microbiology, Louis Pasteur [172] and Robert Koch [155], held sway, and the Dutchmen assumed that beri-beri was caused by a bacterial infection. For two years they toiled to isolate a bacterium and were finally persuaded that they had succeeded. Winkler and Pekelharing returned home, leaving Eijkman to conclude matters before joining them. But Eijkman grew uneasy. At the military hospital that was his base he had been working with chickens afflicted with what appeared to be the same disease, but all his efforts to transmit the supposed infection from sick or dead birds to healthy ones proved inconclusive. Indeed, even those isolated as normal controls fell sick. Nor did minute examination reveal any bacterium or parasite.

Eijkman had begun to wonder whether he and his colleagues might not after all have been on the wrong track, when a chance observation came to his aid: the disease that had rampaged through the fowl population all that summer and autumn abruptly abated. Eijkman looked for any changes in the treatment of the chickens and soon discovered that at the critical period a new cook had arrived at the hospital. The cook was responsible for feeding the chickens, as well as the patients; but he did not want to waste his boiled rice on chickens and so he bought in a consignment of cheap polished rice. Eijkman divided his chickens into two groups: one fed polished, the other unpolished rice. This was the answer: the first group quickly fell sick but could be cured when administered rice polishings, the second group thrived. Eijkman inferred that the chickens were being poisoned by a toxic substance in the rice and that the polishings contained an antidote, but it was Pekelharing who made the correct deduction: the rice polishings contained an 'active principle' which warded off beri-beri. The vitamin was isolated in 1912 by Frederick Gowland Hopkins, a renowned Cambridge biochemist. It was not until 17 years later that Eijkman and Gowland Hopkins (Pekelharing being by then dead) shared the Nobel Prize for their work.

See C. Eijkman, in *Nobel Prize Lectures in Physiology and Medicine 1922–1944* (Elsevier, Amsterdam, 1965).

~

79 Baccy and quanta

Although quantum theory had its origins in Albert Einstein's work on the photoelectric effect (for which, and not for relativity, he received the Nobel

Prize), its indeterminacy was something Einstein could never accept. This was the origin of his famous assertion that the Lord does not play dice. If the Universe was ruled by chance, he said, he would rather be a croupier in a gambling casino than a physicist. Einstein [161] was not alone in his epistemological distaste. In 1913, two future winners of the Nobel Prize, Einstein's then assistant, Otto Stern [6] and Max von Laue [112], while on a walk up the Ütliberg near Zurich, shook hands on an oath, contemptuously dubbed by Wolfgang Pauli [25] the Ütlischwur (an allusion to William Tell's *Rütlischwur*—the oath that led to the union of the Swiss cantons): 'If this nonsense of Bohr's should in the end prove right we will leave physics.' (They of course reneged.)

The tireless efforts of Niels Hendrik David Bohr (1885–1962) to convince Einstein, which continued for decades, were not unlike those of a priest grappling with a heretic for the salvation of his soul. Abraham Pais, friend, disciple, and biographer of Bohr, recalls the following encounter, which took place in Bohr's office.

After we had entered, Bohr asked me to sit down ('I always need an origin for the co-ordinate system') and soon started to pace furiously round the oblong table in the centre of the room. He then asked me if I could put down a few sentences as they would emerge during his pacing. It should be explained that, at such sessions, Bohr never had a full sentence ready. He would often dwell on one word, coax it, implore it, to find the continuation. This could go on for many minutes. At that moment the word was 'Einstein'. There Bohr was, almost running around the table and repeating: 'Einstein . . . Einstein . . .' It would have been a curious sight for someone not familiar with Bohr. After a little while he walked to the window, gazed out, repeating every now and then: 'Einstein . . . Einstein . . .'

At that moment the door opened very softly and Einstein tiptoed in.

He beckoned to me with a finger on his lips to be very quiet, his urchin smile on his face. He was to explain a few minutes later the reason for his behaviour. Einstein was not allowed by his doctor to buy any tobacco. However, the doctor had not forbidden him to steal tobacco, and this was precisely what he set out to do now. Always on tiptoe he made a bee-line for Bohr's tobacco pot, which stood on the table at which I was sitting. Meanwhile Bohr, unaware, was standing at the window, muttering 'Einstein . . . Einstein'. I was at a loss what to do, especially because I had at that moment not the faintest idea what Einstein was up to.

Then Bohr, with a firm 'Einstein', turned around. There they were, face to face, as if Bohr had summoned him forth. It is an understatement to say that for a moment Bohr was speechless. I myself, who had seen it coming, had distinctly felt uncanny for a moment, so I could well understand Bohr's own reaction. A moment later the spell was broken when Einstein explained his mission and soon we were all bursting with laughter.

See Abraham Pais, *Niels Bohr's Times* (Oxford University Press, Oxford, 1991).

~

80 The country doctor, his captive, and the professor

The history of physiology—and, of course, of medicine—is replete with examples of daring self-experimentation [85]. Less often the fearless experimenters have used members of their family as captive guinea-pigs. Edward Jenner [143], the country doctor credited with discovering the means to inoculate against smallpox, was perhaps most famous of all those who engaged in such dubious practices. Here is a later and notably reckless example in which scientific curiosity took precedence over paternal concern. It led in 1894 to a seminal discovery in physiology:

> Dr George Oliver, a physician of Harrogate, employed his winter leisure in experiments on his family, using apparatus of his own devising for clinical measurements. In one such experiment he was applying an instrument for measuring the thickness of the radial artery [in the arm]; and, having given his young son, who deserves a special memorial, an injection of an extract of the suprarenal gland, prepared from material supplied by the local butcher, Oliver thought that he had detected a contraction or, according to some who have transmitted the story, an expansion of the radial artery. Whichever it was, he went up to London to tell Professor Schäfer [Professor of Physiology at University College, later metamorphosed into Sir Edward Sharpey-Schäfer] what he thought he had observed, and found him engaged in an experiment in which the blood pressure of a dog was being recorded; found him, not unnaturally, incredulous about Oliver's story and very impatient at the interruption. But Oliver was in no hurry, and urged only that a dose of his suprarenal extract, which he produced from his pocket, should be injected into a vein when Schäfer's own experiment was finished. And so, just to convince Oliver that it was all nonsense, Schäfer gave the injection, and then stood amazed to see the mercury mounting in the arterial manometer till the recording float was lifted almost out of the distal limb.

Thus the extremely active substance formed in one part of the suprarenal gland, and known as adrenaline, was discovered.

This account is taken from a lecture, 'Accident and opportunism in medical research', by Sir Henry Dale, reprinted in the *British Medical Journal*, ii, 451 (1948).

~

81 Whispers from the void

Late in life Andrei Sakharov, the venerated physicist, begetter of the Soviet hydrogen bomb, and indefatigable dissident, told an interviewer: 'Do you know what I love most in life? It is the radio background emanation—the barely discernible reflection of unknown cosmic processes that ended billions of years ago.' This radiation was discovered, or at least publicized, in 1965, but its existence had been predicted some 20 years before. It was another Russian, the emigré physicist George Gamow (1904–68), who formulated the theory of what became known as the big bang—the moment when the Universe created itself out of nothing. Edwin Hubble at the Mount Wilson Observatory in California had discerned the famous red shift in the light from remote stars, which told him that the Universe is expanding. Working back from the rate of this expansion, Gamow calculated what would have happened at time zero, when the mass of the matter that now makes up the Universe erupted from its point of origin. The creation of all this matter would have been accompanied by the release of an enveloping torrent of radiation. As the Universe inflated so this spread outwards, diminishing in energy density as it thinned.

It was left to Gamow's associates, Ralph Alpher and Robert Herman [137] (the man who resisted Gamow's demand that he change his name to Delter), to calculate what the energy of the radiation should be today. The answer (for engineers and astronomers express the energy of radiation in terms of the temperature of an idealized hot body giving off the same kind of flux) was 2.7 degrees on the Kelvin scale (which measures from the absolute zero of temperature, at which all motion stops). Their paper on the subject was published in a physics, not an astronomy journal, and fell on deaf ears. Jeremy Bernstein, who has written most lucidly on the episode, puts this down to physicists' customary asperity about cosmology—

expressed most candidly in the words of Lev Landau, the Russian physicist [137]: 'Cosmologists are often in error but never in doubt.'

We move now to the laboratories of the Bell Telephone Company in New Jersey, for decades one of the world's foremost centres of discovery and invention, thanks to the company's enlightened practice of engaging the best scientists and allowing them scope to follow their own ideas, even when these bore no evident relation to practical objectives. A discovery of momentous importance to astronomy had already been made at Bell, for in 1929 an engineer, Karl Jansky, was entrusted with the task of tracking down the sources of static in short-wave radio reception so that ways could be sought of suppressing it. Jansky built a sensitive antenna on the roof of the laboratory in Holmdel and soon identified thunderstorms, near and remote, as a major source, but after that there still remained a continuous hiss, varying in intensity on a diurnal cycle. Eventually he found that the hiss came from the centre of the Milky Way, and so prefigured the science of radio astronomy.

Bell did not pursue the subject further, but 30 years later the thoughts of radio engineers turned to satellite communications, and, to make a start, they decided to try bouncing microwave signals (that is to say radiation of wavelengths of a centimetre or so up to a metre) off a weather balloon. To receive the signals a giant horn antenna was constructed, but it was agreed that when it had served its immediate purpose the Bell scientists could use it as a radio telescope for astronomical observations. The interested parties were two physicists, Arno Penzias and Robert Wilson. The intensity of background noise in microwave radiation from known sources had been calculated, but Penzias and Wilson found to their annoyance that the noise level recorded from their antenna was significantly higher. The temperature of the ubiquitous background was about 2.7 degrees. They tried all they knew to eliminate it. First some pigeons, found roosting in the horn, were displaced, together with the resulting encrustation of 'white dielectric deposit'. This did not solve the problem, nor could any other source of noise be discovered: it did not come from the nearby city of New York, nor was it a residuum of the radiation released by nuclear bomb tests.

Penzias and Wilson were defeated, but then fate intervened: one day in 1964, Penzias was chatting on the telephone to an astronomer friend at the Massachusetts Institute of Technology, who inquired how his work was going. Penzias told his plaintive tale and then his friend recalled a conversation with a colleague at the Carnegie Institute in Pittsburgh, who had related that while on a visit to Johns Hopkins University in Baltimore he had chanced to attend a lecture by a young astronomer from Princeton, called P. J. E. Peebles. Now Peebles was a student of Robert Dicke, who had a particular interest in the predicted cosmic background radiation. He had not read the papers of Gamow or of Alpher and Herman, but had covered the same territory himself and had set up an antenna on the roof of his department at Princeton University to see what he might detect. (As a wise commentator on the ways of scientists once observed, 'two months in the laboratory can often save an hour in the library'.) Penzias's friend suggested that he and Dicke might find common ground.

Dicke and Peebles immediately recognized that Penzias and Wilson had unknowingly shot their fox. But Penzias and Wilson were unimpressed, the more so because Wilson had learned what cosmology he knew from Fred Hoyle, the British astronomer, who had propounded the steady-state theory of the Universe and would have none of the big bang—the derisive name he had himself invented for Gamow's conception. At all events, in July 1965 the two groups published their papers in the same journal: that of Penzias and Wilson simply reported their observations and drew no inferences, while Dicke and his colleagues set out the theoretical grounds for equating the radiation with the cosmic microwave background. In 1978, it was Penzias and Wilson who were garlanded with the Nobel Prize.

But, Jeremy Bernstein points out, evidence for the cosmic background radiation actually predated the theory, for in 1941 an astronomer named Andrew McKellar had examined the wavelengths of light arriving from a constellation, which carried the signature of an organic compound, cyanogen. Analysis of the spectrum revealed that the temperature of this gas was 2.3 degrees. In a standard textbook on the spectra of molecules, published some years later, another Nobel Laureate, Gerhard Herzberg, noted this result, remarking that its meaning was questionable. Herzberg had not read the papers by Gamow and by Alpher and Herman, which explained its meaning; nor had they read Herzberg's book.

The conclusion one might draw from the saga of the cosmic background radiation, seen now as an ineluctable vindication of the big bang theory, is that scientists too seldom leave their own comfortable hutches.

A fine account of the episode and its background is in Jeremy Bernstein's collection of writings, *Cranks, Quarks, and the Cosmos* (Basic Books, New York, 1993).

82 The lying stones of Mount Eivelstadt

This cause célèbre occurred in the eighteenth century and brought ridicule on a German scholar at the University of Würzburg. Dr Johann Beringer was not only a member of the faculty there, but was an intimate of the Prince-Bishop, whom he attended as personal physician. Beringer was also a keen palaeontologist and collected fossils. In 1725, some young men of the locality presented him with a collection of finds from a site near the town on Mount Eivelstadt. The specimens were fakes, stones incised with images of a wide variety of modern animals and plants. As the collection grew Beringer's excitement mounted and by 1726 he had published a book, describing his discoveries. They were not confined to plants and animals:

> Here were clear depictions of the sun and moon, of stars and of comets with their fiery tails. And lastly, as the supreme prodigy commanding the reverend admiration of myself and of my fellow examiners, were magnificent tablets engraved in Latin, Arabic and Hebrew characters with the ineffable name of Jehovah.

The site had, of course, been 'salted' by the mischievous students. Finally, Beringer, rooting around on Mount Eivelstadt, came upon the culminating specimen, a stone that bore his own name. The stricken savant demanded an inquiry and it was soon established that the volunteer helpers had been hired by two of Beringer's colleagues at the university, who found him insufferably arrogant and had resolved to puncture his pomposity a little. But, alarmed that the hoax was succeeding all too well, they had tried to warn their victim not to publish his book because the stones might be frauds. Secure in his self-esteem, he did not heed the rather obvious hints. It was said that Beringer devoted much of the rest of his life to collecting up copies of his book (a fate shared more than 200 years later by a Polish professor, who had published a book on genetics just before this science was proscribed by the communist regime, in thrall to the nonsensical doctrines of the Russian charlatan Lysenko [86]).

See *The Lying Stones of Dr Johann Bartholomew Adam Beringer—being his Lithographiae Wurceburgensis*, translated and edited by Melvin E. Jahn and Daniel J. Woolf (University of California Press, Berkeley, 1963).

83 The mind of a mathematician

John (Jáncsi or Johnny to his friends) von Neumann was one of a remarkable group of Hungarian physicists and mathematicians, who emerged
from Budapest in the years after the First World War. His interests were
exceptionally wide: his contributions to theoretical physics, to the mathematical concepts on which the modern computer is based, to many areas
of pure mathematics, to games theory, and even to economics, are prodigious. He participated critically in the Manhattan Project and in a whole
series of other American military undertakings. He supervised the construction at Princeton University of the world's fastest computer in the
years after the Second World War, the Johnniac, of which he quipped: 'I
don't know how really useful this will be. But at any rate it will be possible
to get a lot of credit in Tibet by coding "*Om Mane Padme Hum*" [the
mantra meaning 'oh, thou flower of the lotus'] a hundred million times in
an hour. It will far exceed anything prayer wheels can do.' His friend and
collaborator, Hermann Goldstine, declared that von Neumann was not
human, but a demi-god, who 'had made a detailed study of humans and
could imitate them perfectly'. Johnny von Neumann died in 1957 at the age
of 53.

Abraham Pais, who was on intimate terms with most of the great physicists of the period, wrote of von Neumann:

> In my life I have met men even greater than Johnny, but none as brilliant. He
> shone not only in mathematics but was also fluently multilingual and
> particularly well-versed in history. One of his most remarkable abilities I
> soon came to note was his power of absolute recall.

This attribute is illustrated in the following reminiscence by Herman
Goldstine:

> As far as I could tell, von Neumann was able on once reading a book or
> article to quote it back verbatim; moreover, he could do it years later without
> hesitation. He could also translate it with no diminution in speed from its
> original language into English. On one occasion I tested his ability by asking
> him to tell me how the *Tale of Two Cities* started. Whereupon, without any
> pause, he immediately began to recite the first chapter and continued until
> asked to stop after about ten or fifteen minutes. [Von Neumann was in fact
> not alone among the great mathematicians in his powers of total recall. Three

centuries earlier, Gottfried Leibniz [15] in old age could recite the entire *Aenead*, which he had not read since childhood.] Another time, I watched him lecture on some material written in German about twenty years earlier. In this performance von Neumann even used exactly the same letters and symbols he had in the original. German was his natural language, and it seemed that he conceived his ideas in German and then translated them at lightning speed into English. Frequently I watched him writing and saw him ask occasionally what the English for some German word was.

Von Neumann could also calculate in his head with preternatural speed and precision. Here is Goldstine again:

One time an excellent mathematician stopped into my office to discuss a problem that had been causing him concern. After a rather lengthy and unfruitful discussion, he said he would take home a desk calculator and work out a few special cases that evening. The next day he arrived at the office looking very tired and haggard. On being asked why he triumphantly stated he had worked out five special cases of increasing complexity in the course of a night of work; he had finished at 4.30 in the morning.

Later that morning von Neumann unexpectedly came in on a consulting trip and asked how things were going. Whereupon I brought in my colleague to discuss the problem with von Neumann, who said, 'Let's work out a few special cases'. We agreed, carefully not telling him of the numerical work in the early morning hours. He then put his eyes to the ceiling and in perhaps five minutes worked out in his head four of the previously and laboriously calculated cases! After he had worked about five minutes on the fifth and hardest case, my colleague suddenly announced out loud the final answer. Von Neumann was completely perturbed and quickly went back, at an increasing tempo, to his mental calculations. After perhaps another five minutes he said, 'Yes, that is correct'. Then my colleague fled, and von Neumann spent perhaps another half hour of considerable mental effort trying to understand how anyone could have found a better way to handle the problem. Finally he was let in on the situation and recovered his aplomb.

The quotations are taken from Abraham Pais's book, *The Genius of Science* (Oxford University Press, Oxford, 2000) and Herman Goldstine's *The Computer* (Princeton University Press, 1980).

84 The old melon

Adolf von Baeyer (1835–1917) was a colossus in the great era of organic chemistry in the nineteenth century, one of the founders of the subject, in which German supremacy stood unchallenged. His laboratory in Munich was a Mecca for aspiring chemists from all over the world, one of whom, John Read [23], later professor in Aberdeen, offered some vignettes of life there in his book, *Humour and Humanism in Chemistry*. What follows is taken from Read's review in *Nature* of a memoir about Baeyer and his times by one of his German associates. At the time of the events recorded here Baeyer had entered into a new area of organic chemistry; his affections were for the moment concentrated on two substances, both of them important and versatile intermediates in organic syntheses:

By means of a 'Kunstgriff' [a masterstroke] of which Baeyer was very proud (treatment with sodium amalgam in presence of sodium bicarbonate), the diketone was reduced to quinitol. At the first glimpse of the crystals of the new substance Baeyer ceremoniously raised his hat!

It must be explained here that the Master's famous greenish-black hat plays the part of a perpetual epithet in Prof. Rupe's narrative. As the famous sword pommel to Paracelsus, so the 'alte Melone' [the melon, as a bowler is termed in Germany] to Baeyer; the former was said to contain the vital mercury of the medieval philosophers; the latter certainly enshrined one of the keenest chemical intellects of the modern world . . . Baeyer's head was normally covered. Only in moments of unusual excitement or elation did 'the Chef' [boss] remove his hat: apart from such occasions his shiny pate remained in permanent eclipse.

When, for example, the analysis of the important diacetylquinitol was found to be correct, Baeyer raised his hat in silent exultation. Soon afterwards the first dihydroxybenzene was prepared, by heating dibromohexamethylene with quinoline: Baeyer ran excitedly to and fro in the laboratory, flourishing the 'alte Melone' and exclaiming: 'Jetzt haben wir das erste Terpen, die Stammsubstanz der Terpene!' ['Now we've got the first terpene, the base-substance of the terpenes'—a class of compounds important in nature and the basis of many drugs]. Such is the picture from behind the scenes of the dramatic way in which the Master entered upon his famous investigations of terpenes.

Incidents of this kind may appear to be slight, yet cumulatively they throw a stream of light on the personality of this great chemist. There is no doubt, for example, that at times 'the Chef' was unduly impulsive. One morning he burst into the private laboratory, and, without having lit his cigar (an indication in itself of unusual emotional disturbance), raised the ancient 'Melone' twice, and exclaimed: 'Gentlemen [the audience was composed of Claisen and Brüning], I have just had word from E. [Emil] Fischer that he has brought off the complete synthesis of glucose. This heralds the end of organic chemistry: let's finish off the terpenes, and only the smears ('Schmieren') will be left!' [the derisive term *Schmierchemie* was used by organic chemists to denote physiological chemistry, or biochemistry as it is now known].

Baeyer favoured the use of simple apparatus, and the introduction into his laboratory of any device savouring of complexity had to be undertaken with great tact. The first mechanical stirrers, worked by water turbines, were smuggled in one evening. On the following morning, 'der Alte' beheld them in full working order. For a time he affected to ignore them; then he contemplated them unwillingly, with no air of challenge; next came the first remark, so anxiously awaited: 'Geht denn das?' [Is it working?] 'Jawohl, Herr Professor, ausgezeichnet, die Reduktionen sind schon bald fertig.' [Excellently, the reductions are already nearly done.] The Herr Professor was finally so much impressed that he took the exceptional step of summoning the Frau Professor [as the wives of professors were addressed]. 'Die Lydia', as she was called in the laboratory, stood by the merrily clattering apparatus for a while in silent admiration; then she uttered these unforgettable words: 'damit müsste man gut Mayonnaise machen können'! What a great deal depends on one's point of view.

This could indeed have been the genesis of the food processor.

From John Read's review, *Nature*, **131**, 294 (1933).

~

85 Strong medicine

The chemists of the German pharmaceutical company C. H. Boehringer und Sohn were after a vasoconstrictor (an agent that causes swollen blood vessels to contract) to mitigate the symptoms of the common cold. If such a compound could be found that would pass through the mucous surfaces, it might, when dropped into the nose, shrink the small vessels and unblock the nasal passage. Helmut Stähle had synthesized a series of related compounds (familiar to organic chemists as imidazoline derivatives), which he hoped might serve the purpose. One day in 1962 samples were delivered to the medical director of the company, Dr Wolf. By a happy chance Dr Wolf's secretary, Frau Schwandt, was in the grip of a heavy cold, and feeling that little harm could come from trying a smidgeon of the new substance, which was of a kind judged generally harmless, he instilled a little of a dilute solution into her nose. Frau Schwandt apparently yawned and fell into a deep slumber from which she could not be roused until the next day. There was naturally much alarm; a doctor was summoned and discovered that the human guinea-pig's blood pressure had dropped precipitately. Frau Schwandt happily suffered no lasting harm and the chemical eventually reached the market under the name of Clonidine. It acted, it turned out, on the peripheral nervous system, and found widespread use as a treatment for hypertension and a range of other disorders.

There are, of course, innumerable recorded instances (see also [80] and [134]) of heroic experiments conducted by physiologists, pharmacologists, and doctors on themselves and their colleagues. One such led to the discovery, which could never have come about by design, of a radically new treatment for alcoholism. The chief protagonist in the drama was a pharmacologist, the head of research of a Danish pharmaceutical concern, Erik Jacobsen. The events unfolded during the Second World War.

It was the custom of Jacobsen and his colleagues, including the technicians, to try on themselves all new substances, synthesized for possible use as drugs. Jacobsen and his friend, Jens Hald, were interested in a compound known as disulfiram, which was used in the form of an ointment to treat scabies, a condition caused by a skin parasite, endemic at that period of privation throughout occupied Europe. Hald had the notion, based on what was known of the drug's action, that it might also be useful for killing intestinal parasites. After some encouraging results in experiments on rabbits, which suffered no ill-effects even from massive doses, Jacobsen and Hald themselves took a course of disulfiram pills for some days and concluded that the compound was indeed harmless. Then one day Jacobsen decided to treat himself to a beer with his sandwich lunch, which he consumed while sitting in the library in the company of his colleagues. By the end of the lunch break Jacobsen felt queasy and groggy and his head throbbed. Slowly his symptoms subsided and he was well enough to go back to work. He ruled out food poisoning as the cause of his malaise, for his wife and daughters reported no symptoms from the same lunch. Then some days later he took lunch in a restaurant with the managing director of the company. They drank a convivial aquavit and Jacobsen returned to the laboratory looking, to the alarm of his colleagues, very red in the face, and again his head throbbed and he felt sick. At the end of the week it happened again:

> That Friday, while a fellow pharmacologist gave an informal luncheon talk, Jacobsen sipped a beer and ate the meatball sandwich his wife had made. Shortly afterward, he had another attack and went home early. It was a trip of several miles, and as he weaved on his bicycle through Copenhagen's narrow streets, he wondered: Could it be those meatballs? He asked his daughters what they had eaten for lunch. Meatballs, just like their father. They were fine; the meatballs could not be responsible.

Then one day Jacobsen encountered Hald in the corridor and they discussed the disulfiram experiment: Hald confessed that he had had the same experiences as Jacobsen. Suspicion fell on the disulfiram pills. The two men conducted more extensive trials on themselves and another colleague, and, then, to make doubly sure of their conclusion, Jacobsen took a course of the pills before injecting himself with a small amount of alcohol. The effect was dramatic: Jacobsen's blood pressure fell alarmingly, almost to zero, and he nearly died. It was now clear that alcohol reacted with disulfiram, or rather with a product of its breakdown in the body, to form a highly toxic product. Soon after this daunting experience Jacobsen had a chance visit from a chemist friend, who immediately identified the odour on Jacobsen's breath as acetaldehyde, the first, and toxic, product of oxidation of alcohol, which normally quickly undergoes further oxidation to acetic acid (as when vinegar is formed from wine). It was the build-up of acetaldehyde that was mainly responsible for the disagreeable effects that Jacobsen and Hald had experienced.

The postscript to the story is that Jacobsen adverted to his adventures with disulfiram in a public lecture. He did not know that there was a journalist in the audience and was surprised to see the story turn up next day in the leading Copenhagen newspaper. There it was seen by a psychiatrist, whose speciality was the treatment of alcohol addiction by aversion therapy, an unpleasant and seldom successful measure. The psychiatrist contacted Jacobsen and soon disulfiram was used, as it still is, to treat chronic alcoholics. Jacobsen suggested a name for disulfiram preparations—Antabuse.

Perhaps the most famous exponent of self-experimentation was the biologist, J. B. S. Haldane, celebrated for his work in physiology, genetics, and biochemistry, not forgetting his mathematical dexterity and his knowledge of the Greek and Latin classics; he stood out, besides, for his unwavering belief in communism, his tempestuous relations with the academic establishment, and his relish for conflict. He was one of the few men of his generation who enjoyed the First World War, which he felt privileged to have experienced. Haldane was perhaps unique among physiologists in eschewing the use of animals in research, in favour of experiments on human subjects, most of all himself. He first learned the practice from his father, John Scott Haldane, Professor of Physiology in Oxford, who achieved fame for his work on the effects of gases in mines, which saved many lives; he even once breathed a mixture of air and carbon monoxide until half of the respiratory protein in his blood, the haemoglobin, had been sequestered by carbon monoxide. This might have killed him. When he was still a small boy J.B.S. would accompany his father down mineshafts, serving as pupil, assistant, and, not infrequently, as guinea-pig. Here is his account of one such excursion, when he and his father were lowered in a large bucket and crawled along a narrow shaft:

After a while, we got to a place where the roof was about eight feet high and a man could stand up. One of the party lifted his safety lamp. It filled with blue flame and went out with a pop. If it had been a candle this would have started an explosion, and we should probably have been killed. But of course the flame of the explosion inside the safety lamp was kept in by the wire gauze. The air near the roof was full of methane, or firedamp, which is a gas lighter than air, so the air on the floor was not dangerous.

To demonstrate the effects of breathing firedamp, my father told me to stand up and recite Mark Anthony's speech from Shakespeare's *Julius Caesar*, beginning 'Friends, Roman, Countrymen'. I soon began to pant, and somewhere about 'the noble Brutus' my legs gave way and I collapsed on the floor, where, of course, the air was all right. In this way I learnt that firedamp is lighter than air and not dangerous to breathe.

Haldane's father was a consultant for the Admiralty on matters of diving, and had transformed underwater safety practices and the procedures used for decompression. In 1908, at the age of 15, J.B.S. had already been permitted one dive.

Shortly afterwards there was a sequel when John Scott Haldane was invited to take part in a trial run of a new Admiralty submarine. He needed an assistant and explained to his family that since the ship was on the secret list, his choice was limited. When her husband had become thoroughly worried about the assistant, Mrs Haldane casually asked: 'Why not take Boy [as her son was called by the family]?' 'Is he old enough?' John Scott Haldane replied, turning to his son to ask: 'What's the formula for soda-lime?' J.B.S. rapped out the formula. Shortly afterwards he made his first trip in a submarine.

When the Great War came Haldane joined the Black Watch and hurled himself with heady enthusiasm into the fray as a platoon commander in France. He was several times wounded and undertook a series of unauthorized and foolhardy escapades. Then in 1915 the first gas attacks took the British army completely by surprise. The Chancellor, Lord Haldane, telegraphed his brother in Oxford for advice and J.S. left immediately for France. He discovered that 90,000 gas masks, which were being distributed to the soldiers, were of a type that he believed to be ineffective. He at once sent for his colleague, Professor C. G. Douglas from Oxford, and summoned his son from the trenches. Together with a handful of volunteers the three men took turns to sit in a chamber into which chlorine gas was pumped. J.B.S. wrote:

> We had to compare the effects on ourselves of various quantities, with and without respirators. It stung the eyes and produced a tendency to gasp and cough when breathed. For this reason trained physiologists had to be employed. An ordinary soldier would probably restrain his tendency to gasp, cough and throw himself about if he were working a machine-gun in a battle, but could not do so in a laboratory experiment with nothing to take his mind off his own feelings. An experimental physiologist has more self-control. It was also necessary to see if one could run or work hard in the respirators, so we had a wheel of some kind to turn by hand in the gas chamber, not to mention doing fifty-yard sprints in respirators outside.

There was no lasting damage, Haldane continued, because they all knew when to stop, but he remained 'very short of breath and incapable of running for a month or so'. It was in this state that he returned to his regiment and participated in the battle of Festubert, in which he was twice wounded. Haldane's biographer suggests that the outcome of these few days of experimentation in the gas chamber saved thousands of lives and may have averted an immediate collapse of the front.

Haldane returned to the service of his country just before the Second World War, when a new submarine, the *Thetis*, sank while undergoing trials in the Mersey, taking 99 men, sailors, and civilians to their death. Haldane was invited to investigate the problems associated with the escape gear installed in British submarines. This led to a series of dangerous experiments, involving exposure to high pressures and high carbon dioxide and oxygen concentrations for long periods. Haldane was always elated by danger, hugely enjoyed the work, and may not have been above indulging in a little exhibitionism. One of his assistants was a young naval surgeon, Lieutenant Kenneth Douglas:

> He did take many quite serious risks in my presence on many occasions and this criticism [that he was playing to the gallery], although it may have had a grain of truth, was entirely unfair. On one occasion, he breathed oxygen at 100 feet (4 atmospheres absolute) in a bath surrounded by blocks of ice. Rather foolishly, he suggested that I breathed oxygen as his attendant as well, to allow immediate decompression if necessary. The result of this was that the wet and frozen professor and the young naval doctor both had oxygen poisoning at the same time, and it is only by good fortune that I did not convulse and Haldane did not drown. Again, Haldane convulsed several times in my arms in the wet pressure pot where he was in a diving suit under-water and I was on a platform above him.

Haldane's work resulted in changes in submarine escape techniques and a considerable expansion in the scope of submarine warfare. He had in fact been interested in the effects of carbon dioxide on the body long before and had performed experiments on himself, designed to make his acidity increase grossly by suppressing elimination of the metabolically generated carbon dioxide. He did this by eating three ounces of sodium bicarbonate, but, then, to maintain his acidified state, without actually drinking hydrochloric acid, he disturbed his acid-alkaline balance by consuming an ounce a day of ammonium chloride for several days. The acid poisoning caused shortness of breath, which persisted for some days after the end of the experiment. This result led to a treatment for a condition called tetany in babies, caused by excessive alkalinity and sometimes fatal.

For the story of Clonidine, see H. Stähle's account in *Chronicles of Drug Discovery*, ed. E. S. Bindra and D. Lednicer (Wiley, New York, 1982). The discovery of Antabuse and other tales of self-experimentation are described in *Who Goes First?* by Lawrence K. Altman (Random House, New York, 1987). For the experiences of the redoubtable J. B. S. Haldane, see the excellent biography by Ronald Clark, *J.B.S.* (Hodder and Stoughton, London, 1968), from which the above quotations are taken; Haldane's experiment with carbon dioxide is related in one of his brilliant essays on science, published in the *Daily Worker* and collected under the title, *Possible Worlds* (Chatto and Windus, London, 1927, and frequently reprinted).

86 A Russian tragedy

In the sombre years that followed the First World War and the Revolution, the Soviet regime conceived a new class of savants—the 'peasant' or 'barefoot' scientists, who, toiling in 'hut laboratories', would bring folk wisdom to bear on the problems then besetting Soviet agriculture. The privations grew worse, until in 1929 Stalin enforced the disastrous policy of collectivization of the farms. The traditional methods of cultivation were abandoned, and, aggravated by droughts, famine swept the land, killing an estimated eight million people. Desperate for quick remedies and fearing for their own heads, the administrative *apparatchiks* panicked and embraced every ignoramus and charlatan with a nostrum for improving crop yields.

The most ruthless, tenacious, and plausible of the charlatans was a Ukrainian peasant, Trofim Denisovich Lysenko, who ingratiated himself with Stalin and over the next two decades took an iron grip not only on the country's agriculture but on all of biology. In particular, he denounced the body of biological science practised in the academies, and especially genetics, as a bourgeois, fascist imposture to be mercilessly expunged. Biology in the Soviet Union and its imperium was blighted for a generation and many of the leading practitioners were arrested and shot or left to perish in prison. The most illustrious victim was Nikolai Vavilov, the country's pre-eminent agronomist and, at the time of Lysenko's rise, President of the Lenin Academy of Agricultural Sciences.

Lysenko was, by all accounts, a hypnotic presence, intense and articulate, lean and saturnine in appearance. Here is how he appeared to the writer of an article in *Pravda* at about the time he first came to public notice, and evidently before he had studied the crafts of demagoguery:

> If one is to judge a man by first impression, Lysenko gives one the feeling of a toothache; God give him health, he has a dejected mien. Stingy of words and insignificant of face is he; all one remembers is his sullen look creeping along the earth as if, at very least, he were ready to do someone in. Only once did this barefoot scientist let a smile pass, and that was at mention of Poltava cherry dumplings with sugar and sour cream.

Vavilov was arrested on a plant-gathering expedition to the Ukraine. Here is how it happened:

Vavilov and his companions first went to Kiev. From there they went by car to Lvov and on to Chernovitsy. From there, in three overcrowded cars, Vavilov and a large group of local specialists proceeded towards the foothills to collect and study plants. One of the cars could not negotiate the difficult road and turned back. On the way the occupants met a light car containing men in civilian clothes: 'Where did Vavilov's car go?' asked one of them. 'We need him urgently.' 'The road further on is not good, return with us to Chernovitsy. Vavilov should be back by 6 or 7 p.m., and that would be the fastest way to find him.' 'No, we must find him right away; a telegram came from Moscow; he is being recalled immediately.'

In the evening the other members of the expedition returned without Vavilov. He was taken so fast that his things were left in one of the cars. But late at night three men in civilian clothes came to fetch them. One of the members of the expedition started sorting out the bags piled in the corner of the room, looking for Vavilov's. When it was located it was found to contain a big sheaf of spelt, a half-wild local type of wheat collected by Vavilov. It was later discovered to be a brand-new species. Thus, on his last day of service to his country, 6 August 1940, Vavilov made his last botanic-geographic discovery. And, although it was modest, it still cannot be dropped from the history of science. And few scientists reading of it in a Vavilov memorial volume published in 1960 could have guessed that the date of this find is a date that scientists throughout the world will always recall with bitterness and pain.

Vavilov died in prison of starvation and neglect more than two years later. Lysenko's power grew, and he sought to extend his malign influence to the physical sciences as well. The chemists vacillated but the physicists stood firm. Stalin eventually came to realize that his protégé was not all he had supposed, and the voices of sanity at last became audible; but Stalin's successor, Khrushchev, a son of the soil, was not interested in the scientists' opinions. Aleksandr Nesmeyanov, a chemist and President of the Soviet Academy of Sciences, has told how he and Igor Kurchatov, the illustrious head of the Soviet nuclear bomb project, tried unsuccessfully to remonstrate with the premier.

Once when I. V. Kurchatov and I initiated a conversation about the impossible situation in biology, which was being suppressed by pseudo-science, we decided to be received by Khrushchev and talk to him about it. The meeting did not start in the best way. Kurchatov told Khrushchev about the gains that the United States had derived from hybrid corn and how we were losing out a lot by lacking modern genetics in our science. Khrushchev became agitated and he drew a couple of long ears of corn from his desk. He started waving them in our direction and telling us that this was our corn and that we understood nothing about agriculture. He advised us to stay with our physics and chemistry and keep out of biology. After that, he became visibly bored while we were telling him about the poor state of biology in our country and about Lysenko's mistakes.

On my return from the meeting, I got a phone call from Khrushchev. He told me, 'Comrade Nesmeyanov, hands off Lysenko or else heads will roll.' That was the end of the story, and I was busy with other things. I kept attending the Council of Ministers, and there were more interactions than before, and more unpleasant situations as well. Sometimes, they may have been unintentional but, in other cases, it was unmistakable that Khrushchev meant to interfere in the affairs of the Academy, under the guise of giving instructions for the improvement of our activities. It was becoming more and more clear that he was applying the saying, 'For the watch to go, you have to shake it'. This 'shaking' was Khrushchev's only means of interfering in our affairs though, and he applied it with increasing frequency. Then an incident happened at the end of 1960. Khrushchev hinted at the unsatisfactory performance of the Science Academy, and he said that the reason was that the Academy dealt with little flies. [Khrushchev was referring here to fruit-flies, the Western geneticists' most rewarding object of study, which had been singled out by Lysenko for especial mockery.] I stood up at this point, and to the horror of the Politbureau members, I declared that it was important to investigate these little flies too. It was unheard of and unprecedented to say anything that contradicted Khrushchev's viewpoints, and I added: 'It is possible to replace the President of the Academy by someone better suited to this position, M. V. Kel'dysh, for example'. 'I think so too', snapped back Khrushchev. The meeting then continued. For me, all that was left to do was just 'wait'.

In Stalin's day such an utterance would have been suicidal, but Nesmeyanov came to no harm. As for Lysenko, the fall of Khrushchev finally deprived him of his last all-powerful patron. He was stripped of his titles and authority and spent his last years in a small laboratory in an agricultural institute, discredited and reviled, but spared the fate of his many victims.

The first passage is taken from Zhores A. Medvedev, *The Rise and Fall of T. D. Lysenko* (Columbia University Press, New York, 1969); Nesmeyanov's account comes from an interview with Emiliya G. Perevalova, *Chemical Intelligencer*, 6, 32 (2000).

87 The way of the world

When George I ascended to the throne of England, he gave it as his greatest source of pride that Newton was now numbered among his subjects. Since those times science has become gradually more and more identified with national prestige. This has not been lost on the more politically aware scientific manipulators.

While American scientists during the Cold War became quite skilled in obtaining funds by sounding alarms about Soviet progress, there is evidence that their Soviet counterparts were not at all laggard in exploiting the Cold War to similar purpose. Thus, Representative Melvin Price [chairman of a congressional subcommittee on research and development] relates the following conversation with a Soviet physicist at the Dubna laboratory:

> 'The Dubna Laboratory asked our group when we were there two years ago how we got the money to build our accelerators. We told him the legislative process of getting money on our program. He said, "That is not the way I understand." He said, "I understand you get it by saying the Russians have a 10-billion electron volt synchrotron and we need a 20-billion electron volt synchrotron and that is how you get your money." I said, "There may be something to it." I said, "How do you get your money?" He said, "The same way." ' Price related this tale to John Williams, director of the AEC research division, who commented, 'That is certainly a very true story.'

The synchrotron is a machine designed to accelerate protons to immense speeds (its more recent use has been to propel electrons at close to the speed of light, when they generate high-intensity radiation, useful for a variety of experiments in many areas of science). The device has the form of a circular underground tunnel, several miles across. The Dubna Laboratory is the site of Russian nuclear research and of particle accelerator projects. The Dubna accelerator was at this time so disastrously unsuccessful that its users dubbed it 'the Stalin Memorial Accelerator'.

Robert Wilson, a leading American expert on accelerator design, had a better argument when he gave evidence before a Senate Committee. What, he was asked, would this expensive project do for the defence of the United States? 'Nothing', was the reply, 'but it will make the United States worth defending.'

The exchange between Price and the Soviet scientists is recorded in Daniel S. Greenberg's *The Politics of Pure Science* (New American Library, New York, 1967).

88 Tug of war on the thread of life

The discovery of the structure of DNA by Francis Crick and James Watson was one of the most dramatic chapters in the history of science, made more so in Jim Watson's own exuberant narrative. In 1952, Watson was 24 years old and a visiting research fellow in the Cavendish Laboratory in Cambridge. The experimental information that Watson and Crick had to guide them was exiguous, and they were aware that they were not alone in their pursuit of the structure. The effort at King's College in London was impeded by the mutual antagonism between the protagonists, Maurice Wilkins and Rosalind Franklin; the main competition, as Watson saw it, was from California, where resided the formidable Linus Pauling, by common consent the world's pre-eminent structural chemist. By happy chance Watson was sharing an office with Pauling's son, Peter, who was a graduate student in the Cavendish. The talk, as Watson remembers it, was mainly about girls, but

> A fetching face, however, had nothing to do with the broad grin on Peter's face as he sauntered into the office one afternoon in the middle of December and put his feet up on the desk. In his hand was a letter from the States that he had picked up on his return to Peterhouse for lunch.
>
> It was from his father. In addition to routine family gossip was the long-feared news that Linus now had a structure for DNA. No details were given of what he was up to, and so each time the letter passed between Francis and me the greater was our frustration. Francis then began pacing up and down the room thinking aloud, hoping that in a great intellectual fervour he could reconstruct what Linus might have done. As long as Linus had not told us the answer, we should get equal credit if we announced it at the same time.
>
> Nothing worth while had emerged, though, by the time we walked upstairs to tea and told Max [Perutz] and John [Kendrew] of the letter. Bragg [the director of the laboratory] was in for a moment, but neither of us wanted the perverse joy of informing him that the English labs were once again about to be humiliated by the Americans. As we munched chocolate biscuits, John tried to cheer us up with the possibility of Linus's being wrong. After all, he had never seen Maurice's and Rosy's pictures [the X-ray diffraction photographs from King's College]. Our hearts, however, told us otherwise.

And now the denouement: in February, Pauling completed his paper and sent a copy of the manuscript to Cambridge. Watson was by then in a lather of nervous anticipation.

Two copies, in fact, were dispatched to Cambridge—one to Sir Lawrence [Bragg], the other to Peter. Bragg's response upon receiving it was to put it aside. Not knowing that Peter would also get a copy, he hesitated to take the manuscript down to Max's office. There Francis would see it and set off on another wild-goose chase. Under the present timetable there were only eight months more of Francis's laugh to bear. That is, if his thesis was finished on schedule. Then for a year, if not more, with Crick in exile in Brooklyn [at the Polytechnic Institute, where he was to work], peace and serenity would prevail.

While Sir Lawrence was pondering whether to chance taking Crick's mind off his thesis, Francis and I were poring over the copy that Peter brought in after lunch. Peter's face betrayed something important as he entered the door, and my stomach sank in apprehension at learning that all was lost. Seeing that neither Francis nor I could bear any further suspense, he quickly told us that the model was a three-chain helix with the sugar-phosphate backbone in the centre. This sounded so suspiciously like our aborted effort of last year that immediately I wondered whether we might already have had the credit and glory of a great discovery if Bragg had not held us back. Giving Francis no chance to ask for the manuscript, I pulled it out of Peter's outside coat pocket and began reading. By spending less than a minute with the summary and introduction, I was soon at the figures showing the locations of the essential atoms.

At once I felt something was not right . . .

Pauling's model was inconsistent with the experimental data, available to Watson and Crick but not to Pauling, but, worse, it was chemically improbable. Homer had nodded indeed. Only a matter of weeks later Watson and Crick found a model structure so compelling in every detail that there could be little doubt it was correct.

The humiliation of an English lab to which Watson refers relates to the structure of the polypeptide chain, the string of linked amino acids from which proteins are constructed. It was known from X-ray diffraction pictures that the insoluble protein, keratin (the substance of hair and nails and the outer layer of the skin), had a regular structure, almost certainly some form of helix. Several laboratories had tried to infer what ordered structures a polypeptide chain could adopt, and Bragg, Perutz, and Kendrew had gone into print with conjectures that soon proved to be wide of the mark.

It was Pauling who put them right. In 1948, he was a visiting professor in Oxford, a damp and dispiriting place in the period of post-war austerity. Pauling caught a severe cold, a sinus infection followed, and he took himself to bed. 'The first day I read detective stories', he later wrote, 'and just tried to keep from feeling miserable, and the second day, too. But I got bored with that, so I thought, "Why don't I think about the structure of proteins?"' He took paper, pencil, and ruler and sketched out the linear structure of the polypeptide chain. Pauling had devoted much of his life to determining and interpreting the lengths of chemical bonds between carbon and nitrogen, oxygen and carbon atoms, and so on, and the angles between those bonds; he was therefore able to retrieve these numbers from his capacious memory. He cut out the chain and twisted it around, and, taking into consideration also that the hydrogen atom attached to each nitrogen would form a hydrogen bond (a weak but essential secondary interaction) with an oxygen atom from another repeating unit along the chain, he looked for ways to produce a regular structure.

Soon Pauling found a helical conformation which looked highly convincing; he called his wife and asked her to bring him his slide-rule, so that he could calculate the geometry of his helix. It took 18 peptide units (amino acid residues, as they are called) for the structure to repeat itself and these residues made up five turns of the helix. Pauling was pleased and excited and forgot about his afflictions. He kept quiet about this discovery, which did not instantly seem to fit with experimental data, in particular a critical spacing between structural elements discerned in X-ray diffraction photographs. But Pauling's paper-folding had led him to the structure that became famous as the alpha-helix. Bragg and his colleagues missed it because they restricted themselves to folds of the chain that led to an integral number of amino acid units in every turn of helix. When Pauling came out with the structure, which had 18 units to five twists of the helix, the Cambridge group were mortified.

Max Perutz has described his reaction on seeing Pauling's work in print, while browsing in the library one Saturday morning, only a short time after he and his colleagues had published their own paper. But there remained the X-ray anomaly. Perutz, thunderstruck, as he put it, by Pauling's revelation, had cycled home to lunch in a daze. Then, brooding over the lapse and also the apparent contradiction with the X-ray diffraction pictures, he suddenly recalled a visit he had made to the man who had obtained them, William Astbury in Leeds. Perutz realized with a start that Astbury's set up would have precluded observation of the X-ray spot in the region that Pauling's structure demanded.

In mad excitement, I cycled back to the lab and looked for a horse hair that I had kept tucked away in a drawer. I stuck it on a goniometer head [a device for accurate adjustment of angles] at an angle of 31° to the incident X-ray beam; instead of Astbury's flat plate camera I put a cylindrical film around it that would catch all reflections with Bragg angles [the angles between the line of the incident X-ray beam and the reflected rays corresponding to the various regularities in the structure] of up to 85°.

After a couple of hours, I developed the film, my heart in my mouth. As soon as I put the light on I found a strong reflection at 1.5 Å [1 Å, or Ångström, is the unit of length, equal to a hundred-millionth of a centimetre, in which distances in the inter-atomic range are commonly expressed] spacing, exactly as demanded by Pauling and Corey's α-helix.

On Monday morning Perutz confronted his director, Sir Lawrence Bragg, with both the chagrin and the triumph. When Bragg asked how the idea of the experiment had come to him, Perutz replied that it was all because their failure to see what Pauling had seen had put him into a rage. Bragg's rejoinder was, 'I wish I had made you angry earlier!' This was the title that Perutz chose for the book in which he relates the story.

James D. Watson's version of the discovery of the DNA structure comes from his now-classic book, *The Double Helix* (Weidenfeld and Nicolson, London, 1968) and Max Perutz's reminiscences are in *I Wish I'd Made You Angry Earlier* (Cold Spring Harbor Laboratory Press/Oxford University Press, Oxford, 1998). See also an interview with Linus Pauling by I. Hargittai in *Chemical Intelligencer*, 4, 34 (1996).

~

89 The trivial and profound

Richard Feynman insisted that physics to him was play—that he took up a problem for no better reason than that it presented an intellectual challenge or tickled his curiosity. His first academic appointment was at Cornell University, where he arrived in 1945, a youthful professor, often mistaken for a student. Here he shows how unpredictable the outcome of research can be.

Within a week [of his arrival at Cornell] I was in the cafeteria and some guy, fooling around, throws a plate in the air. As the plate went up in the air I saw it wobble, and I noticed the red medallion of Cornell on the plate going around. It was pretty obvious to me that the medallion went around faster than the wobbling.

I had nothing to do, so I start to figure out the motion of the revolving plate. I discover that when the angle is very slight, the medallion rotates twice as fast as the wobble rate—two to one. It came out of a complicated equation! Then I thought, 'Is there some way I can see in a more fundamental way, by looking at the forces or the dynamics, why it's two to one?'

I don't remember how I did it, but I ultimately worked out what the motion of the mass particles is, and how all the accelerations balance to make it come out two to one.

I still remember going to Hans Bethe [62] and saying, 'Hey, Hans! I noticed something interesting. Here the plate goes around so, and the reason it's two to one is . . . ' and I showed him the accelerations.

He says, 'Feynman, that's pretty interesting, but what's the importance of it? Why are you doing it?'

'Hah!' I say. 'There's no importance whatsoever. I'm just doing it for the fun of it.' His reaction didn't discourage me; I had made up my mind I was going to enjoy physics and do whatever I liked.

I went on to work out equations of wobbles. Then I thought about how electron orbits start to move in relativity. Then there's the Dirac equation in electrodynamics. And then quantum electrodynamics. And before I knew it (it was a very short time) I was 'playing'—working, really—with the same old problem that I loved so much, that I had stopped working on when I went to Los Alamos [to work on the atom bomb]: my thesis-type problems; all those old-fashioned, wonderful things.

It was effortless. It was easy to play with these things. It was like uncorking a bottle: Everything flowed out effortlessly. I almost tried to resist it! There was no importance to what I was doing, but ultimately there was. The diagrams and the whole business that I got the Nobel Prize for came from that piddling around with the wobbling plate.

So Feynman progressed from the wobbling plate to the most rarified frontiers of theoretical physics. His biographers paint a picture that differs in only one respect: Feynman was in reality unhappy when he arrived at Cornell, for the inspiration about what he was going to do had up till then eluded him.

Feynman was a theoretician without peer, but his occasional audacious forays into experimental science were not uniformly successful. His illustrious mentor at Princeton, John Archibald Wheeler, recalls the outcome of an experimental initiative, sparked by a discussion of the kind of seemingly elementary question that so often exercises the most profound minds in physics:

> Was it a problem in the junior course in mechanics that started us thinking about the familiar lawn sprinkler? Shaped like a swastika, it shoots out four jets of water. The recoil drives the sprinkler arms round and round. But where does the recoil act? Doesn't it act at the point where the stream of water suddenly changes direction from straight out to straight transverse? But suppose the arm sucks water instead of squirting it out. Surely, we said to each other, there is an identical change in direction and therefore an identical reaction. Surely the sprinkler will again turn round when water in the arms is sucked in rather than being shot out. Oh no, it won't. Oh yes, it will. We had a great time trying out both sides of this question on our colleagues. As the days went by, more and more colleagues up and down the corridor took positions. The debate grew more animated. No argument of theory was strong enough to still the disagreements. The situation called for an experiment.
>
> Feynman made a six-inch miniature lawn sprinkler out of glass tubing and hung it from a flexible tube of rubber. He checked that it worked OK as a sprinkler. Then he wangled the whole dangling gadget through the throat of a great glass carboy filled with water. He got this outfit set up on the floor of the cyclotron lab, where there was a handy compressed-air outlet. He ran the compressed air in through a second hole in the cork at the top of the carboy. Ha! A little tremor as the pressure was first applied, as water first began to run backward through the miniature lawn sprinkler. But, as the flow continued there was no reaction. Then increase the air pressure. Get more backward flow of water. Again a momentary tremor at the start of this maneuver but no continuing torque. OK, more pressure. And more! Boom! The glass container exploded. Water and fragments of glass went all over the cyclotron room. From that time onward Feynman was banished from the lab.

Whether the Princeton physicists came to any conclusion about the proposition Feynman had sought to test Wheeler does not divulge.

The first passage comes from Feynman's oral memoirs, edited by Edward Hutchings, *Surely You're Joking, Mr Feynman: Adventures of a Curious Character* (Norton, New York, 1985); J. A. Wheeler's reminiscence is from '*Most of the Good Stuff*': *Memories of Richard Feynman*, ed. Laurie M. Brown and John S. Rigden (American Institute of Physics, New York, 1993).

90 Phlogiston consigned to flames

Antoine Laurent Lavoisier (1743–94), '*d'immortelle mémoire*', is generally seen as the founder of modern chemistry. He introduced the principle of accurate measurement, especially by weighing, of the reagents and products of chemical reactions; it was this fastidious adherence to quantitative evaluation, rather than mere observation, that led him to most of his great discoveries. He was a man of conspicuous vanity, hauteur, and no little concupiscence. Rich by birth, he had married an even richer and intelligent and beautiful wife, but it was his membership of the rapacious *Fermiers Généraux,* the association of tax-tenants, who levied tributes on a wide range of goods, that brought him eventually to the scaffold.

Lavoisier was not above laying claim to the work of others and he seldom acknowledged the efforts of his contemporaries. Nevertheless, it was he who (among many other achievements) defined the difference between elements and compounds and understood (even if he shared the discovery with Joseph Priestley in England and Karl Wilhelm Scheele in Sweden) the significance of oxygen—which he named from the Greek, meaning 'acid generator' (a misnomer, of course, perpetuated in current German usage as *Sauerstoff*, or sour substance). Lavoisier formulated the principle of conservation of matter and put paid to the phlogiston theory, which had dominated chemistry for a half century.

Phlogiston was the brain-child of Georg-Ernst Stahl in Germany; it was an imponderable fluid, which permeated combustible substances, and when they burned was released in a swirling motion, manifested as flame. And so, Priestley—a tenacious devotee of the theory to the day of his death—inferred that when substances were burned in air, phlogiston was lost, leaving behind an inert residue that would not support further combustion or, indeed, life; this gas (nitrogen) he designated 'dephlogisticated air'. But Lavoisier showed that substances burned in air or oxygen actually gained weight in predictable measure and some (the red oxide of mercury, for instance, to which he had been introduced by Priestley) could be made to yield up their oxygen again. Lavoisier exulted in his victory over Priestley and organized a curious entertainment in one of the famous *soirées* at his Paris house, attended by the quality of the city.

His vanity was such that it often made him appear ridiculous. For example, in 1789 shortly after the fall of the Bastille, Lavoisier devised a mock-trial of the phlogiston theory. He invited a large and distinguished party, and enacted this trial in front of them. Lavoisier and a few others presided over the judicial bench, and the charge was read out by a handsome young man who presented himself under the name of 'Oxygen'. Then the defendant, a very old and haggard man who was masked to look like Stahl, read out his plea. The court then gave its judgement and sentenced the phlogiston theory to death by burning, whereupon Lavoisier's wife, dressed in the white robe of a priestess, ceremonially threw Stahl's book on a bonfire.

Lavoisier did not last long after this ludicrous episode. When the Jacobins came to power he was arrested, tried and, according to the published accounts, sent to the guillotine under the rubric: *La République n'a pas besoin des savants* (but there is now doubt about the truth of this hallowed article of belief). Lavoisier's contemporary, the mathematician Joseph-Louis Lagrange, observed that 'it needed but a moment to sever that head, and perhaps a century will not be long enough to produce another like it'. Witnesses of the mass execution in which Lavoisier died testified that he bore himself nobly. One remarked: 'I do not know whether I saw the last and carefully played role of an actor, or whether my judgement of him before was wrong, and a really great man has died.'

Lavoisier, it should be added, was not the only man of science to fall victim to the Revolution. A distinguished astronomer, Jean Sylvain Bailly, who mapped the trajectories of the moons of Jupiter, was accused of complicity in the attack by the militia on a peacefully demonstrating crowd in the Champ de Mars in 1791. As senior deputy for Paris he may, indeed, have shared responsibility for the massacre. He was, in any event, sent to the guillotine. Another academician who died in prison, probably by poison, while awaiting execution, was a mathematician, the Marquis de Condorcet, and many lesser luminaries also perished.

The description of Lavoisier's masque is taken from *History of Analytical Chemistry* by Ferenc Szabadváry (Gordon and Breach, London, 1960); see also Bernadette Bensaude-Vincent's notable biography, *Lavoisier* (Flammarion, Paris, 1993).

91 The errant compass

Hans Christian Oersted (1777–1851) was Professor of Physics at the University of Copenhagen when he tried a lecture demonstration that changed the course of physics. Oersted was fascinated by magnetism and he did not hold with the prevailing view that magnetism and electricity were unrelated phenomena—fluids, as Ampère [17] had called them. Rather, he thought, they were forces radiated by all substances, which might well interfere with each other, and, indeed, it was well established that a compass in a ship, struck by lightning, would sometimes reverse its polarity. He conjectured, therefore, that a thin wire carrying an electric current might, when placed beside a compass needle, cause it to deflect.

For reasons never made clear Oersted chose to test his hypothesis before an audience at one of his lectures instead of first trying it out for himself. He had, he afterwards confessed, hesitated, for prudence suggested that the experiment might not, after all, work and he would look foolish; but on an impulse he went ahead all the same. Oersted sent the current through his thin platinum wire until it glowed. The compass was lying directly beneath, but before he could move it alongside the wire, the needle deflected. The effect was small and probably invisible to the audience, which manifested apathy, but Oersted was astonished: the current flowing in the direction of the needle's axis had deflected it *sideways*? How could this be?

Oersted cogitated for three months before taking to the laboratory. Eventually, after much experimentation, he was driven to infer that the electric current was generating a magnetic force, and he formulated the rule, which became famous, that an electric current engenders a magnetic force at right angles to its direction of flow. Twelve years later Michael Faraday in England and Joseph Henry in America showed the converse— that a changing magnetic field induces a flow of electricity in a nearby circuit. It remained for James Clerk Maxwell [44] to explain it all.

See, for instance, Hans Christian Oersted in *Dictionary of Scientific Biography*, ed. C. C. Gillespie (Scribner, New York, 1980).

~

92 Liberation by fire

Miriam Rothschild, the self-taught and distinguished zoologist, who has written captivatingly about her passions—notably in the book, *Flukes, Fleas and Cuckoos* (1952)—has told how she was liberated from the tyranny of her early studies on trematodes, which are microscopic parasitic worms. It was the beginning of the Second World War and Miriam Rothschild was on the staff of the Marine Biological Station in Plymouth. As a volunteer air-raid warden, she had suggested that fire-fighting equipment be installed in the laboratory. The director had demurred: there would be no bombing of cities in this war, he opined, and, even if there were, Plymouth would be spared, for everyone knew that the oil storage tanks were empty, the docks of negligible importance, and the city was anyway remote from the flight path of German bombers. The air raid duly ensued, the storage tanks, which were not empty at all, were set ablaze, and the laboratory, too, was hit.

When dawn came and there was light enough to see—for of course we had no artificial light working—I staggered to my room to assess the damage. An incredible sight met my eyes. The door had gone and the room appeared to be empty, except for a huge pile of tiny splinters of glass on the floor and, picking its way delicately among the debris, the sole survivor: my tame red-shank.

Where were my notebooks and manuscripts? Where were the labelled drawings? Where were the cultures of intermediate hosts [of the parasitic worms], the infected gobies, the hundreds of isolated infected snails? Where were the microscope, the Cambridge Rocker, the Camera Lucida, the watch glasses, the finger bowls, the tubes, the shelving, the jars? Gone. Seven years' work had vanished, pulverized with a ton of glass.

For three days I felt nothing except for a vague backache. I was stunned. A blank.

A German reconnaissance plane slipped in over the still-burning oil tanks, wheeled through the pall of smoke and vanished unmolested. Would we have another raid immediately? The fire surely provided a perfect visible target. Nothing happened.

Next morning I found that my redshank had died, possibly from delayed shock or some internal injury due to the blast. Had she died in great pain? I felt deeply disturbed as I saw her lying quietly among the slivers of broken glass—an unequivocal indictment of the human race. I grieved for her.

The following day I was seized by a sense of meaningless excitement and light-headedness. Without realizing it I had gradually become an appendage of my trematode life cycles. At the time I had no assistant, which meant that I could not afford a days' illness, let alone a vacation or a free weekend. There were all those intermediate hosts to tend and feed and rear; all those snails to nurture; scores of beautiful ephemeral cercariae [the parasite larvae] to count, draw and describe; all those shells to measure; all those twinkling flame-cell patterns to unravel, and my flock of seagull hosts to coax from egg to adult. It had meant a remorseless 16-hour day. Why, I even saw cercariae in the clouds and flame cells beating in my dreams.

Now, all at once, I was free.

I packed my bags and left Plymouth, never to return as a research scientist. I did not know that butterflies and fields of flowers were to be exchanged for free-swimming cercariae and the turbulent Atlantic Ocean. But temporarily, at least, the German Air Force had liberated me.

From Miriam Rothschild, in *The Scientist*, July 1987; reproduced in *From Creation to Chaos: Classic Writings in Science,* ed. Bernard Dixon (Blackwell, Oxford, 1989).

~

93 How small is small?

Throughout the 1930s Leo Szilard brooded on the prospects of a nuclear chain reaction [20] and the possibility, therefore, of an atomic bomb, and from time to time whipped himself into a paroxysm of apprehension. In 1939, he met Isidor Rabi [21] in Washington to discuss his fears. Rabi told Szilard that the same thought had presented itself to the great Italian physicist Enrico Fermi [29], also by then in the United States, who, however, had shown no inclination to pursue the matter. Szilard insisted on at once calling on Fermi.

'Fermi was not in,' Szilard later recalled, 'so I told Rabi to please talk to Fermi and say these things ought to be kept secret because it's very likely that if neutrons are emitted [from the fission of uranium], this may lead to a chain reaction, and this may lead to the construction of bombs. A few days later I went again to see Rabi, and I said to him, "Did you talk to Fermi?" Rabi said, "Yes, I did." I said, "What did Fermi say?" Rabi said, "Fermi said 'Nuts!'" So I said, "Why did he say 'Nuts!'?" and Rabi said, "Well, I don't know, but he is in and we can ask him." So we went over to Fermi's office, and Rabi said to Fermi, "Look, Fermi, I told you what Szilard thought and you said 'Nuts!' and Szilard wants to know why you said 'Nuts!'" So Fermi said, "Well, there is a remote possibility that neutrons may be emitted in the fission of uranium and then of course a chain reaction can be made." Rabi said, "What do you mean by 'remote possibility'?" and Fermi said, "Well, ten percent". Rabi said, "Ten percent is not a remote possibility if it means that we may die of it. If I have pneumonia and the doctor tells me that there is a remote possibility that I might die, and it's ten percent, I get excited about it."'

From this meeting on, Szilard realized how differently he and Fermi could view the same scientific evidence. 'We both wanted to be conservative,' Szilard recalled later, 'but Fermi thought that the conservative thing was to play down the possibility that this may happen, and I thought the conservative thing was to assume that it would happen and take the necessary precautions.'

As we know, Szilard was soon proved right. Szilard's biographers believe that the difference in outlook of the two men reflects their outlook on life: 'Put simply, science was Fermi's life, whereas for Szilard science was an endeavour ineluctably united with politics and personal sensitivities.' The personalities of the two men were indeed scarcely reconcilable. When, for instance, Fermi was supervising the construction of the first atomic pile, in which a chain reaction was to be tested, in a squash court in Chicago, he urged his workforce on. In his shirtsleeves, he joined in the labour of heaving heavy graphite blocks into the assembly. But, much to Fermi's annoyance, Szilard, who was averse to physical exertion of any kind, could not be induced to participate. This episode created a lasting rift.

See *Genius in the Shadows: A Biography of Leo Szilard, the Man Behind the Bomb* by William Lanouette and Bela Szilard (Scribner, New York, 1993); *Leo Szilard: His Version of the Facts*, ed. Spencer Weart and Gertrude Weiss Szilard (MIT Press, Cambridge, Mass., 1978); *Rabi: Scientist and Citizen* by John S. Rigden (Basic Books, New York, 1987).

~

94 Seeing sparks

In German a radio transmitter is a sender of sparks, for the discovery of radio waves was indeed associated with the appearance of sparks. In 1886, Heinrich Hertz, whose name is enshrined in the unit of frequency (the number of cycles per second of an electromagnetic oscillation), was a young professor at the Technical University of Karlsruhe, a backwater of learning, where he had to deliver lectures on such matters as meteorology for farmers. He tried his best, with minimal resources and little optimism, to get a programme of research going. His interest was in electromagnetic radiation and especially in the exploration of Maxwell's theory [44]. In the summer of that year Hertz married and on the day of his great discovery, 1 November 1886, his wife, who took an interest in his work, was at his side.

Hertz had modified an induction coil to generate sparks of high intensity across a gap between two small spheres on the ends of metal rods. This was a common enough set up for demonstration experiments on electric discharges, but Hertz had modified the apparatus to study the effect of changing the configuration of the circuit: he used long rods with larger spheres at the far ends to act as condensers for storing charge. He had an arrangement for regulating the spark gap and a shunt (a conductor of variable resistance) with which he could vary the voltage across the gap. When he reduced the resistance to zero, so as to short out the spark gap altogether, he was astonished to observe that weak sparks still crossed. On the bench, near his apparatus, lay another coil of metal, the ends terminating in spheres separated by the distance of a spark gap. As Hertz worked the induction coil he was startled to see—or perhaps his wife did—not only the brilliant discharge between the spark gap that it was feeding, but tiny, weak sparks crossing the spark gap in the inert coil some distance away. It was the merest chance, and, as Hertz later wrote, 'It would have been impossible to arrive at these phenomena by the aid of theory alone'.

Hertz realized that this strange and unaccountable observation portended something quite new. It did not take him long to discover that it was an oscillation of the current in the primary spark gap that his detector was picking up and he measured this frequency with a rotating mirror—a primitive stroboscope. Hertz's investigations showed him that it was not an induction effect that he was observing, as he had at first supposed, but that a radiation was traversing the room and finding the detector coil. It had a very long wavelength, but travelled at the velocity of light. Here was the genesis of radio and all that flowed from it. Hertz did not live to see the technological revolution that followed from his discovery: he was rewarded by an appointment to a better position, the chair of physics at the University of Bonn, and there he succumbed at 36 to blood poisoning. Here is what he wrote to his parents not long before he died:

> If anything should really befall me, you are not to mourn; rather you must be proud a little and consider that I am among the specially elect destined to live for a short while and yet to live enough. I did not desire to choose this fate, but since it has overtaken me, I must be content; and if the choice had been left to me, perhaps I should have chosen it myself.

This is reminiscent of the words of Enrico Fermi [29], who died prematurely some 70 years later: he did not mind too much, he averred, because most of what he was capable of achieving he had already achieved.

Hertz's observation of radio waves is an example of simultaneous discovery, so common an occurrence throughout the history of science. That same year Oliver Lodge in England also found evidence of such an electromagnetic radiation. But instead of writing a paper he departed on a leisurely climbing holiday in the Alps, intending to prepare his work for publication when he got back. It was too late: when Lodge returned to London he found news of Hertz's paper awaiting him. The genial Lodge seems to have been surprisingly little put out.

See the memoir of Heinrich Hertz by his sister, Johanna Hertz: *Heinrich Hertz: Memoirs, Letters, Diaries* (San Francisco Press, San Francisco, 1977); also *Heinrich Hertz: A Short Life* by Charles Susskind (San Francisco Press, San Francisco, 1995).

~

95 A Victorian tragedy, a twentieth-century sequel

Philip Gosse was a nineteenth-century biologist, a respected figure, a Fellow of the Royal Society, who had struggled earnestly to come to terms with Darwin's theory. He was also an outstanding popularizer of natural history in his writings and on the lecture podium. A severe figure of unbending Victorian rectitude, he had no time for frivolity and little even for his family. An entry in his diary reads: 'Received green swallow from Jamaica. E. delivered of a son.'

As a Christian fundamentalist, a member of the austere Plymouth Brethren sect, Gosse was deeply troubled by the evident contradiction between the fossil record and biblical chronology. After years of agonizing, he arrived at a solution to this problem and some others thrown up by scientific discoveries, and in 1884 he published the fruits of his lucubrations in a book, which he called *Omphalos*, the ancient Greek for the navel. For one of Gosse's preoccupations had been the thorny problem of whether Adam, who, unlike his descendants, was not born of woman, possessed this anatomical feature. The essence of Gosse's theory was that God's creation had been designed to embody the *appearance* of pre-existence. There was thus a true, or 'diachronic' time, and, in addition, God's spurious, or 'prochronic' timescale, in which reposed the fossils that Gosse had so single-mindedly studied. The story of how the book was received is told by

Gosse's son, Edmund, the novelist and man of letters (and Philip's only child, whose emergence into the world had received such a cryptic diary entry).

The theory, coarsely enough, and to my Father's great indignation, was defined by a hasty press as being this—that God hid the fossils in the rocks in order to tempt geologists into infidelity. In truth, it was the logical and inevitable conclusion of accepting, literally, the doctrine of a sudden act of creation; it emphasised the fact that any breach in the circular course of nature could be conceived only on the supposition that the object created bore false witness to past processes, which had never taken place. For instance, Adam would certainly possess hair and teeth and bones in a condition which it must have taken many years to accomplish, yet he was created full-grown yesterday. He would certainly—though Sir Thomas Browne denied it—display an *omphalos*, yet no umbilical cord had ever attached him to a mother.

Never was a book cast upon the waters with greater anticipations of success than was this curious, this obstinate, this fanatical volume. My Father lived in a fever of suspense, waiting for the tremendous issue. This 'Omphalos' of his, he thought, was to bring all the turmoil of scientific speculation to a close, fling geology into the arms of Scripture, and make the lion eat grass with the lamb. It was not surprising, he admitted, that there had been experienced an ever-increasing discord between the facts which geology brings to light and the direct statements of the early chapters of 'Genesis'. Nobody was to blame for that. My Father, and my Father alone, possessed the secret of the enigma; he alone held the key which could smoothly open the lock of geological mystery. He offered it, with a glowing gesture, to atheists and Christians alike. This was to be the universal panacea; this the system of intellectual therapeutics which could not but heal all the maladies of the age. But, alas! atheists and Christians alike looked at it and laughed, and threw it away.

In the course of that dismal winter, as the post began to bring in private letters, few and chilly, and public reviews, many and scornful, my Father looked in vain for the approval of the churches, and in vain for the acquiescence of the scientific societies, and in vain for the gratitude of those 'thousands of thinking persons', which he had rashly assured himself of receiving. As his reconciliation of Scripture statements and geological deductions was welcomed nowhere; as Darwin continued silent, and the youthful Huxley was scornful, and even Charles Kingsley, from whom my Father had expected the most instant appreciation, wrote that he could not 'give up the painful and slow conclusion of five and twenty years' study of geology, and believe that

God has written on the rocks one enormous and superfluous lie',—as all this happened or failed to happen, a gloom, cold and dismal, descended upon our morning tea cups. It was what the poets mean by an 'inspissated' gloom; it thickened day by day, as hope and self-confidence evaporated in thin clouds of disappointment. My Father was not prepared for such a fate. He had been the spoiled darling of the public, the constant favourite of the press, and now, like the dark angels of old

> so huge a rout
> Encumbered him with ruin.

He could not recover from amazement at having offended everybody by an enterprise which he had undertaken in the cause of universal reconciliation.

The picture of a devious God, intent on hoaxing mankind, makes a strange reappearance a century on. The Abbé Georges Lemaître (1894–1966) was not only a Catholic priest but also a respected theoretical physicist. His main interest was in cosmology, and it was he who first explicitly formulated the concept of the expanding Universe, which led to the big bang theory of its formation [81]. Lemaître, a Belgian, came to lecture at the University of Göttingen, where Victor Weisskopf, a distinguished Austrian physicist, who emigrated to the United States shortly before the Second World War, was a student. Lemaître's subject was the Earth's age [16]; this and others had calculated from the abundance of certain elements, the products of radioactive decay of parent elements with very long half-lives (the measure of how long the atoms of a radioactive element endure before transformation into the daughter element) [149].

Abbé Lemaître told us that such investigations had revealed that the earth was about 4.5 billion years old.

When we were sitting with him after his talk, someone asked him whether he believed in the Bible. He said, 'Yes, every word is true'. But, we continued, how could he tell us the earth is 4.5 billion years old, if the Bible says it is about 5,800 years old? He said, I suppose with tongue in cheek, 'That is no contradiction'. 'How come?' we nearly shouted. He explained that God made the earth 5,800 years ago with all the radioactive substances, the fossils, and other indications of an older age. He did this to tempt humankind and to test its belief in the Bible. Then we asked, 'Why are you so interested in finding out the age of the earth if it is not the actual age?' And he answered, 'Just to convince myself that God did not make a single mistake'.

The passage about Philip Gosse comes from Edmund Gosse, *Father and Son: A Study of Two Temperaments* (Heinemann, London, 1907, and many times reprinted). Victor Weisskopf's reminiscences are recounted in *The Joy of Insight: Passions of a Physicist* (Basic Books, New York, 1991).

~

96   A visit to the Führer

Max Planck was a tragic figure. When Adolf Hitler seized power in Germany Planck was the most respected and influential scientist in the country. It was he who had sparked the revolution in physics with the discovery that radiant energy is quantized; exists, that is to say, in the form of defined packets, or quanta. Planck, an ascetic figure, the last of a line of Lutheran pastors, became a friend of Albert Einstein, with whom he played chamber music. His eldest son fell in battle in the First World War, his beloved twin daughters both died in childbirth, and his younger son was executed in the last weeks of the Second World War, accused of participation in the plot on Hitler's life the year before.

When Hitler became Chancellor and promulgated the racial (Nuremberg) laws, Planck was President of the Kaiser-Wilhelm-Gesellschaft (now the Max-Planck-Gesellschaft)—the organization set up, with the blessing of the Kaiser, to promote scientific progress through its chain of research institutes, encompassing all branches of science. Dismayed at the dismissal of the Jews, among whom were many of his friends, Planck found himself in a painful dilemma. A public protest might have resulted in his own ejection from his position of power within the academic establishment, and he therefore saw it as his duty—misguidedly, as some of his more upright colleagues, and especially the expelled Jews, felt—to keep quiet, cleave to his presidency, and try to protect what could be salvaged of German physics. Einstein for one could not forgive him and never communicated with him again. But in May of 1933, when the Jewish exodus was beginning, Planck sought an audience with Hitler and supposedly tried to remonstrate. Here is Planck's account of the interview, set down 14 years after the event:

> After Hitler seized power, I, as the President of the Kaiser-Wilhelm-Gesellschaft, had the obligation to pay my respects to the Führer. I thought I might use this opportunity to put in a word in favour of my Jewish colleague, Fritz Haber, without whose process for the conversion of atmospheric nitro-

gen into ammonia the last war would have been lost from the outset. [Haber, a fervently patriotic baptized Jew, a Nobel Laureate, and the architect of the chemical warfare programme during the Great War, was dispossessed of all he held dear and hounded out of Germany.] Hitler replied with the words: 'I have nothing against the Jews. But the Jews are all communists, and these are my enemies, against whom my entire struggle is directed.' To my observation that there are different kinds of Jews, both valuable for mankind and value-less, among the first of which are old families with the highest German culture, and that one must surely distinguish between them, he replied: 'That is not correct. A Jew is a Jew; all Jews hang together like burrs. Where there is a Jew other Jews of all kinds immediately gather. It should have been the obligation of the Jews themselves to draw a line between the different types. They have not done this and that is why I must proceed equally against all Jews.' Upon my remark that it would be self-destructive if valuable Jews were forced to emigrate, because we urgently require their scientific labours and their value would otherwise accrue to other countries, he declined further comment, lapsed into generalities and finally concluded: 'They say I some-times suffer from weak nerves. This is a slander. I have nerves of steel.' Thereupon he violently slapped his knees, spoke ever faster and whipped himself into such a rage that I had no option but to fall silent and take my leave.

Some doubts have, alas, been cast on the accuracy of Planck's account. He was, when he set it down in 1947, 89 years old, had just emerged from a life-threatening illness, and, indeed, died a few months later. The events that occurred during the most tortured period of his life may have become muddled in his mind. At all events, his friends recalled his reports of the visit immediately after the event rather differently. They doubted whether Planck had tried to broach the dangerous question of the Jews and thought that Planck had in any case not found Hitler in a mood to listen to an age-ing intellectual, on whom he quickly turned his back. There are photo-graphs of the hapless Planck standing on the Swastika-draped podium at a session of the Prussian Academy of Sciences. He was seen to raise his right arm slowly, let it drop, and finally raise it mournfully in the Nazi salute.

Planck's account of his encounter with Hitler was published (in German) in *Physikalische Blätter*, **3**, 143 (1947). For a biography of Planck see J. L. Heilbron, *The Dilemmas of an Upright Man* (University of California Press, Berkeley, 1986).

97 Butterfly in Beijing

The phenomenon of chaos—the emergence of patterns out of random-ness—has in recent years touched almost every area of science. Irregularities in physical and biological (indeed, even in economic) processes were always regarded as defying theoretical analysis and were accordingly shunned by theoreticians. Turbulence in fluid flow was one practical problem, which had troubled both engineers and physiologists, and physicists had long been irked by the seemingly random transitions between steady and sporadic flow of water from a dribbling tap. The ideas behind chaos theory had been hazily prefigured in earlier years, but the beginnings of the subject can properly be dated to 1961, and the place, MIT, the Massachusetts Institute of Technology. Edward Lorenz was a meteor-ologist, trained as a mathematician; his interest was long-range weather forecasting, and he had recognized early on that any system of equations aimed at simulating the change with time of a weather pattern would become tractable only with the advent of the high-speed computer. Lorenz had acquired one of the first commercially available machines and he had written a crude program for the change in a weather pattern, based on 12 equations. His computer disgorged an endless series of successive weather maps.

Lorenz supposed, as did everyone else, that the evolution of weather would be deterministic—that the state of the weather at any instant would uniquely determine its state at any later time; so the accuracy of a forecast would depend only on the precision with which the state at the start could be defined. Lorenz's computer generated forecasts in the form of a numerical read-out, which could then be converted into graphical form. The revelation came one day when Lorenz decided to look more closely at one part of the output, and so, to save time, he re-started the computation at some point in the middle of the previous run. Then he went for coffee.

On his return Lorenz was surprised to see that the forecast from his program had deviated substantially from the previous result. But then he recalled a difference between the two experiments. He had given the machine its starting values with less precision the second time than the first: instead of 0.506127, for instance, for one of the features that defined the weather he had punched in 0.506. But the difference was only one part in 5,000, far less than Lorenz could conceive might affect the outcome. One part in 5,000 would amount to no more than an infinitesimal puff of air.

Lorenz might well have assumed that his computer was behaving erratically. Instead, he pursued his observation and found that the mathematical phenomenon was real: however small the difference between the starting values, the predictions would diverge until, after a while, all similarity between them had vanished. Here is how James Gleick puts it in his book on chaos:

> But for reasons of mathematical intuition that his colleagues would begin to understand only later, Lorenz felt a jolt: something was philosophically out of joint. The practical import could be staggering. Although the equations were gross parodies of the earth's weather, he had a faith that they captured the essence of the real atmosphere. That first day he decided that long-range weather forecasting must be doomed.

'I realized', Lorenz concluded, 'that *any* physical system that behaved nonperiodically would be unpredictable.' His conclusion held up when, years later, a vastly more powerful computer was programmed to model the weather with not 12 but no less than half a million equations. So was born the 'butterfly effect': the beat of a butterfly's wing in Beijing would be enough to change the weather in New York a month later.

Edward Lorenz did not stop at that, however. He discovered much simpler systems of equations that generated divergence, in accordance with what became known as the 'sensitive-dependence-on-initial-conditions' effect. His intuition told him that the deviations in the results seen in repeated cycles of computation should recur, that a pattern of change should emerge, and so indeed it proved. The fluctuating magnitudes of a variable, when plotted out in a three-dimensional graph, would distribute themselves around a focus, which became known as 'the Lorenz attractor'. Such pictures have now entered into the repertoire of graphic designers. This, then, was how the phenomenon of chaos came to light, but Lorenz published his results in meteorological journals, which scientists in other disciplines did not read, and it took years for the importance of his observations to permeate into the many areas in which they are now a commonplace. These include flow of liquids in tides, waves and pipes (not least arteries and veins), the beating of the heart, the fluctuations of animal populations, and many more.

Gleick quotes the words of a physicist: 'Relativity eliminated the Newtonian illusion of absolute space and time; quantum theory eliminated the Newtonian dream of a controllable measurement process; and chaos eliminates the Laplacian fantasy [145] of deterministic predictability.' This is a much more profound truth than the observation of a psychiatrist, Ernest Jones, that man's psyche had sustained only three grievous blows, delivered by Galileo, by Darwin, and by Freud.

See James Gleick's outstanding book, *Chaos—Making a New Science* (Viking, New York, 1987; Heinemann, London, 1988).

~

98 Cook knows best

We owe the concept of radiocarbon dating to the chemist Willard Libby (1908–80). Carbon, which is a constituent of all the compounds that participate in the processes of life, contains a small proportion of a radioactive isotope [149]. When an organism dies, metabolism ceases and there is no more turnover of all these carbon compounds; so the radioactive isotope is not replenished and decays away. Consequently the amount of radioactivity detected in the substance of the deceased animal or plant (wood or cotton, for instance) gives a measure of how long it has been dead. This technique revolutionized the practice of archaeology and won Libby the Nobel Prize in 1960. The following recollection comes from the American biochemist, Daniel Koshland, at the time a research student.

I remember one Saturday afternoon when Frank Westheimer [a distinguished chemist and Koshland's research supervisor] burst into the laboratory and said, 'Come right away. We need you at a conference.' I followed along dutifully to find Frank, Bill Libby, George Whelan, two other professors, and some assorted graduate students and postdocs assembled in a room. The problem put to us was that Libby wanted to know how to ash a penguin. Someone had told Libby that he should have a verified modern sample of carbon composition to compare with his ancient samples of carbon dating and that he should accumulate animals from the North Pole, South Pole, Equator, etc.

The Penguin had been flown in from the Antarctic and we were charged with converting all the carbon in the flesh, beak, claws, feathers, etc. to CO_2. The group started with obvious answers such as fuming sulfuric acid, aqua regia [a mixture of nitric and hydrochloric acid], fuming nitric acid, chromate solutions and so on. Each suggestion was discarded on the recommendation of someone whose experience showed it couldn't do the job. Finally, in frustration, the group dispersed for dinner. Several days later I happened to meet Libby, and I asked what had been decided. Libby said no chemical solution had been found, but he had mentioned the problem to his wife. She pointed out that all body materials were synthesized from a common source, and she therefore suggested that we cook the penguin and collect the grease, which of course could be easily oxidized to CO_2. We followed her advice and the problem was solved. Both this imaginative course and the exchange of ideas among professors and students over the course of several hours are typical examples of what made the atmosphere at Chicago so exciting at that time.

From D. E. Koshland, *Annual Reviews of Biochemistry*, **65**, 1 (1996).

~

99 Chemistry in the kitchen: the discovery of nitrocellulose

Nitrocellulose, which became known as gun-cotton, was discovered (as also was ozone) by a German chemist, Christian Friedrich Schönbein (1799–1868). He was appointed professor at the University of Basel in 1829 and the story, often repeated (though its exact veracity has been questioned), goes that he made his discovery at home in the kitchen.

The university laboratory closed daily for lunch, but Schönbein, an eager and impatient experimenter, would sometimes continue his work at home. On one such occasion, it is said, a flask in which he was heating a nitration mixture of nitric and sulphuric acids, broke and the corrosive liquid spilled on his wife's work surface. Terrified of the displeasure any obvious defacement might cause, he seized the first object that came to hand to mop up the spill. It was his wife's cotton apron, which he hastily washed under the tap and hung up to dry near the stove. A smokeless conflagration and explosion ensued and the apron vanished, leaving no trace behind.

The episode may have jeopardized Schönbein's domestic tranquillity, but it soon led to fame and fortune. He was invited to demonstrate the new explosive at the Woolwich Arsenal in London and he took the occasion to present Queen Victoria and Prince Albert with a brace of pheasants, the first ever shot with cartridges filled with gun-cotton.

See, for example, G. I. Brown's *The Big Bang: A History of Explosives* (Sutton, Stroud, 1998).

100 The living fossil

In 1938, three days before Christmas, the young curator of the small East London Museum on the eastern seaboard of South Africa was assembling a display of fossils. Her name was Marjorie Courtenay-Latimer. That morning her work was interrupted by a telephone call from the harbour, where a trawler, laden with a catch that included sharks and other fish, had just docked. The trawlermen had supplied Miss Courtenay-Latimer with fish of many species for the museum's collection. She now had a backlog of specimens from earlier catches still to be mounted and she wanted to get her fossil display finished before the Christmas holiday. She was not therefore anxious to accumulate more fish. 'But I thought of how good everyone at Irvin and Johnson had been to me, and it being so near Christmas, I thought the least I could do would be to go down to the docks to wish them the compliments of the season.'

So Miss Courtenay-Latimer climbed onboard the trawler and eyed the pile of fish, sponges, seaweed, and other detritus, which lay on the deck. There, projecting out of the heap, was a strange blue fin. 'I picked away the layers of slime to reveal the most beautiful fish I had ever seen. It was five foot long, a pale, mauvy blue with faint flecks of whitish spots; it had an iridescent silver-blue-green sheen all over. It was covered in hard scales, and it had four limb-like fins and a strange little puppy-dog tail. It was such a beautiful fish—more like a China ornament—but I didn't know what it was.' Thus Miss Courtenay-Latimer, recollecting 60 years on how the first coelacanth had revealed itself to her. The trawler captain, too, had been struck by its exotic allure; it had snapped at his hand when he examined it in the net, and at first he had thought of throwing it back to let it live.

Miss Courtenay-Latimer wanted it for her museum, but preservation presented a problem. The local cold-store owner would not accommodate the creature for fear the hints of putrefaction, already perceptible, would contaminate his stock, neither would the guardian of the mortuary help. The town's chemists could find only a little formalin, which proved insufficient for preservation; so as a last resort, as the fish began to exude oil, it was taken to a taxidermist. Miss Courtenay-Latimer was convinced she had happened on something remarkable, and it struck her that it looked like—but surely couldn't be, for it had been alive—a fossil fish. She had at once written, with a drawing of her find, to Dr J. L. B. Smith, chemistry lecturer but icthyologist by vocation, at Rhodes University in Grahamstown. She knew Smith, who was honorary curator of fishes for the small natural history museums around the coast. But Smith was away and found the letter only two weeks later, by when the rotting body parts of the fish had (to the subsequent dismay of Smith and biologists around the world) been thrown away.

When Smith received the letter he gazed in perplexity at the drawing. 'I did not know', he wrote later, 'any fish of our own, or indeed of any seas like that; it looked more like a lizard.'

> And then a bomb seemed to burst in my brain, and beyond that sketch and the paper of the letter, I was looking at a series of fishy creatures that flashed up as on a screen, fishes no longer here, fishes that had lived in dim past ages gone, and of which only fragmentary remains in rocks are known.

When Smith arrived in East London and saw and caressed the fish, he turned to its discoverer and declared, 'Lass, this discovery will be on the lips of every scientist in the world'. The coelacanth-like fishes had originated some 400 million years ago and had adapted and survived unsuspected into our own era. Smith named the species *Latimeria chalumnae* in honour of Marjorie Courtenay-Latimer. It was to be another five years, the Second World War intervening, before a second specimen was brought to the surface. When Smith received the word from the Comoros Island, it was only through a personal appeal to the Prime Minister, Dr Malan, that an aircraft of the South African Air Force was procured to fetch the fish before it could decay.

The story of the coelacanth has been well told by Samantha Weinberg in *A Fish Caught in Time: The Search for the Coelacanth* (Fourth Estate, London, 1999), and J. L. B. Smith's own account, *Old Fourlegs: The Story of the Coelacanth* (Longmans, Green, London, 1956) is something of a classic.

101 The sound of physics

Richard Feynman [89] is one of the few scientists since Einstein to have impressed himself on the public consciousness. This is in large measure the result of the two books of transcribed reminiscences and *obiter dicta*, which caught the popular imagination. Here is an extract from one of them, which gives a hint of the ebullient spirit that informed his science and his life.

Feynman was always careless with engagements and seldom wrote down addresses or telephone numbers. In 1957, he had been invited to take part in a conference on gravity, to be held at the University of North Carolina.

I landed at the airport a day late for the conference (I couldn't make it the first day), and I went out to where the taxis were. I said to the dispatcher, 'I'd like to go to the University of North Carolina.'

'Which do you mean,' he said, 'the State University of North Carolina at Raleigh, or the University of North Carolina at Chapel Hill?'

Needless to say, I hadn't the slightest idea. 'Where are they?' I asked, figuring that one must be near the other.

'One's north of here, and the other's south of here, about the same distance.'

I had nothing with me that showed which one it was, and there was nobody else going to the conference a day late like I was.

That gave me an idea. 'Listen', I said to the dispatcher. 'The main meeting began yesterday, so there were a whole lot of guys going to the meeting who must have come through here yesterday. Let me describe them to you: They would have had their heads kind of in the air, and they would be talking to each other, not paying attention to where they were going, saying things to each other, like "G-mu-nu. G-mu-nu".'

His face lit up. 'Ah, yes', he said. 'You mean Chapel Hill!' He called the next taxi waiting in line. 'Take this man to the university at Chapel Hill.'

'Thank you', I said, and I went to the conference.

From Richard Feynman, ed. Edward Hutchings, *Surely You're Joking Mr Feynman! Adventures of a Curious Character* (Norton, New York, 1985).

~

102 Grand Guignol

The spontaneous-generation controversy dates back to the very birth of biological science. It smouldered for centuries and occasionally flared up in impassioned disputes. By about the seventeenth century it had acquired a religious dimension, for the notion that life could arise spontaneously from inert matter was contrary to Christian teaching, which held that God had personally assembled a specimen of each species that it might procreate and duly die. The argument was settled for practical purposes by Louis Pasteur in the late nineteenth century, although dissenting voices were not finally silenced until another 50 years had passed.

Early in the nineteenth century the debate focused in large measure on parasites, such as tapeworms, which, common sense dictated, could originate only in the putrefying matter inside the intestine. Then, in the fourth and fifth decades of the century, the transfer of parasites or their eggs between hosts of different species was demonstrated. In 1854, Friedrich Küchenmeister (1820–90), a doctor in Dresden, who had developed a particular interest in parasites, motivated in large part by a religious ardour to understand God's purpose, uncovered the life cycle of the tapeworm. His method was peculiarly gruesome.

Küchenmeister had studied the so-called bladder-worms, found in pigs, cows, and some other animals. The name derived from the bubbles in which the creatures encyst themselves. These had been discerned in muscles, they appeared to cause no problems to the animal, and when examined under the microscope showed similarities to the head of the tapeworm. Küchenmeister took bladder-worms from the muscles of several species of animals and fed them to other animals, which, when dissected after a sufficient interval, were indeed found to harbour tapeworms. He also suspected that human tapeworms came from ingested pork; evidence for this was the high incidence of tapeworms in the local pork butchers and their families.

In 1855, determined to test his hypothesis, Küchenmeister had a brainwave: he asked permission to try an experiment on a condemned criminal. Some days before the man's execution Küchenmeister had, in those less fastidious days, noticed that the pork he was enjoying at dinner contained cooked bladder-worms. Hastening to the butcher he procured some meat from the same carcass. He teased out some bladder-worms and mixed them into a lukewarm soup and a black pudding, both of which were offered to the condemned felon (without what would now be called informed consent). The man consumed two helpings and three days later met the hangman. Küchenmeister opened the corpse, examined the viscera, and found young, developing tapeworms attached to the intestinal wall. Five years on, Küchenmeister was offered another chance to confirm his result. This time he was given access to the prisoner four months before the execution. The outcome was gratifying: when anatomized, the felon's intestine yielded up a tapeworm five feet long. The findings, important as they were, caused widespread revulsion in the community of biologists. A reviewer of the published account, quoting Wordsworth, called Küchenmeister

> One that would peep and botanize
> Upon his mother's grave.

Other, more scrupulous seekers after truth have inflicted worse abuses on their own bodies, as when in 1767 John Hunter, celebrated Scottish anatomist and surgeon to George III, injected pus from the sores of a gonorrhoea patient into his own penis to determine how the disease was transmitted. Hunter was unlucky, because his patient evidently had syphilis as well, and never recovered his health. Moreover, he drew the false conclusion that syphilis and gonorrhoea were aspects of one disease and so set back venereology by many years. It was only a century later that the agents of the venereal diseases were identified and the final proof was again obtained only by a suspension of ethical principle. In 1885, a German bacteriologist, Ernst von Bumm, cultured the bacterium of gonorrhoea, and, to make sure, he inoculated a healthy woman with a culture. Four years later Albert Neisser in Breslau, while searching for the spirochaete that causes syphilis, injected four healthy people with syphilitic secretions. This did, indeed, cause a public outcry when it became known and Neisser was very properly censured and fined.

But Neisser's offence was less heinous than the callous Alabama syphilis study, which continued for 40 years in the twentieth century: large numbers of patients, all black (for syphilis was commonly regarded in the southern states as a 'negro disease' and was held to have a genetic component) were given only placebos, so that the progress of the malady could be followed until death supervened.

Details of Friedrich Küchenmeister's discoveries are in W. D Foster, *A History of Parasitology* (Livingstone, Edinburgh, 1965); see also *Parasite Rex: Inside the Bizarre World of Nature's Most Dangerous Creatures* by Carl Zimmer (The Free Press, New York, 2000). For a history of self-experimentation, see Lawrence K. Altman: *Who Goes First?* (Random House, New York, 1987).

103 The mathematical wallpaper

Sophie (or Sofya) Kovalevsky was a mathematician of genius. Her name crops up in today's textbooks in the Cauchy–Kovalevsky theorem of differential equations, and she also made notable contributions to mechanics and to physics, especially to the theory of light propagation through crystalline solids. Her life is the stuff of romantic fiction.

Sophie Kovalevsky was born in 1850 into the Russian nobility, the daughter of General of Artillery Korvin-Krukovsky. It was an uncle who first sparked her interest in mathematics:

more than anything else he loved to communicate the things he had succeeded in reading and learning in the course of his long life.

It was during such conversations that I first had occasion to hear about certain mathematical concepts which made a very powerful impression on me. Uncle spoke about 'squaring the circle', about the asymptote—that straight line which the curve constantly approaches without ever reaching it—and about many other things which were quite unintelligible to me and yet seemed mysterious and at the same time deeply attractive. And to all this, reinforcing even more strongly the impact of these mathematical terms, fate added another and quite accidental event.

Before our move to the country from Kaluga, the whole house was repainted and papered. The wallpaper had been ordered from Petersburg, but the quantity needed was not estimated quite accurately, so that paper was lacking for one room.

At first it was intended to order extra wallpaper from Petersburg, but through rustic laxity and characteristic Russian inertia the whole matter was postponed indefinitely, as often happens in such situations. Meanwhile time passed, and while everyone was intending and deciding and disposing, the redecoration of the rest of the house was finished.

Finally it was decided that it simply wasn't worth going to all the trouble of sending a special messenger to the capital, five hundred versts away, for a single roll of wallpaper. Considering that all the other rooms were in order, the nursery might very well manage without special paper. One could simply paste ordinary paper on the walls, particularly since our Polibino attic was filled with stacks of old newspapers accumulated over many years, lying there in total disuse.

By happy accident, however, it turned out that there in the attic, in the same pile with the old newspapers and other rubbish, were stored the lithographed lecture notes of Academician Ostrogradsky's course on differential and integral calculus, which my father had attended as a very young army officer. And it was these sheets which were utilized to paper the walls of my nursery.

I was then about eleven years old. As I looked at the nursery walls one day, I noticed that certain things were shown on them which I had already heard mentioned by Uncle. Since I was in any case quite electrified by the things he told me, I began scrutinizing the walls very attentively. It amused me to examine these sheets, yellowed by time, all speckled over with some kind of hieroglyphics whose meaning escaped me completely but which, I felt, must signify something very wise and interesting. And I would stand by the wall for hours on end, reading and rereading what was written there.

I have to admit that I could make no sense of any of it at all then, and yet something seemed to lure me on toward this occupation. As a result of my sustained scrutiny I learned many of the writings by heart, and some of the formulas (in their purely external form) stayed in my memory and left a deep trace there. I remember particularly that on the sheet of paper which happened to be in the most prominent place on the wall, there was an explanation of the concepts of infinitely small quantities and of limit. The depth of that impression was evidenced several years later, when I was taking lessons from Professor A. N. Strannolyubsky in Petersburg. In explaining those very concepts he was astounded at the speed with which I assimilated them, and he said, 'You have understood them as though you knew them in advance'. And, in fact, much of the material had long been familiar to me from a formal standpoint.

Sophie's father, she noted in her memoirs, 'harboured a strong prejudice against learned women' and decided to put a stop to his daughter's mathematical studies with her tutor, a man, in any case, of limited capacity. So

Since I was under my governess's strict surveillance all day long, I was forced to practice some cunning in this matter. At bedtime I used to put the book [Bourdon's *Algebra Course*, which she had managed to procure through her tutor] under my pillow and then, when everyone was asleep [the dragon of an English governess behind a screen in the same room], I would read the night through under the dim light of the icon-lamp or the night lamp.

Then came another stroke of good fortune: a neighbouring landowner was a physics professor and one day he brought to the house his new elementary physics text. Sophie pounced on it and soon came up against trigonometrical functions, which she had not encountered before. She was baffled and her tutor could not help. Wrestling with it all alone, she succeeded in working out for herself the meaning of a sine and how to calculate it. When she told Professor Tyrtov how much she had understood of his book, he evidently responded with an indulgent smile.

> But when I told him the means I had used to explain the trigonometric formulas he completely changed his tone. He went straight to my father, heatedly arguing the necessity of providing me with the most serious kind of instruction, and even comparing me to Pascal.

The result was a grudging agreement that she should be tutored by the aforementioned Professor Strannolyubsky, a mathematician at the naval academy in St Petersburg, who also quickly recognized her as a prodigy.

But Sophie's interests were not limited to mathematics. She was passionate about literature and she and her sister befriended Dostoevsky. It is thought that Dostoevsky modelled the characters of Aglia and Alexandra in *The Idiot* on Sophie and her sister, whom he briefly courted. Sophie entered into a stormy marriage with Vladimir Kovalevsky, who was to become Professor of Palaeontology at St Petersburg University, and after some years gave birth to a daughter. Marriage and escape from the suffocating atmosphere of the parental home had allowed her to travel and at the age of 20 she approached the august German mathematician Karl Weierstrass. Weierstrass was then an ageing bachelor and probably something of a misogynist. In response to Sophie's request for his help he set her a test, consisting of problems so difficult that he must have felt confident of shaking off her unwanted attentions. Matters turned out differently, and Weierstrass quickly perceived that he was in the presence of an exceptional talent.

Weierstrass became Sophie's tutor, counsellor, and friend. Under his guidance she flowered and in due time presented her doctoral thesis to the University of Göttingen, based on three papers, two in pure mathematics and one in theoretical astronomy. She thereupon returned to her husband in Russia and for seven years, to Weierstrass's dismay, seemed to have abandoned mathematics. At the end of this period she separated from her husband and accepted an invitation to Stockholm from a noted Swedish mathematician, Gösta Mittag-Leffler, whom Weierstrass had sent to seek her out in Russia. In Stockholm, she recovered her mathematical interests and rose to the position of Professor of Mathematics—the first woman to occupy a professorial chair in a European university, for it was not until 17 years later that a similar mark of recognition was grudgingly bestowed on

Marie Curie [9]. Mittag-Leffler had to work hard to secure the appointment for Kovalevskaya. Most of the country's mathematicians supported her candidature, but objections came from other quarters; indeed, August Strindberg, the ill-natured playwright, called her 'a monstrosity', a freak of nature. She continued to make important contributions to mathematics, and for a paper in mechanics ('On the rotation of a solid body around a fixed point') she received a highly coveted prize from the French Academy of Sciences, which in fact was doubled, by reason of 'the quite extraordinary service' that her paper had rendered to theoretical physics.

Meanwhile Sophie, or Sonya, as she became known in Sweden, had started again to write. Her restless nature had evidently reasserted itself and she appeared to be forsaking mathematics once more, this time for her second love, literature. During her time in Stockholm she published several novels, a play, and articles in Swedish literary magazines, and was full of plans for new literary projects. But in the winter of 1891 Sophie Kovalevsky died of pleurisy, aged 41, with the words 'too much happiness' on her lips.

The extracts above are from Sofya Kovalevskaya's *A Russian Childhood*, translated by Beatrice Stillman (Springer-Verlag, New York, 1978). See also the biography of Sophie Kovalevsky's life, *The Little Sparrow: A Portrait of Sophia Kovalevsky* by D. H. Kennedy (Ohio University Press, Athens, Ohio, 1983), as well as a chapter, reflecting on this book, in Jeremy Bernstein's collection of essays, *Cranks, Quarks and the Cosmos* (Basic Books, New York, 1993) and the obituary by Sophie Kovalevsky's friend, the intrepid scientist and anarchist Prince Peter Kropotkin, *Nature*, **41**, 375 (1891).

104 Out of the mouths of poets

One of the more alarming medical therapies, loosely based on physiological theorizing, was introduced by a formidable Viennese psychiatrist, Julius Wagner-Jauregg (1857–1940). He did not hold with the Freudian philosophy, which in Vienna admitted of little dissent, and was an early enthusiast for pharmacological intervention. He had, he thought, seen an improvement in the condition of some of his disturbed patients after they recovered from feverish infections. This led him to a hypothesis (which later proved wrong) of how an increased body temperature might redound on the brain. He decided to try his theory on his patients, whom he proceeded to infect with streptococci, staphylococci, and then with tuberculosis. The results of these terrifying experiments boosted his confidence in the approach, and he now went one step further: he injected blood from a victim of fulminating malaria into a syphilitic patient in an advanced state of paresis. There was, Wagner-Jauregg delightedly reported, a notable improvement in the patient's mental state.

The method caught on; Wagner-Jauregg was rewarded with the Nobel Prize in 1927. The treatment had wide currency during the 1920s and 30s, but it is not clear how many patients died of the cure. Fortunately better drugs emerged in the next decades. The patients might, indeed, have been better served by the nineteenth-century regime of sudden precipitation into an icy sea.

Wagner-Jauregg's inspiration was prefigured in a curious manner:

One day in 1927 the great Hofrat Julius Wagner-Jauregg of Vienna was sitting in a compartment of a railway carriage in Sweden, waiting for the train to take him to Stockholm, where he was to receive the Nobel Prize for Medicine. He had won the Prize for his discovery of the treatment of the insane by raising their temperatures (actually by giving them fevers in the form of malaria). As he was waiting for the train to start, a lady stepped into the compartment and sat opposite him. They fell into conversation, and it turned out that the lady too was on her way to the Royal Palace in Stockholm, and that she too was going to receive the Nobel Prize. Hers was to be the Prize for Literature, for she was the Sardinian poet Grazia Deledda. She had written a romance about a young man who was insane, and in his insanity stumbled about the Macedonian marshes, and so got soaked, and so got a high fever, and so was cured of his insanity.

The story is from an article by a well-known biochemist, W. E. van Heyningen, in *Trends in Biochemical Sciences*, N177, August 1979.

~

105 Venomous vermin

The story of typhus has been told by Ludwik Gross, the man who—against the unquestioned wisdom of the time (which rejected any notion of a link between viruses and cancers)—discovered the first leukaemia virus. He carried out this research in such time as he could spare from his duties as a busy hospital doctor in New York. As a young man, Gross worked at the Pasteur Institute in Paris and there in 1934 had met Charles Nicolle, recipient of the Nobel Prize in 1928 for his discovery that typhus was carried by lice (as had been conjectured for some time). Nicolle headed the Pasteur Institute of Tunis during the first decade of the twentieth century, and there, during a typhus epidemic, he was struck by a curious circumstance: the townspeople were going down with the disease in their homes and in the streets, but no one ever became infected in a hospital, seething with typhus victims. Patients were, of course, washed and clad in hospital garb, and Nicolle deduced that infection lurked in dirty clothing and probably, therefore, in lice. He injected a monkey with the blood of a typhus patient and, when the animal fell sick, collected lice from its body and applied them to another monkey, which also then developed typhus. Nicolle further showed that the lice that harboured the infective agent, clearly a bacterium, turned red and died. Their powdery droppings, Nicolle found, would engender typhus when rubbed on the skin. It took some 30 years to develop an effective vaccine.

Gross again had the story from the horse's mouth: in 1936 he called on Dr Rudolf Weigl in Lwów in Poland [now Lvov, and in the Ukraine], who had studied typhus and the related disease, trench fever, since his days in a military bacteriological laboratory during the First World War.

Dr Weigl had a very modest laboratory in which he worked with his wife and a few assistants. Here are the results of his studies. A normal healthy louse is relatively free from microbes. When the louse ingests a drop of blood from a patient suffering from typhus, it becomes red and dies after a few weeks. In the meantime the Rickettsias [the *Rickettsia prowazeki* bacteria, named after two American researchers, Howard T. Ricketts and Stanislas von Prowazek, who both fell victim to typhus, caught in the laboratory] multiply by the millions in the intestinal tracts of the infected lice. Millions of live Rickettsias are then excreted in the feces of such lice, in the form of a dark powder, which is highly infectious. Dr Weigl transmitted the Rickettsias from louse to louse by making a watery suspension from their infected intestines and infecting healthy lice through miniature enemas [!]. In the course of these experiments, several members of his staff became infected with typhus and died. Dr Weigl himself became infected with typhus, but recovered. The lice had to be fed daily with human blood. Dr Weigl and his wife fed healthy lice kept in small boxes, similar to match boxes, except that one side of these boxes consisted of a very dense screen, through which the lice could prick the skin and ingest blood. Dr Weigl developed typhus a second time while trying to feed infected lice located on his skin, and he learned from this that one can develop typhus twice, although this was probably an exceptional occurrence. Over 100 lice were needed to produce a single dose of vaccine. Subsequently this number was reduced to 30. Nevertheless, production of very large doses of vaccine by this method, which was very complicated and dangerous for laboratory technicians, presented tremendous difficulties.

Some two years later an American researcher found that the rickettsia bacteria could be grown in fertilized chicken eggs and could then be killed and extracted to produce a safe vaccine.

Weigl's work was quickly assimilated into the teaching of the Polish medical schools and led to at least one remarkable coup when the country was subjugated by the Germans a few years later. More than five million Poles died during the brutal occupation and more than half a million were deported to Germany and elsewhere as slave labourers. Two young doctors, Stanislav Matulewicz and Eugeniusz Łazowski, thought up a ruse that saved hundreds of their compatriots in the area, not far from Warsaw, where they worked. They had learned that rickettsia share an antigenic feature with some strains of the harmless proteus bacterium. This means that antibodies generated in the blood of those infected with typhus recognize not only the typhus microbe, but also proteus. Because an infestation with proteus is hardly a likely occurrence, this cross-reactivity, as it is called, came to serve as the standard diagnostic test for typhus: if the serum of a typhus suspect reacted with proteus bacteria grown in the laboratory (that is to say, it agglutinated or clumped the bacteria [59]), this was regarded as proof enough of typhus.

The advantage of this, the Weil–Felix reaction, was, of course, that laboratory workers did not need to expose themselves to the lethal rickettsia. The two doctors knew, moreover, that the Germans had a dread of typhus, which had ravaged the population of their country, and especially the army, during and after the First World War. So newly arrived prisoners in concentration camps were quarantined to avert outbreaks of the disease. The Gestapo doctors in Auschwitz would kill all inmates with any symptoms that could presage a typhus infection. On one day alone in Auschwitz 746 prisoners were killed for this reason.

It occurred to Matulewicz that a person inoculated with proteus would give a positive Weil–Felix reaction and be assumed to have typhus. The first trial was on a slave labourer, who had briefly returned from Germany and was prepared to risk death to avoid being sent back. The ruse worked: a blood sample was sent to the German state laboratory and a telegram soon arrived confirming typhus. Thereupon the two doctors propagated a spurious typhus epidemic by a programme of inoculation. The Germans were on their guard against any tricks, such as substituting infected for healthy blood samples, but here, of course, they could draw the blood themselves to make the diagnosis. A senior army doctor and two assistants were sent to ascertain what was happening, but the two Poles entertained the senior officer with an abundance of vodka and he was content to leave the tour of inspection to his juniors. They were shown decrepit barns in which typhus patients were supposed to have died and they thought it prudent not to venture further. For the benefit of their superior, the Poles brought in an old man dying of pneumonia, whose distressed state was attributed to typhus. The Germans were totally fooled and the population of Rozvadow and the surrounding countryside were spared their malign attentions.

See Ludwik Gross, *Proceedings of the National Academy of Sciences, USA*, **93**, 10539 (1996) and, for an account of the proteus deception, J. D. C. Bennett and L. Tyszczuk, *British Medical Journal*, **301**, 1471 (1990); also Bernard Dixon's book, *Power Unseen: How Microbes Rule the World* (Oxford University Press, Oxford, 1994).

~

106 A principle misapplied

R. V. Jones, the physicist, has related the following cautionary tale. Outside the Clarendon Laboratory in Oxford he

came across a dirty beaker full of water just when I happened to have a pistol in my hand [Jones does not enlarge on this—perhaps unusual—circumstance]. Almost without thinking I fired, and was surprised at the spectacular way in which the beaker disappeared. I had, of course [sic!], fired at beakers before; but they had merely broken, and not shattered into small fragments. Following Rutherford's precept [presumably to test conscientiously the reproducibility of experimental observations] I repeated the experiment and obtained the same result: it was the presence of the water which caused the difference in behaviour. Years later, after the War, I found myself having to lecture to a large elementary class at Aberdeen, teaching hydrostatics *ab initio*. Right at the beginning came the definitions—a gas having little resistance to change of volume but a liquid having great resistance. I thought that I would drive the definitions home by repeating for the class my experiments with the pistol, for one can look at them from the point of view of the beaker, thus suddenly challenged to accommodate not only the liquid that it held before the bullet entered it, but also the bullet. It cannot accommodate the extra volume with the speed demanded, and so it shatters.

The experiment became duly public in Aberdeen, and inspired the local Territorial contingent of the Royal Engineers, who used sometimes to parade on Sundays to practise demolition. One task that fell to them, or, more accurately, refused to fall to them, was the demolition of a tall chimney at the local paper works. There are various standard procedures for this exercise, one of the oldest being to remove some of the bricks of one side, and to replace them by wooden struts. This process is carried out so far as to remove the bricks from rather more than half-way round the base of the chimney and to a height comparable with the radius. A fire is then lit in the chimney, to burn through the struts and cause the chimney to fall.

The Royal Engineers, however, decided this time to exploit the incompressibility of water as demonstrated by my experiment. Their plan was to stop up the bottom of the chimney, fill it with water to a height of 6 ft or so, and simulate the bullet by firing an explosive charge under the water. Since diversion on the Sabbath was rare in Aberdeen, the exercise collected a large audience and the charge was duly fired. It succeeded so well that it failed completely. What happened was that, as with the beaker, every brick in contact with the water flew outwards, leaving a slightly shortened chimney with a beautifully level-trimmed bottom 6 ft up in the air. The whole structure then dropped nicely into the old foundation, remaining upright and intact—and presenting the Sappers with an exquisite problem.

R. V. Jones, 'Impotence and achievement in physics and technology', *Nature*, **207**, 120 (1965).

∼

107 A message from space

Much of what we now know of the nature of the Universe comes from radio astronomy. The great parabolic receiver dishes, gathering in signals from the void, which are now a familiar feature of the rural scenery, are a product of an accidental discovery, an offshoot of the struggle for survival in the early years of the Second World War. True, it had been foreshadowed some 10 years earlier by the observation made by Karl Jansky (1905–50), while searching for the causes of atmospherics in radio reception at the Bell Telephone Laboratories in the United States [81]. Jansky had found that the noise intensity fluctuated on a diurnal cycle, and his instinct had led him to measure the intervals between the noise maxima very precisely. He was rewarded by the discovery that they occurred once every 23 hours and 56 minutes—the period of the Earth's rotation relative to the stars. The source of the intruding signals, then, must be outside the Solar System and appeared, indeed, to come from the Milky Way. This was confirmed by an amateur cosmologist, Grote Reber, who built a parabolic receiver in his garden in Wheaton, Illinois to track the origin of Jansky's signals. The work attracted little attention from the astronomers and there the matter rested for a decade.

J. S. Hey, a physicist, was recruited in 1942 into war work. His task was to improve the erratic radar system then operated by the British Army. The physicists were engaged in a contest of jamming and countermeasures with the Germans. Hey warmed to this work.

The sound basic training, the urgent demands of defence, and the invigorating research environment, all fed my enthusiasm.

During 1941, the enemy made increasing endeavours to jam radar operations. The War Office became anxious lest their radar devices, particularly vulnerable to airborne jamming, might be rendered useless. On 12 February, 1942, the passage of the German warships, Scharnhorst and Gneisenau through the English Channel, slipping by almost unnoticed until it was too late to muster any effective attack on them, to the accompaniment of radar jamming from the French coast, resulted in a drastic reappraisal of the jamming menace. The War Office decided to augment its efforts to contend with radar jamming and sought assistance from the Army Operational Research Group in dealing with this awkward problem. The investigation of jamming is hardly an appealing subject for a scientist, on the face of it a negative, irksome task. Nevertheless, the challenge had to be met and I readily accepted the invitation to be responsible for analysing Army radar jamming, and advising on anti-jamming measures. I cooperated with Army staff officers in devising instructions for radar operators and organising an immediate reporting system. A mobile J-watch laboratory was strategically sited on the cliffs at Dover and manned by a member of my team. I had a peculiar role as a civilian scientist holding a key position in an Army system and the work proved not dull but exciting with my advice often being sought urgently by Anti-Aircraft Command and by the War Office.

On 27 and 28 February 1942, a remarkable series of reports from many parts of the country described the daytime occurrence of severe noise jamming experienced by anti-aircraft radar working at wavelengths between 4 and 8 m[etres], and of such intensity as to render radar operation impossible. Fortunately no air raids were in progress but alarm was widespread at the incidence of this new form of jamming, all wondering what it might portend. Recognising that the directions of maximum interference recorded by the operators appeared to follow the Sun, I immediately telephoned the Royal Observatory at Greenwich to inquire whether there was any unusual solar activity, and was informed that although we were within two years of the minimum of the sunspot cycle, an exceptionally active sunspot was in transit across the solar disc and situated on the central meridian on 28 February. [The sunspots come into view as the Sun rotates; they are strongly magnetic, whereas the magnetic strength of the Sun as a whole is weak.] It was clear to me that the Sun must be radiating electromagnetic waves directly—for how else could the coincidence in direction be explained—and that the active sunspot region was the likely source. I knew that magnetron valves generated centimetric radio waves [the radiation bounced back from aircraft in radar detection] from the motion of electrons in kilogauss magnetic fields [the gauss is the unit of magnetic field strength], and I thought, why should it not be possible for a sunspot region, with its vast reservoir of energy and known emission of corpuscular streams of ions and electrons [37] in a magnetic field of the order of 100 gauss, to generate metre wave radiation.

When I wrote a paper giving details of the event, my director B. F. J. Schonland [Sir Basil Schonland, a South-African physicist who became director of the Atomic Energy Research Establishment at Harwell] recalled Jansky's discovery of galactic radio noise with which I had been hitherto unacquainted. What was surprising, however, was that several radio scientists, experienced in ionospheric and communications research, were sceptical of my conclusion. They found it hard to believe that such powerful radio outbursts had escaped notice in previous decades of radio research. It seemed almost an effrontery for a comparative novice in the field to be presenting a paper on an energetic solar radio phenomenon.

The discovery of the intense radio emission from the Sun had some features in common with Jansky's discovery of cosmic radio noise. Both were examples of observations for one purpose leading to hitherto unknown phenomena. In both instances the aim had been to study types of interference limiting the effectiveness of practical systems.

Hey's work, and related results obtained independently, a little later, at the Bell Telephone Laboratories, had to await publication until after the war. Hey suggests that the failure of earlier researchers to detect such obvious and intense emission from the Sun at times of sunspot activity, which, he says, 'almost clamoured to be observed', stemmed from their adherence to current dogma: none so blind as will not see. Only a single amateur astronomer came close in 1938.

From J. S. Hey, *The Evolution of Radio Astronomy* (Elek Science, London, 1973).

108 The emperor's chessboard

According to legend, a Chinese Emperor asked a sage what reward he would require in return for an important service. The sage named his price: nothing more than some rice, two grains to be placed on the first square of a chessboard, four on the second, eight on the third, and so on. A modest demand, the Emperor thought, and happily agreed; but he had failed to grasp the principle of geometric progressions. The entire rice crop of the empire would have had to go on a single square, long before the sixty-fourth was reached.

It was the same simple calculation, so obvious to anyone conversant with numbers, that led to what has been arguably the most important technological innovation of our time. It won Kary Mullis, the maverick American biologist, the Nobel Prize in 1993. Here is how he recalls the blinding moment of revelation, an experience granted to only a few scientists in a working lifetime.

> One Friday evening later in the spring [of 1983] I was driving to Mendocino County [in California] with a chemist friend. She was asleep. Every weekend I went north to my cabin and on the way sat still for three hours in the car; I like midnight driving; my hands occupied, my mind free. On that night I was thinking about my proposed sequencing experiment.

Mullis, who worked for Cetus, an infant biotechnology company, was pondering an idea that might improve the efficiency of the procedure used in determining the sequence of nucleotides in DNA. The nucleotides are the links that make up the long chains of DNA. They are of four kinds, abbreviated A, C, G, and T. The 'sequence' of the DNA is the order in which these units occur along the chain. The two strands of the famous double helix [88] have 'complementary' sequences: every A faces, and is linked to, a T in the opposite chain, and every C is linked to a G. The sequencing process makes use of an enzyme, which in nature copies DNA during cell division. To start the enzyme (DNA polymerase) running along the chain and doing its work one needs a so-called primer. This is a short segment of a DNA strand, readily synthesized in the laboratory, and complementary to the start of the segment of DNA to be sequenced. If, thought Mullis, one were to bracket the desired segment with two primers, one on either strand of the double helix (made up of two strands running in opposite directions), molecules of the enzyme would proceed along the DNA both ways and the sequences of both strands could be determined simultaneously. This would give an important internal check on the correctness of the answer, for if the sequence of one strand is known that of its complementary strand can immediately be inferred. (The scheme would not, as it happens, have worked.)

It was then that inspiration struck: with the two primers at opposite ends, the segment in between would be copied by the enzyme. Suppose now one separated the two strands of the newly formed DNA—easily done by heating—then if there were a sufficiency of primer molecules in the mixture, the enzyme would start again on the new strands. From two copies would come four, from four, on the next cycle, eight, and so on. The snag was only that at the temperature required to separate the DNA strands the enzyme would be inactivated and a fresh dose would have to be added each time. That nuisance, though, could be overcome by using an enzyme preparation from a thermophilic bacterium—one that flourishes in hot springs and contains heat-resistant proteins. Mullis continues:

> The idea of repeating a procedure over and over again might have seemed unacceptably dreary. I had been spending a lot of time writing computer programs, however, and had become familiar with reiterative loops—procedures in which a mathematical operation is repeatedly applied to the products of earlier reiterations. That experience had taught me how powerful reiterative exponential growth processes are. The DNA replication procedure I had imagined would be just such a process. Excited, I started running over powers of two in my head: 2, 4, 8, 16, 32 . . . I remembered vaguely that two to the tenth power was about 1,000 and that therefore two to the twentieth power was around a million. I stopped the car at a turnoff overlooking Anderson Valley. From the glove compartment I pulled a pencil and paper. I needed to check my calculations. Jennifer, my sleepy passenger, objected groggily to the delay and the light but I exclaimed that I had discovered something fantastic. Unimpressed, she went back to sleep. I confirmed that two to the twentieth power was over a million and drove on.

On Monday morning Mullis, bubbling with excitement, told his colleagues at the Cetus Corporation about his brainchild, for which he coined the name, polymerase chain reaction, or PCR; but they remained obdurately unimpressed—that is, until the method was shown to work.

This, at least, is Mullis's version of the story, but it does not accord too closely with the recollections of others. Mullis's erratic performance in the laboratory and his abrasive and hyperbolic style had not endeared him to his colleagues, and there were some in the company who felt it would be best if they could rid it of his obtrusive presence. His poor credit at the time may have partly explained why his presentation of his PCR idea at an internal seminar met with a cool reception. But there was an additional reason: one of Mullis's associates at Cetus has remarked that the oddest aspect of the history of PCR is that it was not developed with a special problem in mind. It could have been useful for the modest undertaking that Mullis had in hand, but its wider implications were not perceived until it had been made to work. And then the applications began to crowd in. The technology was brought to fruition by a team of Cetus researchers. It made Cetus one of the leading American biotech companies and changed the face of biology and of the biotechnology, pharmaceutical, and agricultural industries. Every biological laboratory now contains automated devices for amplifying DNA by PCR. PCR allows workable quantities of DNA to be generated from samples of no more than a few molecules—for instance, in a blood stain or semen [135]. To most biologists it seems in retrospect scarcely comprehensible that the idea occurred only to Mullis and not to them.

For a commentary on the discovery of PCR, which broadly accepts Mullis's version, see Walter Bodmer and Robin McKie, *The Book of Man: The Quest to Discover our Genetic Heritage* (Little, Brown, London, 1994), but for a more detailed and balanced account see the book by Paul Rabinow, *Making PCR: A Story of Biochemistry* (University of Chicago Press, Chicago, 1996).

109 A modest appraisal

Johann Heinrich Lambert was a German polymath, born in 1728 in Alsace in humble circumstances. He was essentially self-taught and his interests encompassed physics, mathematics, and chemistry. Settling in Berlin, he attracted the attention of Frederick the Great, who bestowed many favours on him. But Lambert's ego was insatiable.

> Lambert was very conceited and there are many anecdotes recording this. One of these records how he was very worried because the king was very slow in making his appointment to the membership of the Academy [the Royal Academy of Sciences of Berlin]. His friend Achard tried to cheer him by saying that he was certain that the king would make the appointment very soon.
>
> I am not a bit impatient,—replied Lambert—because this is a matter for his own glory. It would be a discredit to his reign in the eyes of posterity if he did not appoint me.
>
> His appointment was eventually made and King Frederick the Great, speaking to him at a reception asked him which of the various branches of science he was most expert at, to which Lambert replied shyly 'With all of them'. 'Thus you are an eminent mathematician also?' asked the king. 'Yes, Sir.' 'Who was your master in this science?' 'I myself, Sir.' 'That means, that you are a second Pascal?' 'At least, Sir', replied Lambert. After Lambert had left, the king commented that it seemed he had appointed a great fool to the Academy.

The king's evaluation was too hasty, for Lambert's achievements are by no means negligible. His work on geometry has its place in the history of mathematics, his contributions to astronomy were appreciable, and his name is commemorated in a basic principle of light absorption, the Lambert–Beer law.

The quotation is from Ferenc Szabadváry's *History of Analytical Chemistry* (Gordon and Breach, London, 1960).

~

110 The little green men who weren't

In 1967, Jocelyn Bell was a research student in Cambridge, her supervisor the astronomer Anthony Hewish. His interest at the time was quasars (quasi-stellar radio sources), very intense sources of radio emission and other radiation, discovered in 1963. They were, and to some extent remain, mysterious, but are now thought to be black holes—objects so heavy that they collapse inwards on themselves and, because of the density of matter in their interior, trap electromagnetic radiation as demanded by the General Theory of Relativity. Because they are, in effect, point sources of radiation, they twinkle like any stars, by reason of disturbances in the Earth's atmosphere, which deflect the radiation this way and that before it reaches the telescope.

The magnitude of the radiowave twinkle would, Hewish thought, give a measure of the size of the object. He devised an array of radio detectors, spread across an area of more than four acres, which could measure intensity differences (twinkles) over time periods of fractions of a second. Jocelyn Bell, while making these measurements, experimented with the time periods and was astonished to find, when she inspected the recorder chart paper one morning, that the quasar source she was observing gave rise to a pulse of intensity every 1.34 seconds. Her first guess was that an output from some kind of machinery was imposing itself on the radio signals; but then she realized that the quasar source, with its strange periodicity, entered the field of view of the telescope every 23 hours and 56 minutes, the period of rotation of Earth relative to the stars [107]. Surely any man-made device would have been set to a 24-hour cycle? Radio telescopes, on the other hand, were run on sidereal time, but no emission from any near at hand could be detected. What could be sending out pulses from the void with a regularity of one part in 10 million? Surely these must be signals from extraterrestrial intelligences! The source was accordingly designated LGM1 (for Little Green Men).

Alas, this startling conclusion prevailed for only a few days. It was discarded when Jocelyn Bell, following a further search, found three more sources of pulsating radiation in different parts of the sky. It was two other astronomers, Thomas Gold and Franco Pacini, who came up with an explanation: the pulsars, as they were now called, were rotating neutron stars—tiny dead stars, made up of neutrons, crushed together, the products of an implosion of a rapidly cooling dying star. Only some 10 kilometres across, they could rotate with periods in the region of a second, shooting out radiation like the revolving beam of a lighthouse. Later it was found that the rotation of pulsars slows as they age, and the rate of diminution can be used to calculate the time when they were formed. Most strikingly, the age of a pulsar in one nebula (the Crab) was estimated to be about 1,000 years, and an explosion (a supernova, as it is called) in the Crab nebula was recorded by Chinese and Japanese astronomers in AD 1054.

The discovery of the first pulsar secured a Nobel Prize for Anthony Hewish in 1974, but not for Jocelyn Bell, who had made the original observations. This was denounced by some—the astronomer Fred Hoyle among them—as a scandal (though it was not the view of the generous Jocelyn Bell Burnell).

The story of the discovery of pulsars has been often told. A good summary and explanation appears in David Leverington's book, *A History of Astronomy from 1890 to the Present* (Springer-Verlag, London, 1995).

~

111 The virtues of squalor

The life and achievements of Alexander Fleming (1881–1955) have been obfuscated by the legends that gathered around him, even before his death. Fleming made two important but accidental discoveries, the second of which initiated a new era in medicine.

Fleming passed most of his working life in a dingy laboratory in St Mary's Hospital close by Paddington railway station in London. His superior for much of that time was the redoubtable Professor Colonel Sir Almroth Wright—the model for Sir Colenso Ridgeon in George Bernard Shaw's play, *The Doctor's Dilemma*. Wright believed that the only valid weapon against bacterial infections (and many other medical problems) was immunization, and research on chemical intervention, which in the foregoing years had, thanks to the work of Paul Ehrlich [155] in Germany, already saved many lives, was strongly discouraged. It was indeed the Inoculation Department over which Wright so majestically presided. The methods favoured by Wright were highly conventional, even old-fashioned. The circumstances surrounding Fleming's discovery of lyso-zyme, an enzyme that dissolves the cell walls of some kinds of bacteria, were recorded many years after the event (which occurred in 1921) by V. D. Allison, at the time a young research worker in Fleming's laboratory:

> Early on, Fleming began to tease me about my excessive tidiness in the laboratory. At the end of each day's work I cleaned my bench, put it in order for the next day and discarded tubes and culture plates for which I had no further use [all good practice in bacteriology, in which contamination with alien bacteria is a mark of professional negligence]. He, for his part, kept his cultures . . . for two or three weeks until his bench was overcrowded with 40 or 50 cultures. He would then discard them, first of all looking at them indi-vidually to see whether anything interesting or unusual had developed. I took

his teasing in the spirit in which it was given. However, the sequel was to prove how right he was, for if he had been as tidy as he thought I was, he would never have made his two great discoveries—lysozyme and penicillin.

Discarding his cultures one evening, he examined one for some time, showed it to me and said 'This is interesting'. The plate was one on which he had cultured mucus from his nose some two weeks earlier, when suffering from a cold. The plate was covered with golden-yellow colonies of bacteria, obviously harmless contaminants deriving from the air or dust of the laboratory, or blown in through the window from the air in Praed Street. The remarkable feature of this plate was that in the vicinity of the blob of nasal mucus there were no bacteria; further away another zone in which the bacteria had grown but had become translucent, glassy and lifeless in appearance; beyond this again were the fully grown, typical opaque colonies. Obviously something had diffused from the nasal mucus to prevent the germs from growing near the mucus, and beyond this zone to kill and dissolve bacteria already grown.

Fleming's next step was to test the effect of nasal mucus on the germ, but this time he prepared an opaque, yellow suspension of the germs in saline and added some nasal mucus to it. To our surprise the opaque suspension became in the space of less than two minutes as clear as water . . . . it was an astonishing and thrilling moment [and] the beginning of an investigation which occupied us for the next few years.

Fleming had apparently believed, and recorded in his laboratory notebook, that the bacteria came from his nose. This is far less likely than Allison's interpretation. It also gave rise to the story that a drop from his running nose had landed on the agar plate, as he was applying a culture of bacteria. Whatever the origin of the bacteria on the plate, the culture was labelled A. F. (for Fleming) coccus (the genus of bacteria) and used in subsequent experiments on the mysterious lytic (cell-dissolving) agent. Allison and Fleming looked for the activity in other bodily fluids, in animals and in plants, and found that it was widely distributed; tears were a good source, but the most abundant was egg white. Fleming suspected that the 'factor' might be an enzyme, but did nothing to prove it. It was in Howard Florey's laboratory in Oxford that the protein, lysozyme, was isolated. It did not prove clinically useful, for it is rapidly eliminated from the circulation and pathogenic bacteria are, or rapidly become, resistant to its action.

Fleming's second fortuitous discovery proved vastly more important and was the product of a freakish stroke of luck. Yet it made little impact at the time, even on Fleming himself. What happened was this: early in 1928

Fleming was moved to investigate a supposed relation between the virulence of strains of *Staphylococcus* bacteria and the colour of the colonies which they formed on agar plates. With a research student, D. M. Pryce, he collected samples from all manner of infections—carbuncles and boils, abscesses, skin lesions, and throat infections—and plated them onto nutrient gels. In the summer Pryce left and was replaced by another student, whom Fleming instructed to carry on with the work while he departed for his annual family holiday in Scotland. As usual, he left a stack of plates, bearing cultures, in the corner of the laboratory.

Soon after Fleming's return at the beginning of September Pryce came by to ask how the work was going. Fleming, ever amiable, went to a tray containing discarded culture plates, soaking in lysol—the disinfectant used to sterilize the glass plates in preparation for washing and recycling. (Today's culture plates are plastic and disposable.) Some of the plates in the large pile were not immersed and were still dry, and it was a handful of these that Fleming picked up to show Pryce. Suddenly he noticed something that had escaped him on the plate he was in the act of handing to Pryce. 'That's funny', he muttered, and pointed to a small excrescence of mould that had grown on the agar gel: the bacterial colonies in its vicinity had vanished. Was the mould another source of lysozyme?

Fleming was intrigued and showed the plate to several of his colleagues, who reacted with uniform apathy. But Fleming did investigate further. He picked up the spot of mould with a sterile wire loop and cultured it. Samples of the culture again inhibited the growth of *Staphylococcus*, but not of several other species of bacteria. Fleming showed the mould to the resident mycologist, who pronounced it to be *Penicillium rubrum*. Many more moulds were tested, most of which were without activity. What was encouraging was that the original mould was not toxic: Fleming's research student was induced to eat some and reported that it tasted like Stilton cheese and caused no ill-effects. Animals injected with a filtrate of the mould culture also remained healthy. Then Fleming administered some to the same student, who was suffering from a persistent infection of his nasal atrium, but the results were inconclusive.

There was only sporadic further interest in the mould extract, now called penicillin, until, in 1938, Howard Florey in Oxford stepped into the field. He had developed an interest in lysozyme and recruited Ernst Chain, an emigré biochemist, to join in a study of its properties. After some time, Florey and Chain decided to extend their investigations to other natural bactericidal compounds, which they supposed would generally also be proteins. They found Fleming's paper on penicillin, published nine years earlier, and thought his mould extract might be of interest. Both were adamant that they had not even contemplated possible medical uses. 'I don't think it ever crossed our minds about suffering humanity,' was how Florey put it, 'this was an interesting scientific exercise.'

Penicillin turned out, of course, not to be a protein at all, but Florey and Chain, joined now by another able biochemist, Norman Heatley, made rapid progress in its isolation. The first pure sample was used to treat mice injected with a virulent strain of *Streptococcus*, and Florey and Heatley sat and watched through a Saturday night as the control group fell sick and the penicillin-treated animals frisked in their cage. Florey recalled years later: 'I must confess that it was one of the more exciting moments when we found in the morning that all the untreated mice were dead and all the penicillin-treated ones were alive.' It seemed, he said, like a miracle. A human patient with sepsis, an Oxford policeman, responded dramatically when treated with penicillin, but he died when the small stock of material ran out.

In 1940, air raids were beginning to ravage the cities of Britain and the fear of 'septic wards' overflowing with wounded civilians and soldiers had concentrated the minds of the research establishment. After the retreat from Dunkirk a spectre reared up before the eyes of the Oxford group: what if there were an invasion and cultures and extracts were seized by the Germans? 'Then and there', Norman Heatley reminisced, 'everyone at Oxford smeared the fungus in the linings of our coats', to be retrieved when it was safe.

The therapeutic promise of penicillin was now evident, and the need to prepare and stockpile quantities of the material was pressing. But to scale up production to industrial levels American help was needed, and Florey and Heatley travelled to the United States—rushing across New York with the precious mould in a taxi to get it into a refrigerator before the 90-degree summer heat killed it—to set up the programme. Within months penicillin, now relatively widely available, had changed the practice of clinical medicine.

The story of the isolation of penicillin by the Oxford group—a tour de force of biochemical and chemical ingenuity—has been often told, but, although Florey, Fleming, and Chain shared the Nobel Prize, the myth that the accomplishment was Fleming's alone has not entirely faded. Fleming, though within his narrow limits a very capable experimenter, was not a scientist of Florey's calibre. He himself seems to have been well aware of his limitations. An eminent scientific contemporary recalled that he

> told me often that he didn't deserve the Nobel Prize, and I had to bite my teeth not to agree with him. He wasn't putting on an act, he really meant it, at least around 1945/6. At the same time he would tell me that he couldn't help enjoying his undeserved fame, and I liked him for that. I don't know whether he took a different line with laymen, but if he'd have liked to pretend to be a great scientist with me and others of his scientific colleagues, he had the sense to know that none of us were any more impressed with him than he was himself.

How the mantle of greatness came to descend on Fleming is not altogether clear. That his name and no other became associated in popular perception, and especially in the press, with the discovery of antibiotics stirred the resentment of Florey and his colleagues. Fleming's obtrusive second wife, Amalia, undoubtedly played a part in spreading the myth, but the authoritative biographer of Fleming and Florey, Gwyn Macfarlane, lays the responsibility, first, on Sir Almroth Wright, who claimed the discovery for Fleming (and incidentally the Inoculation Department of St Mary's) in an extravagant letter to *The Times*, and, secondly, on the Dean of St Mary's Hospital Medical School, the egregious Charles McMoran Wilson, Lord Moran, who sought to annex the greatest share of the glory for his institution.

Known to his colleagues Corkscrew Charlie, by reason of his well-earned reputation for deviousness, Moran had served as Winston Churchill's personal physician during and after the Second World War; he had drawn much opprobrium on himself by publishing intimate details of Churchill's maladies and so breaching the contract of confidentiality between doctor and patient. While returning from the Tehran Conference with Stalin and Roosevelt in 1944, Churchill had been struck down with pneumonia. The military doctor in Cairo, where the Prime Minister was treated, urged an immediate recourse to penicillin, but Moran, who was probably totally ignorant of its power, would not countenance it; Churchill was treated with a sulphonamide drug and recovered. But it was later put about, and evidently not denied by the wily Moran, that it was penicillin which had miraculously and providentially saved Churchill's life. Amends to Florey and Chain were, at all events, made by the Nobel Prize committee.

The success of penicillin initiated a scramble to discover other antibiotics. Several thousand are now known, but most have toxic side-effects, and, while often useful in research, have not found application in the clinic.

Among the most potent of the antibiotics are the cephalosporins, discovered as early as 1945 by Giuseppe Brotzu, who held the chair of bacteriology at the University of Cagliari in Sardinia. Brotzu noticed that the sea in the vicinity of the town was, notwithstanding the presence of a sewage outfall, strangely free of pathogenic bacteria. Brotzu had read about penicillin and began to wonder whether some micro-organism in the sewage might not be producing an antibiotic. The intrepid professor descended into the sewer pipe and collected samples of the effluvium. When cultured they revealed the presence of a mould, *Cephalosporium acremonium*, which did indeed secrete a substance active against several kinds of pathogen. When tried on patients with staphylococcal infections it proved moderately effective.

Brotzu could not excite the interest of the pharmaceutical industry and so contented himself with publishing the results in a Sardinian journal. This did not, of course, alert the world to his discovery but he sent a copy to a British acquaintance, a doctor who had worked in Cagliari. Thence it found its way to the Medical Research Council in London and soon Edward Abraham and Guy Newton in Florey's institute in Oxford embarked on a survey of the *Cephalosporium* moulds. The outcome was the isolation from one of them of cephalosporin C, which became one of the most useful antibiotics, active against a range of pathogens, including staphylococci that had acquired resistance against penicillin.

For a full account of Alexander Fleming's life and discoveries, on which the above description is based, see Gwyn Macfarlane's magisterial biography, *Alexander Fleming: The Man and the Myth* (Chatto and Windus, London, 1984); note also Macfarlane's equally engrossing and definitive biography of Howard Florey. For other antibiotic discoveries, including that of the cephalosporins, see Bernard Dixon's entertaining book, *Power Unseen: How Microbes Rule the World* (Oxford University Press, Oxford, 1994).

~

112 Hevesy's subterfuge

When, soon after Hitler's accession to power in 1933, the racial laws were promulgated, James Franck (1882–1964), physicist and Nobel Laureate, resolved to leave Germany immediately, even though, as a *Frontkämpfer* in the First World War, he was still at that stage exempt. Fearing that his gold Nobel Prize medal would be confiscated, he entrusted it to a friend, Niels Bohr [79], in Copenhagen. Max von Laue (1879–1960), the most upright and courageous of the German scientists who remained in Berlin and taught the forbidden science of relativity throughout the Nazi period (he told Einstein that he always assured students that the original papers had been written in Hebrew), had the same concern. Bohr debated about what he should do with the three gold medals—Franck's, Laue's, and his own— and conferred with a colleague, Georg von Hevesy (a Hungarian physicist, who pioneered the use of radioactive isotopes [149] in biology and medicine). A desperate situation, they decided, demanded desperate measures. Here, in Hevesy's words, is what they resolved:

> I found Bohr worrying about Max von Laue's Nobel medal, which Laue had sent to Copenhagen for safe-keeping. In Hitler's empire it was almost a capital offence to send gold out of the country and, Laue's name being engraved on the medal, the discovery of this by the invading forces would have had very serious consequences for him. I suggested we should bury the medal, but Bohr did not like this idea as the medal might be unearthed. I decided to dissolve it. While the invasion forces marched in the streets of Copenhagen, I was busy dissolving Laue's and also James Franck's medals.

And so the medals were consigned to aqua regia (the mixture of hydrochloric and nitric acids that dissolves gold by converting it into its nitrate). Confident that the war would in time be won and he would eventually return to his beloved city and institute, Bohr left the jar containing the dissolved medals on a shelf in his laboratory. Shortly after, Bohr was whisked away to Sweden by fishing-boat (and thence clandestinely by air to England). And, indeed, when he returned in 1945, there the jar still was, disregarded by the occupiers. Bohr had the gold recovered and the Nobel Foundation was pleased to recast it into two commemorative medals.

Scientists escaping from Germany at this time used many other ingenious subterfuges to safeguard some at least of their assets and to evade the law that forbade them from taking money or valuables out of the country. The chemist Hermann Mark, for example, used what money he had to buy platinum wire, which he fashioned into coathangers; these passed undetected through the hostile customs inspection.

See Georg von Hevesy, *Adventures in Radioisotope Research*, Vol. 1 (Pergamon, New York, 1962).

113 Crystal clear

The particle-wave duality was perhaps the most startling outcome of the conceptual revolution that swept through physics during the first three decades of the twentieth century. That photons—the packets, or quanta, of light—behaved in some respects as particles, in others as waves, had already been noted when the Prince (later duc) Louis de Broglie conjectured that the same might be true of other particles, such as electrons. Sitting in an eyrie in the Eiffel Tower as a meteorological observer during the First World War, de Broglie had time to ruminate on these matters and to develop an equation relating the momentum of the particle to its wavelength. This formed the substance of his doctoral dissertation, which he submitted to the University of Paris in 1924 and which earned him the Nobel Prize five years later.

The criterion for wave propagation is interference: when two waves meet their combined intensity will be reinforced where their crests superimpose; and, conversely, where the crests of one coincide with the troughs of the other they will cancel. Waves reflected from a regular array—a lattice—of objects, separated from each other by a distance comparable to the wavelength will form a so-called diffraction pattern. This is the basis of X-ray crystallography, in which the positions and intensities of interference events arising from X-ray waves reflected by regularly spaced atoms in a crystal lattice are used to deduce the precise three-dimensional disposition of these atoms. The prediction of the wave nature of electrons was put to the test, more or less accidentally, by G. P. Thomson (son of the illustrious 'J. J.' [73], and later himself Sir George) and, by virtue of a spectacular fluke of good fortune, by Clinton Davisson and Lester Germer, working in the laboratories of the Western Electric Company in New York.

Western Electric (which later became the Bell Telephone Laboratories) had been engaged in a long and expensive legal wrangle with the General Electric Company over a patent for a vacuum tube. Western had bought the patent from its inventor, Lee de Forest, in the expectation that it could be used to improve long-distance telephone communications. The issue hinged on whether their tube differed substantially from the device constructed by Irving Langmuir at General Electric; this, it was claimed, was a true vacuum tube, whereas de Forest's tube depended for its function on the presence in the interior of a little air. Davisson was instructed to study, in relation to the patent, the emission of electrons from a hot metal target inside a vacuum tube when struck by a beam of positively charged particles; but no sooner had he started than he stumbled on a new phenomenon. He had realized that, with some small modifications, the tube could also be used to bombard the target with electrons, which, as was known, elicited emission of secondary electrons of much lower energy. Davisson found that when the tube was operated in this mode he could detect also some high-energy electrons projected back from the target. These, he realized, were a few of the primary electrons that the target had reflected.

Davisson was reminded of Rutherford's experiments with alpha particles [20], which had given such striking insights into the structure of the atom: when fired at a thin metal foil, most of the particles passed straight through, implying that the atoms of the metal were mainly empty space, but some, having struck the nucleus, bounced straight back. Rutherford was astounded. It was, he said, 'as if you fired a 15-inch shell at a piece of tissue paper and it came back and hit you'. (This result put paid to J. J. Thomson's 'plum-pudding' model of the atom, which envisaged the electrons as embedded in a matrix of positive charge. It upheld the Bohr–Rutherford planetary model [27].) Davisson wondered whether the reflected electrons might not yield new information about internal energy levels in the atoms of the target, and he persuaded his employers to let him pursue the matter. His plan was to examine the reflection of electrons by different metals.

The revelation came in 1925, when Davisson and his assistant, Germer, were busy with their electron bombardment experiments and a container of liquid air exploded in the laboratory, shattering all around it, including their vacuum tube. The surface of the hot target, consisting of crystalline nickel, was exposed to air and quickly acquired a coating of nickel oxide. Davisson and Germer rebuilt their tube around the same nickel target, which they heated to a high temperature in the vacuum to eliminate the oxide layer, and again began their laborious experiments. But this time, when they examined the distribution in space of the reflected electrons, they found a new and quite different result: the intensity of the reflected electrons showed strong maxima at sharply defined angles to the target. Baffled, the experimenters broke open the tube and examined the metal. The heat treatment, it turned out, had converted the tiny nickel crystals into large crystalline plates. It took a visit to the annual meeting of the British Association for the Advancement of Science in Oxford to bring home to Davisson what the result signified—that the electrons reflected from the crystal lattice of the nickel were interfering like waves. Davisson was actually vacationing with his wife when he dropped in on the conference, and was startled to hear one of his own earlier experiments (with a platinum target) being cited by the lecturer, Max Born [73], as evidence for the existence of electron waves. Davisson spent the return sea voyage grappling with the new theory of wave-mechanics. Back in the laboratory he and Germer began a search for peaks of intensity at the theoretically predicted angles and after, much effort, found them.

G. P. Thomson arrived at the same point independently and by a different route in 1927, when he was Professor of Physics at the University of Aberdeen. With a colleague he had set up an experiment to bombard a thin metal foil, mounted in a vacuum tube, with an electron beam. Unexpectedly he found the intensity of the electrons passing through the foil to display a pattern of rings, unmistakably interference fringes. Thomson and Davisson duly shared the 1937 Nobel Prize for Physics.

See C. Davisson and L. H. Germer, *Physical Review*, 30, 705 (1927), and the account of their work by Richard K. Gehrenbeck in *History of Physics*, ed. Spencer R. Weart and Melba Phillips (American Institute of Physics, New York, 1985).

~

114 Little brown dog

During Queen Victoria's reign a powerful anti-vivisection movement sprang up in Britain, strongly represented in both Houses of Parliament. Physiologists, such as Claude Bernard and Charles Richet [150] in France, and Michael Foster and Burdon Sanderson in Britain, attracted the hatred of animal-lovers. It was undoubtedly true that many callous, and often unnecessary experiments were conducted on living animals, especially in France, where no legal restraints existed. Claude Bernard [138], the greatest physiologist of all, was a target of violent abuse (even from his own family), and his teacher, François Magendie, even more so. Anti-vivisectionists from England infiltrated Magendie's lecture demonstrations and reported on the cruel and repellent scenes that they had witnessed. A Member of Parliament, Henry Labouchère, recalled the cacophony of screams from the experimental animals that assailed his ears in the corridors of the Medical

School in Paris, and the reaction of the doorman when he remarked on it: 'Que voulez-vous? C'est la science.' Magendie would often be heard to address the struggling dog, strapped down on the table before him: 'Tais-toi, pauvre bête!'

In Britain, the Home Office introduced legislation on the use of animals in research, and the anti-vivisection campaign, in Parliament and the country, was well funded and well organized. It reached its climax with the 'brown dog affair' of 1907. This cause célèbre was initiated by two young Swedish women, who enrolled as medical students in the London School of Medicine for Women, having already witnessed and been distressed by animal experiments in France. They attended demonstrations in physiology classes at University College, but abandoned their studies after a year. They had, however, kept a diary, in which they meticulously recorded their observations, and in April of 1903 they presented it to Stephen Coleridge, a barrister and officer of the National Anti-Vivisection Society.

Coleridge's attention was caught by one case in particular, that of a dog, which had been subjected to a demonstration experiment at University College by a lecturer, Dr William Bayliss. The two ladies had managed to take a close look at the animal just before it was brought into the lecture theatre and had observed half-healed operating scars, one still closed with forceps, on its abdomen. Now the Cruelty to Animals Act forbade the use of an animal for more than one experiment (though this could involve two operations), yet here was the brown dog, tightly muzzled and strapped to the table, while Bayliss opened its neck to expose the salivary glands. The animal, according to the two ladies, had struggled piteously, 'violently and purposefully', and during the half-hour of the demonstration experiment, when Bayliss had attempted to measure the salivary pressure, it had thus been fully conscious. There had, moreover, been no sign or smell of an anaesthetic.

Coleridge relayed this inflammatory intelligence to the audience at a large and indignant public meeting, not without some embellishment. The speech was reported in a national newspaper and questions were asked in the House of Commons. Bayliss, who had been named as the malefactor, instructed his solicitor, who demanded a public retraction and apology from Coleridge. When Coleridge refused, a writ was issued and the trial, which lasted for four days, began at the Law Courts in the Strand on 11 November 1903. The public gallery was packed and rowdy.

The first witness was Ernest Starling, Professor of Physiology at University College (famous, with Bayliss, for his work on cardiac physiology). He testified that he had used the brown dog (small, according to him, but, to Coleridge, large) to study the mechanisms of pancreatic disorders, including diabetes. He had opened the abdomen and ligated a pancreatic duct. Two months later, on the day of Bayliss's demonstration, he performed a follow-up internal examination to assess the consequences of the first operation. Having satisfied himself that all had gone well, Starling passed the anaesthetized animal over to Bayliss for his lecture demonstration on secretion.

Starling had infringed the Cruelty to Animals Act, but stated in his defence that he had done this to avoid sacrificing another dog. Bayliss asserted, and several students who had been present at his demonstration confirmed, that the dog had not struggled, it had merely twitched. Had it struggled he could not have performed the dissection. It had been thoroughly anaesthetized, first with a morphine injection and then with the standard mixture of chloroform, alcohol, and ether, delivered through a pipe that ran under the bench into a tube inserted into the dog's trachea; this had evidently been hidden from the accusers. The demonstration, which was intended to show that the salivary pressure was independent of blood pressure, had failed: Bayliss was unable to achieve electrical stimulation of the nerve controlling the salivary glands and, after trying vainly for a half-hour, had given up. The laboratory technician had then handed the dog to a student, Henry Dale [36], who had taken the pancreas for dissection and killed the animal by a knife in the heart.

Counsel for Bayliss was Rufus Isaacs (later Marquess of Reading and Viceroy of India), who made the most of the lacunae in the Defence's case. The judge summed up even-handedly, but it did not take the jury long to find for the plaintiff. Bayliss was awarded damages of £2,000, with £3,000 costs, leaving Coleridge with a bill amounting in today's terms to some £250,000—soon raised by sympathizers with his movement. With a fine sense of irony, Bayliss donated his winnings to his college for use in physiological research. The fund still exists and is probably sometimes used to buy animals for research.

Reacting against this setback to their cause, a group of anti-vivisectionists, led by Louise Lind-af-Hageby, one of the two Swedish women who had initiated the affair, resolved to erect a memorial to the brown dog and a symbol of their cause. A well-known sculptor was commissioned to produce a bronze image of the dog, to surmount a drinking fountain and horse-trough of granite. After two initial rejections a compliant local council was found: the London Borough of Battersea was at the time a proletarian, socialist stronghold, with several political crusaders among its inhabitants, sympathetic to the anti-vivisectionist cause. Even the local hospital, which eschewed the use of animals, was known to the local populace as 'The Antivivy'. The drinking-fountain was duly erected near Battersea Park. It bore on its base the inscription:

> *In memory of the brown terrier dog done to death in the laboratories of University College in February of 1903, after having endured vivisection extending over more than two months and having been handed over from one vivisector to another till death came to his release. Also in memory of the 232 dogs* [an exaggeration] *vivisected at the same place during the year 1902. Men and women of England, how long shall these things be?*

The statue was unveiled on 15 September 1906 and became at once a focus of angry debate. The bronze dog resisted a determined attack one night by a group of students. They were caught by the police, brought before a magistrate, and pleading guilty to malicious damage, were fined. There followed two years of intermittent protest meetings, rallies, riots, and arrests throughout London. Eventually Battersea Council tired of the endless wrangles; after various compromise solutions had failed to find favour a motion was passed to do away with this tiresome source of contention, and the statue quietly vanished one night in March 1910.

For the whole story see the slim volume by Peter Mason, *The Brown Dog Affair* (Two Sevens Publishing, London, 1997).

~

115 Friends and enemies

A. V. (Archibald Vivian) Hill (1886–1977) was a physiologist who won a Nobel Prize in 1922 for his work on the energetics of muscle. He was also, in the words of his protégé (and later Nobel Laureate), Sir Bernard Katz, 'the most naturally upright man I have ever known'. Hill strove all his life to right political wrongs—to curb the excesses of chauvinism that disfigured relations between intellectuals in the opposing countries during and after the First World War, and to give succour to scholars displaced from Nazi

Germany before the Second. A story from his collection of essays and memoirs recalls some of the absurdities that sprang from the nationalistic passions engendered by the Great War.

It was known since the early years of the twentieth century that the metabolic end-product of muscle activity is lactic acid, which accumulates in large amount with fatigue. What was not then fully understood was how the muscle recovers, and what, in particular, happens to all the lactic acid. Was it eliminated by oxidation or was it recycled through metabolic reactions into the carbohydrate from which it had been formed?

> J. K. Parnas [a well-known German biochemist] had come to Cambridge in 1914, shortly before the war, in order, he hoped, to decide the question by thermal measurements . . . He concluded first, that lactic acid of fatigue is completely burnt and not rebuilt, and, second, that about half the energy so liberated is stored as potential energy in muscle.
>
> Parnas's experiments, for so difficult an investigation, were made in a very short time and were ended by the outbreak of war, so that I had no opportunity of discussing them with him. He, as a German citizen, was interned (and later repatriated) while I was in the Army . . . At any rate his conclusions were wrong. However, they were communicated to the Physiological Society and published in greater detail in German.

These incorrect conclusions misled researchers in the field for some time, until Otto Meyerhof, a giant in the history of biochemistry, looked into the matter again after the war. By 1920 he had a complete answer: the disappearance of lactic acid, the oxygen consumed, the fraction (most) of lactic acid reconverted into carbohydrate, made up a self-consistent and satisfying scheme that soon found its way into the textbooks. But meanwhile:

> In July 1920 an 'International' Congress of Physiology was to be held in Paris, from which 'enemy' scientists were to be excluded. In March 1920, Meyerhof had sent his results for publication in *Pflügers Archiv* [a German journal, at the time required reading for physiologists] and he complained to me bitterly in a letter that he was not to be allowed to attend the Congress and report them, whereas Parnas, no longer a German 'enemy' but now [in consequence of the Versailles Treaty] a Polish 'ally', was intending to read a paper on his contrary findings of 1914, based on much less critical evidence that he had obtained. Yet Parnas in Cambridge from Strassburg [Strasbourg, then still German] in 1914, had been an open and vigorous supporter of German militarism, which Meyerhof had always deplored. In any event, however,

neither of them went to the Congress, for Parnas was cut off in Warsaw by the Russian armies which had invaded Poland in July and it was left to the Congress in Edinburgh in 1923 under the presidency of [the British physiologist] Sharpey-Schäfer [80] (who himself had lost two sons in the war) to become properly international again.

The story comes from A. V. Hill's *Trails and Trials in Physiology* (Edward Arnold, London, 1965).

116 The maestro's gaffe

Erwin Schrödinger (1887–1961) was hailed as a mathematical and scientific genius from the beginning of his student years in Vienna. He was one of the small band of theoretical physicists who wrought a revolution in the perception of matter and the Universe in the years just before and after the First World War. Schrödinger's most celebrated monument is the invention of wave mechanics, a description of the behaviour of fundamental particles, which he, with others, later showed to be another formulation of quantum mechanics. In 1927, he was elected, as successor to Max Planck, to a professorial chair at Berlin University, then a Mecca of theoretical physics. There he became closely associated with Albert Einstein, a friendship that endured intermittently, though with stormy vicissitudes, until a few years before Einstein's death in 1955.

Schrödinger was not a wholly admirable character. He was physically fearless, as his exploits in the Austrian army during the Great War attested, but was wanting in moral courage or scruples in his private life. In 1933, appalled by the excesses of the Nazis, he negotiated a position for himself in Oxford, through the good offices of Frederick Lindemann (later Lord Cherwell) [57]. Lindemann had helped set up the Academic Assistance Council to give shelter to Jewish academics, displaced from their positions in Germany, and he was surprised at Schrödinger's overtures, for Schrödinger was not a Jew. His demands, moreover, were considerable and he asked in particular that a position should be found for Artur March, then an associate professor in Innsbruck, to serve as his assistant. This was duly arranged, as was the generous bonus of a fellowship at Magdalen College.

But it soon transpired that Schrödinger, ever in the grip of a powerful priapic urge, was determinedly pursuing March's wife. When Lindemann, a rather prudish bachelor, found out, he was outraged: it was not the assistant that Schrödinger was after, it was his wife. 'We ought to get rid of the bounder', he told his colleagues. Schrödinger did not in any case care for the stag culture of the Magdalen high table, and to universal disgust took himself off to a professorial chair at the University of Graz in his native Austria. The *Anschluss*, when Austria was conjoined with Germany, soon followed and Schrödinger, whose political views were known, became suddenly persona non grata. At this point he performed his most shameful act of self-interest: he wrote an open letter to the university senate, published in all the leading newspapers, in which he made a grovelling apology for his earlier mistakes, affirmed his exultation at the union of his beloved country with Germany and called for joyous submission to the will of the Führer. Einstein's reaction can be imagined. Nor did his self-abasement profit Schrödinger, for it was evidently thought to lack sincerity, and he was dismissed from his position and banished from the campus.

Schrödinger was now in deep trouble. Lindemann was not disposed to offer the hand of friendship once more; but then help came from an unexpected quarter. Éamon de Valéra, head of the now independent Republic of Eire, had retained his affection for his early calling as a mathematician, and was making plans for the foundation of an Institute of Advanced Studies in Dublin. Schrödinger, by then beatified with a Nobel Prize, was invited to be its first director. He accepted with alacrity and late in 1939 took up his position, after a brief stay at the University of Ghent. He and his family were happy in Ireland and he entered into more than one amorous liaison during his time there, living for some years in a *ménage à trois* with his wife and his Irish mistress, who bore him a child. His theoretical work was moderately productive and it was in Dublin that he wrote his best-selling monograph, which redirected the careers of many physicists, *What is Life?* But it was also in Dublin that he again alienated Einstein, with whom he had in the meantime been reconciled.

Schrödinger, like Einstein, had long been preoccupied with the vision of a unified field theory—an expansion of the General Theory of Relativity that would embrace both gravitational and electromagnetic forces—for he had an almost mystical belief in the unity of nature. He entered into an animated correspondence on this and other matters with Einstein, and was delighted, on informing Einstein of a new mathematical trick of which he was very proud, to be told that he was '*ein raffinierter Gauner*'—a cunning rogue.

But Schrödinger now overreached himself. Intoxicated with his theory, and apparently dreaming even of a second Nobel Prize, he presented his paper on his latest cogitations before a distinguished audience, which included de Valéra, at the Royal Irish Academy in January 1947. His scheme was based on what he believed to be a new formulation of geometry, applied to relativistic space-time; but it proved in reality to be a very modest modification of a line pursued years earlier by Einstein and by Arthur Eddington, which Einstein had abandoned as fruitless. Schrödinger's will-o'-the-wisp was well publicized. The *Irish Press* reported that 'Twenty persons heard and saw history being made in the world of physics yesterday as they sat in the lecture hall of the Royal Irish Academy, Dublin, and heard Dr Erwin Schrödinger . . . ' Schrödinger, the article reported, had 'disappeared through the snowy traffic on his veteran bicycle, before he could be questioned further', but the reporter had caught up with him at his home in a Dublin suburb, and there had learned that the theory was a generalization of which Einstein's Theory of Relativity was now merely a special case. Asked whether he was confident in his conclusions, Schrödinger replied 'I believe I am right. I shall look an awful fool if I am wrong'. And so, indeed, it proved. News of the event quickly found its way to the United States, and Einstein and other leading physicists received copies through the *New York Times* with requests for comment. Einstein responded in measured but devastating terms: the theory could be judged

only on the basis of its mathematical-formal qualities, but not from the point of view of 'truth' (i.e. agreement with the facts of experience). Even from this point of view I can see no special advantages over the theoretical possibilities known before, rather the opposite. As an incidental remark I want to stress the following. It seems undesirable to me to present such preliminary attempts to the public in any form. It is even worse when the impression is created that one is dealing with definite discoveries concerning physical reality. Such communiqués given in sensational terms give the lay public misleading ideas about the character of research. The reader gets the impression that every five minutes there is a revolution in science, somewhat like the coup d'état in some of the smaller unstable republics. In reality one has in theoretical sciences a process of development to which the best brains of successive generations add by untiring labor, and so slowly lead to a deeper conception of the laws of nature. Honest reporting should do justice to this character of scientific work.

These comments were disseminated by radio and newspapers around the world, together with Schrödinger's observation of how he would appear were he wrong. Even before he had seen this crushing reaction Schrödinger had written an apologetic letter to Einstein, avowing that he had found it necessary to puff himself up in order to improve his situation (and especially his salary) at the institute. Einstein replied brusquely, with an explanation of why he believed that the theory represented no significant advance, and thereafter the correspondence between the two men ceased.

Einstein was further alienated when Schrödinger resigned from his position in Dublin—a base act of ingratitude, as he saw it—and accepted the Chair of Physics at the University of Vienna. It was in Vienna that he died, laden with all the honours that his country (and Germany) could bestow. Among his papers his biographer found a folder marked *Die Einstein Schweinerei*, an untranslatable word, implying a dirty, discreditable business.

The story and quotations are from the magisterial biography by Walter Moore, *Schrödinger: Life and Thought* (Cambridge University Press, Cambridge, 1989).

~

117 Hybrid vigour

Science has generated its own vocabulary, loosely rooted in the classical languages. But neologisms today are less scrupulously coined than in the era when the ancient languages were universally taught. Jacques Barzun, the American scholar, has recalled the lament of a university president, deploring the introduction of the Bachelor of Science degree: it would not, he said, ensure that the students knew science, but it would certainly ensure that they knew no Latin. Mixed derivations, in those more learned days, were anathema. An Oxford historian was supposed to have remarked when television arrived that no good would ever come of an invention the name of which was half in Latin and half in Greek.

In old age, Johann Wolfgang von Goethe (1749–1832), who wrote and theorized widely about science (and is now remembered for his elaborate, but incorrect, theory of colour vision), was reported to have had the following illuminating exchange with his disciple, Johann Peter Eckermann: the latter related to his master one day that he had been present at the demonstration of a remarkable new invention. It was a steam carriage, or *automobile*, which could propel itself without horses. Goethe pondered long over this remarkable apparition, and presently summoned Eckermann again. It was surely, he pronounced, Eckermann's little joke, for if such a contrivance had indeed been invented it could not have been given so grotesque a name: it would have been called an *autokineticon*, or otherwise perhaps an *ipsomobile*.

An echo of these purist scruples found its way into the columns of *Nature* 70 years after Goethe's death, when one Sir Courtenay Boyle was reported to have deplored in an article in *Macmillan's Magazine* that barbarous usages had crept into the language; he did not care for the word *motor*, and was even more incensed by the Latin–Greek hybrid, *automotor* (an early form, perhaps, of automobile?). They should be replaced, he urged, by *kion* and *autokion*.

There have been occasional assaults on scientific vocabulary from altogether less fastidious motives than respect for the classical niceties. German nationalism, for instance, spawned a movement to expunge all but Germanic stems from the language. And so a *Telefon* became a *Fernsprecher*, and during the Third Reich there was a movement to construct a wholly Teutonic vocabulary for the physical sciences. This included such risible compounds as *Haarröhrchenkraft*, or hair-tubule-power, for capillarity; *Verschluckung*—engulfment or swallowing—for absorption; and so on. Chemistry itself was to be *Scheidekunst*, or separation-craft. The biolgists of the time also brought forward their own abortions—*Schmarotzer*, or sponger, for parasite; *Umweltlehre*, or whole-world-teaching, for ecology; and many more. As may be supposed, they found little favour even in the inflamed humour of the time and place.

See *Nature*, **63**, 474 (1901).

~

118 Buffon's balls of fire

Georges-Louis Leclerc, comte de Buffon (1707–88) was a scholar of prodigious accomplishment. He is now remembered above all for his contributions to anatomy and zoological classification, but his interests embraced all of science, and his *Histoire naturelle*, published in 44 volumes, was an enduringly influential monument of scholarship. Buffon was rich, self-indulgent and imperious of demeanour. His intellectual audacity and self-assurance often led him into futile controversies, such as his long-running disputation with Thomas Jefferson and other American scholars; for he was convinced that a retarded state of evolution prevailed in North and South America.

Buffon had formed the idea that the American climate was damp and unhealthy, and that this had militated against the emergence of new species and had debauched those that already existed. This, he maintained, was evident from a comparison of the plants and animals common to the Americas and Europe, humankind included. In these views he was abetted by other French scholars, notably the Abbé Raynal and Corneille de Pauw. De Pauw wrote that much of America was covered by 'putrid and death-dealing waters' under a blanket of 'fogs of poisonous salts'. Insects and venomous reptiles were huge and hideous; syphilis was an American disease, which corrupted both man and animals, and could be caught by merely breathing the pestilential air. Jefferson resolved to rebut these Gallic aspersions on his native land and to tackle Buffon, much the most respected of the defamers. He collected specimens of native animals—the skin and bones of a moose, the antlers and skull of a stag, the horns of a caribou, and more—and presented them to Buffon in Paris. He also compiled a meteorological analysis of the climate of Virginia, compared to that of Paris, to the disadvantage of the latter. When they met, Jefferson and Buffon hit it off well enough, and Buffon eventually conceded in a letter that he might not have been entirely right. Jefferson was not satisfied and continued with his mission of refutation of Buffon's calumnies for years after the great man's death.

Jefferson recalled the scene at a dinner, given in Paris a few years later by Benjamin Franklin [47] for some French guests and a party of American visitors. Of the company was the Abbé Raynal, who began, as he often did, to expound his theory of the degeneracy of all things American, including the people, and 'urged it with his usual eloquence'.

> The Doctor [Franklin] at length noticing the accidental stature and position of his guests, at table, 'Come', says he, 'M. l'Abbé, let us try this question by the fact before us. We are here one half Americans, and one half French, and it happens that the Americans have placed themselves on one side of the table, and our French friends are on the other. Let both parties rise, and we will see on which side nature has degenerated.' It happened that his American

guests were Carmichael, Harmer, Humphreys, and others of the finest stature and form; while those of the other side were remarkably diminutive, and the Abbé himself particularly, was a mere shrimp. He parried the appeal, however, by a complimentary admission of exceptions among which the Doctor himself was a conspicuous one.

One of the formative influences on Buffon's thinking was the work of Isaac Newton, which he read in the original English (when not in Latin). For years he pondered deeply the question of the Earth's age. He thought that Bishop Ussher's calculation from the generations of the biblical dynasties, that the planet was created in 4004 BC, was absurd, and he began to ruminate on how the origin of the Solar System might be dated. He had formed a theory that the planets had been ejected from the Sun by a collision with a comet: fragments of molten material had coalesced in space, cooled and solidified. This theory had a corollary: Earth, derived from a molten globule, spinning about its axis, should be distended in the equatorial plane. The discovery that such was indeed the case must have strengthened Buffon in his conviction that he was on the right track.

Buffon knew the rate of cooling of a white-hot ball of iron and he calculated that for a sphere the size of Earth to cool to the ambient temperature would take a minimum of 50,000 years. This did not take him as far as he wanted to go for it seemed likely that Earth would have reached its present temperature long ago. But his published cogitations were sufficient to excite the ire of the Church, for his scheme contradicted not merely Bishop Ussher but Genesis itself. He was denounced, though in relatively emollient terms, because the Church had learned its lesson with Galileo, and was invited to answer the charges of the theology faculty of the Sorbonne. Happily the first volume of *Histoire naturelle*, in which the argument appeared, was not burned. Rather, a compromise was reached, whereby Buffon would retract his conclusions in the forthcoming second volume. 'It is better', he explained later, 'to be humble than hanged.'

There Buffon's researches on the Earth's age rested for several years, flickering into life from time to time with the appearance of some new discovery (including a calculation, based on the rate of recession of the oceans, which yielded the startling conclusion that the planet was two billion years old). And then came the revelation that the planet was warmed by an internal source of heat, for a French scientist had reported that the temperature was higher at the bottom than at the top of a deep shaft. Moreover, the radiant heat from the Sun appeared insufficient to account for the temperature rise in summer. Perhaps, then, the planet was after all still cooling? If so an estimate of its age was again possible.

Buffon decided to make accurate measurements on rates of cooling of iron balls and extrapolate their diameters to that of Earth. First he confirmed Newton's conjecture that the rate of cooling of a sphere was directly proportional to its diameter. For this he measured the time that it took for a white-hot ball to cool to the point at which it could just be touched, and then to room temperature. To make the measurements he employed a number of young women, whose soft skin would afford the highest level of sensitivity to temperature differences. After iron he tried a series of other materials—several metals, and then clay, marble, glass, and limestone, which cooled more rapidly than the metals. Next he made a correction for the heat that the planet was receiving from the Sun while it cooled, and thus arrived at his final conclusion: Earth was 74,832 years old. He inferred from the temperatures at different times when life might have begun and animals and finally people appeared on the scene. Buffon published his results in a book, which was received with much interest. The theologians again objected, but Buffon, by now a septuagenarian, no longer cared. He made a perfunctory apology, but obdurately refused to publish a retraction.

In the event, Buffon became dissatisfied with his estimate. The fossil evidence [95] pointed to a much greater age, perhaps millions of years, he thought; but he drew no further conclusions, although he continued to brood on the matter until his death. Suffering from kidney stones, and in continuous pain, he rejected the services of a surgeon. His funeral cortège drew many thousands of Parisians into the streets to pay their last homage to the greatest savant of his time. The debate about the Earth's age swung back and forth for another 200 years, until the many contradictions were finally resolved by the discovery of radioactivity [16].

For the story of Buffon's life see *Buffon: A Life in Natural History* by Jacques Roger (Cornell University Press, Ithaca, NY, 1997), and his work on the Earth's age is described in an absorbing book by Martin Gorst, *Aeons: The Search for the Beginning of Time* (Fourth Estate, London, 2001), from which the above account is mainly taken. For the debate between Buffon and Jefferson, see *Thomas Jefferson: Scientist* by Edward T. Martin (Collier, New York, 1961).

119 Science *in extremis* and the phosphorescent toothpaste

With resilience and determination enough, science has been known to thrive even in a prisoner-of-war camp [35]. James Chadwick (1891–1974), who was to win a Nobel Prize for the discovery of the neutron, kept boredom and frustration at bay in a German prison camp during the First World War by experimenting in an improvised laboratory. Chadwick was born into the working class in the north of England and suffered in his early years from a crippling shyness. But his talents were recognized by a schoolmaster and he won a scholarship to what is now the University of Manchester. There he came to the notice of the recently appointed professor of physics, Ernest Rutherford [16], whom he later followed to the Cavendish Laboratory in Cambridge.

One of Rutherford's cleverest associates was Hans Geiger (who gave his name to the Geiger counter, still used to detect radioactive emissions). When Geiger returned to his native Germany, Chadwick arranged to spend a year with him in Berlin. That year was 1914. Chadwick was given some bad advice by the local branch of the travel agent, Thomas Cook, and in consequence was forced to endure nearly five years of internment under conditions of great privation in a makeshift camp on the race course at Ruhleben, outside Berlin. In time a group of the prisoners formed a Science Circle and, tiring of lecturing to each other, asked their captors for space to set up a laboratory. In the autumn of 1915 they were given part of a loft in a stable block. The temperature touched –10 °C in winter and rose to 37 °C at the height of summer, but the prisoners persisted. Lamps fuelled with animal fat gave light and a little heat.

Few chemicals were available and poisonous substances were barred. But Chadwick found a source of radioactivity: a toothpaste, popular in Germany at this time, was advertised as radioactive. It was sold by the Auer company, its 'active ingredient' presumably a by-product of the manufacture of the incandescent gas mantles for which the company was famous. Posters showed a young woman, displaying a mouthful of coruscating teeth. What manner of diseases the paste may have caused is not recorded; certainly, in those early days of radioactivity, the emanations were widely regarded to possess health-enhancing powers. Indeed, a highly radioactive potion marketed in the United States as a tonic is thought to have claimed many victims. Chadwick, at all events, procured large quantities of the toothpaste through the indulgence of prison guards. He constructed an electroscope for detecting electric charge from wood and tin foil and began experiments. The source of radioactivity in the toothpaste resembled no radioactive element that Chadwick could identify, but it later turned out to contain a highly dangerous element, thorium.

After another year the camp authorities agreed to install an electricity main, albeit at the prisoners' expense, and this broadened the horizons of Chadwick's research. A chemist in the group told him about liquid crystals, of which he then had no knowledge, and he resolved to study their response to a magnetic field. He constructed an electromagnet from a piece of iron and some copper wire, provided for him by the guards, but before the project was completed there arrived in the camp the latest volume of the British Chemical Society's annual reviews. There Chadwick found that the problem had already been solved. By then the German officers responsible for the running of the camp had become very accommodating and with their help, and that of the sympathetic Max Planck [96] and of an official of a prisoners' help organisation, a wider range of materials was made available. A German publisher provided 200 technical books, but, to Chadwick's disgust, the Foreign Office in London would not allow an elementary inorganic chemistry textbook to be sent to the camp for fear it might afford useful information to the enemy.

In 1917, better laboratory accommodation was offered and more advanced apparatus was constructed, including a burner, fired by rancid butter, and fed with air blown by mouth through a nozzle, which served as a blowtorch for glassblowing. By this and other ingeniously improvised means Chadwick and his colleagues built an apparatus to study the photochemical reaction of chlorine with carbon dioxide; and the prisoners also began a study of a mysterious phenomenon, the ionization (generation of charge-bearing particles) of the air at the surface of phosphorus. The results from the laboratory at Ruhleben camp were of limited value, but their work evidently kept the prisoners absorbed and allowed Chadwick to develop his ideas and learn from his colleagues. Best of all, he taught physics to a young cadet from the Royal Military Academy in Woolwich, who had been similarly trapped while visiting Germany: Charles D. Ellis became Chadwick's most valued collaborator in Cambridge and an author of one of the classic works of twentieth-century physics, which bore the names, Rutherford, Chadwick, and Ellis. Most importantly, Chadwick and Ellis were spared from the dangers of the Great War, which cost the lives of so many of their contemporaries—notably the most brilliant of all Rutherford's young protégés, H. J. G. Moseley, felled by a sniper's bullet in the ill-starred Dardanelles campaign.

After an illustrious career at the Cavendish Laboratory, Chadwick was appointed Professor of Physics at the University of Liverpool, where he established a highly effective research group. During the Second World War he emerged as an influential figure in the Manhattan Project. At this stage of his life he was discovered to possess unsuspected administrative and diplomatic skills. After the war his contribution to the creation of the atomic bomb preyed on his mind and he confessed that he remained dependent on sleeping pills ever after. His last position before he retired was as master of a Cambridge college.

During the Second World War, 20 years after Chadwick's ordeal, French prisoners of war at a camp in Edelbach in Germany (Oflag XVII) established a 'university', with rather greater success than Chadwick and his colleagues had enjoyed. There were among them several geologists, and here, in the words of a report in *Nature*, is what they achieved:

> Not content with lectures, the geologists made a thorough examination of the area—only 400 metres square—enclosed within the barbed wire. No stone was left unturned, and trenches and secret tunnels provided many critical exposures. A microscope was constructed in the camp and equipped with polarizers [required for the study of crystals] from piled cover glasses. Thin sections [so thin as to be the transparent enough for microscopy] were mounted with a mixture of violin wax and edible fat. Only the determination of certain untwinned felspars [a class of crystalline minerals] remained to be completed on the return to France.

The results represented a very considerable advance in geological science. They showed, the article concludes, that

> quartz and orthoclase were remarkably plastic during the physico-chemical conditions that attended their formation, and that, in consequence, the granite formed by the transformation of pre-existing rocks could readily become intrusive. It follows that to prove that a granite is intrusive does not prove that it has ever been in a liquid condition.
>
> The memoir is full of important observations and stimulating suggestions, and should be read by all workers in the field of plutonic geology.

Rita Levi-Montalcini was not a prisoner of war, but, as a member of the large Piedmontese Jewish community, she was in hiding from the fascist zealots, eager to implement Mussolini's racial manifesto. Effectively imprisoned in her parents' flat in Turin, she converted her mother's kitchen into a laboratory, and with only occasional furtive visits from her former professor (who was also Jewish) to encourage her, she laid the foundations of her life's work on embryology. The experimental material was fertilized hen's eggs, procured from a nearby farm. The remains of the eggs, after each experiment on the developing embryos had been concluded, were turned into omelettes. When Italy surrendered Levi-Montalcini was at last free to communicate her results to the outside world. An invitation to Washington University St Louis ensued, and her researches there led her in due course to Stockholm to share the Nobel Prize in 1986.

James Chadwick's experiences in the prison camp are described in *The Neutron and the Bomb: A Biography of Sir James Chadwick* by Andrew Brown (Oxford University Press, Oxford, 1997). The account of the work of the French prisoners of war is in *Nature* 163, 967 (1949). Rita Levi-Montalcini tells of her wartime experiences in *In Praise of Imperfection: My Life and Work* (Basic Books, New York, 1988).

~

120 Their Lordships kick a football

The discovery of buckminsterfullerene in 1985 was for chemists the event of the decade, perhaps of the previous several decades. Carbon is the most intensively studied of the elements, for its compounds are what constitute organic chemistry and, indeed, the chemistry of life. It distinguishes itself from other elements in its capacity to form long and complex chains in an endless variety of configurations. The carbon atom is quadrivalent, forms bonds—that is to say, to four other atoms, which may or may not be carbon. These bonds are symmetrically disposed pointing from the carbon atom towards the corners of a regular tetrahedron, a pyramid with triangular faces and four apices. Carbon in its elementary state will form a crystal made up of atoms linked together in this manner, and that crystal is diamond. There is another form of elementary carbon, in which three bonds point towards the corners of an equilateral triangle, with another, weaker bond pointing out of the plane of the triangle. In this form the carbon atoms give rise to a planar structure of fused hexagons like a honeycomb, and multiple layers of this kind stack up on one another. That form of carbon is graphite, and the sliding of the layers over one another accounts for its properties as a lubricant. All this has been known for well over a century. Imagine then the astonishment, and in some circles derisive disbelief, when the discovery of a completely new form of elementary carbon hit the headlines in 1985.

The new state of carbon was first observed in outer space by spectroscopic analysis [70]—two species, the more abundant with weights corresponding exactly to a cluster of 60 carbon atoms, C_{60}, and the lesser component to 70 carbon atoms, C_{70}. What was magic about these numbers of atoms? The answer is that they can be linked to form a closed round shell, with flat facets for sides. But, as geometricians have always known (and the chemists took unconscionably long to realize), fused hexagons alone cannot form a closed shell; they must be regularly interspersed with pentagons—exactly like a soccer ball with its hexagonal and pentagonal facets, or one of the architect Buckminster Fuller's geodesic domes (because generally too large, a less exact model). Clearly the angles that the carbons make in a pentagon differ from those in a hexagon, but so long as the pentagons are surrounded by hexagons the strain induced by the distortion is small. The C_{60}, with its 60 carbon atoms at the vertices of the polygon, has the least distortion, and nearly all other numbers are totally precluded by the much larger strain energy that their structure would imply. Oddly, very large shells are no longer spherical, but flattened on one side.

Two teams of researchers managed, after four or five years of struggle, to mimic in the laboratory the conditions in outer space in which the footballs form, and buckminsterfullerene became available for study in bulk. It soon became apparent that it had some remarkable properties, which could open new horizons, most obviously in lubrication and in superconductivity—the phenomenon of essentially zero electrical resistance [177]. The smallest new revelations about buckminsterfullerenes, the study of which rejuvenated chemistry in a time of need, were publicized with unbridled hyperbole and soon found resonance beyond the walls of the academies. On 10 December 1991, the word was uttered in the chamber of the House of Lords in London and reported in the parliamentary record, *Hansard*, as follows:

Lord Errol of Hale asked Her Majesty's Government:

What steps they are taking to encourage the use of Buckminsterfullerene in science and industry.

The Parliamentary Under-Secretary of State, Department of Trade and Industry (Lord Reay): My Lords, the Government have been following with interest the emergence of Buckminsterfullerene and support research currently being undertaken at Sussex University through the Science and Engineering Research Centre [he means Council]. However, it must be left to the judgment of firms whether they wish to pursue research into commercial applications of Buckminsterfullerene and other fullerenes.

Lord Erroll of Hale: My Lords, I thank my noble friend for his answer, which is good so far as it goes. Can he not offer more substantial support in this country for the development of this exciting new form of carbon? It is already being manufactured in no less than three factories in the United States.

Lord Reay: My Lords, as I said, the Government continue to fund academic research into Buckminsterfullerenes at Sussex University. Many grants have been made available since 1986 which have gone towards that research. SERC also supports a number of researchers investigating the theoretical aspects of chemical bonding relating to fullerenes. The Government funding for collaborative research between industry and the academic world into the commercial applications of Buckminsterfullerenes may be available also under the Link [a scheme whereby private investment in research can in some cases be matched by Government funding] or other schemes.

Baroness Seear: My Lords, forgive my ignorance, but can the noble Lord say whether this thing is animal, vegetable or mineral?

Lord Reay: My Lords, I am glad the noble Baroness asked that question. I can say that Buckminsterfullerene is a molecule composed of 60 carbon atoms known to chemists as C60. Those atoms form a closed cage made up of 12 pentagons and 20 hexagons that fit together like the surface of a football.

Lord Williams of Elvel: My Lords, is the noble Lord aware, in supplementing his Answer, that the football-shaped carbon molecule is also known, for some extraordinary reason, as 'Bucky ball'? It created a considerable stir within the scientific community. As the British Technology Group either has been, or is shortly to be, privatised, is this not a case that should be taken up by the privatised BTG and promoted as a British invention? [This stretches the history of the origins of buckminsterfullerene considerably, although Harry Kroto of Sussex University was one of the scientists who shared the Nobel Prize for its discovery.]

Lord Reay: My Lords, the privatised BTG will be free to take that decision. We do not feel that it is for the Government to say whether or not buckminsterfullerenes have commercial usages, nor whether companies should become involved. It must be up to them.

Lord Renton: My Lords, is it the shape of a rugger football or a soccer football?

Lord Reay: My Lords, I believe it is the shape of a soccer football. Professor Kroto, whose group played a significant part in the development of Buckminsterfullerenes, described it as bearing the same relationship to a football as a football does to the earth. In other words, it is an extremely small molecule [!].

Lord Campbell of Alloway: My Lords, what does it do?

Lord Reay: My Lords, it is thought that it may have several possible uses; for batteries, as a lubricant or as a semi-conductor [a superconductor perhaps?] All that is speculation. It may turn out to have no uses at all.

Earl Russell: My Lords, can one say that it does nothing in particular and does it very well? [This is a laboured allusion to Gilbert and Sullivan's *Iolanthe*: Gilbert applied the quip to the House of Lords.]

Lord Reay: My Lords, that may well be the case.

Lord Callaghan of Cardiff: My Lords, where does the name come from?

Lord Reay: My Lords, it is named after the American engineer and architect, Buckminster Fuller, who developed the geodesic dome, which bears a close resemblance to the structure of the molecule.

The intellectual content of this exchange is not atypical of the level of scientific discourse in both Houses of Parliament. It is reminiscent of a dissertation by a Minister of the Crown in a public discussion about dangerous breeds of dogs: dogs, he instructed his audience, had, according to the expert opinion he had sought on the matter, no DNA. Or consider the intervention by a veteran Member of Parliament when the freezing of embryos came up for debate in the House of Commons: as a housewife, she declared, she well knew how difficult it was to make a pie with pastry that had been deep-frozen for more than six months. QED.

The colloquy in the House of Lords is reproduced in Hugh Aldersey-Williams's excellent book about buckminsterfullerene, *The Most Beautiful Molecule* (Wiley, New York, 1995).

121 The power of incantations

Bruce Frederick Cummings was an English amateur marine biologist, who desperately hankered after an academic career and was thwarted by poverty and ill-health; he died of tuberculosis in 1919 aged 30. Under the pseudonym W. N. P. Barbellion, he wrote a classic memoir, with the title *The Journal of a Disappointed Man*. It is full of acute observation, couched in limpid prose, but the tone is sour with resentment at the airs of the overweening academic establishment of the day.

Here is an example. The scene is a meeting of the Entomological Society and Professor Edward Poulton is one of the emerging breed of biologists, in thrall to the new genetics. Cummings is sharply aware of the gulf that divides Poulton and his like from the scientifically backward beetle-hunters. 'There were', he writes, 'a great many Scarabees present who exhibited to one another poor little pinned insects in collecting boxes', but

> It was really a one-man show, Professor Poulton, a man of very considerable attainments, being present, and shouting with a raucous voice in a way that must have scared some of the timid, unassuming collectors of our country's butterflies and moths. Like a great powerful sheep-dog, he got up and barked, 'Mendelian characters', or 'Germ plasm', what time the obedient flock ran together and bleated a pitiful applause. I suppose, having frequently heard these and similar phrases fall from the lips of the great man at these reunions, they have come to regard them as symbols of a ritual which they think it pious to accept without any question. So every time the Professor says, 'Allelomorph', or some such phrase, they cross themselves and never venture to ask him what the hell it is all about.

The vocabulary was that of the new biology: Mendelian characters were the inherited appearances of an organism, conferred by given genes (phenotypes, as we would now say), while 'germ plasm' was an expression coined by a German biologist, August Weismann, to denote the hereditary substance passed on, supposedly unchanged, in the specialized cells (germ cells, or gametes) that unite in sexual reproduction; this prefigures the concept of the gene. 'Allelomorph' is an obsolete term for an allele—a member of the pair of genes for a given protein that an offspring receives from the parents.

From W. N. P. Barbellion (B. F. Cummings), *The Journal of a Disappointed Man* (Chatto and Windus, London, 1919).

122 Hoax!

The Piltdown skull is probably the most famous and successful hoax in the history of science. It broke upon the turbulent and divided sodality of anthropologists on eighteenth December 1912 at a meeting in London of the Geological Society. The affair had been incubating for nearly four years, ever since a prominent amateur archaeologist, Charles Dawson, had come by some fragments of a human skull. They had been unearthed by workmen in a gravel pit at Piltdown in Sussex. Dawson, who had always hoped that the Sussex Downs would yield up prehistoric human remains, eagerly sifted through the spoil at the pit and found more fragments of ancient, deeply stained bone, together with worked flints and animal remains. Excited, he alerted his friend, Arthur Smith Woodward, Keeper of Palae-ontology at the Natural History Museum (then still an arm of the British Museum) in London, and a young Frenchman, whom he had befriended while digging in Sussex. This was none other than Father Pierre Teilhard de Chardin, who was to become a cult figure 50 years on, through his mystical conceptions of the noosphere and the omega point, elaborated in his book *The Phenomenon of Man*, which sought to reconcile biblical teaching with evolution.

Teilhard de Chardin had come to Sussex to study in a Jesuit college and was, like Dawson, an enthusiastic amateur archaeologist. It was not long before the three men found other treasures, in particular pieces of the lower jaw, stained like the skull and containing two teeth. The jaw was simian in aspect, but the teeth were abraded like those found in early human jaws but never in those of apes. More fragments of skulls soon came to light, in the company of similar jawbones. They belonged, Dawson and Smith Woodward proclaimed, to the earliest man, or 'missing link'. The name that they gave him was *Eoanthropos*, the man of the dawn. Their report was met with mingled elation and scepticism, but within the British scientific establishment it was the sanguine view that prevailed. Smith Woodward and the leading anatomists, Arthur Keith and Grafton Elliot Smith, rebuffed all sceptics with sarcastic hauteur, and over the next few years all three were rewarded with knighthoods for their distinguished work—for it was a cause of some national pride that the primal man was English. Later finds at Piltdown seemed to confirm their claims and won round several distinguished sceptics, notably the doyen of American anthropologists, Henry Fairfield Osborn.

But the doubts persisted, though it was not until some 40 years later that they broke surface in publications by Kenneth Oakley, a geologist and Keeper of Anthropology at the Natural History Museum, and anthropologist Joseph Weiner. Oakley had access to the original specimens (denied to most scholars, who were allowed to examine only casts) and had applied a chemical test. Radiocarbon dating [98] had not yet emerged as a tractable method and Oakley instead measured the fluorine content of the bones. Buried bones absorb fluoride from their surroundings and its concentration in bone gives a measure of age. The Piltdown bones, it transpired, were (in anthropological terms) modern, perhaps recent burials amidst the ancient gravel. A little later Oakley began to consider the alternative hypothesis—that the remains had been planted. Further examination, detailed in 1953 by Weiner, Oakley, and anatomist Wilfrid Le Gros Clark, revealed that the bones had been stained with potassium dichromate to make them look ancient, and the teeth had been crudely filed with a modern implement, evidently an iron file, for there were specks of iron embedded in the surface. The cranium was that of a man, the jawbone of an orang-utan. Who, then, had perpetrated such an outrageous fraud?

Smith Woodward would have been aware of trouble in store, but in 1948, on his deathbed, he dictated the text of a book, *The Earliest Englishman*, in which he affirmed the authenticity of the Piltdown find. The debate was joined by a succession of scholars and amateur sleuths. Initially suspicion fell on Dawson, the eager and ambitious, but not notably competent, amateur. He had died in 1916 and no evidence of his guilt was ever found. He would, it was widely felt, have been better cast as gullible victim than as miscreant. Other candidates were brought forward—W. J. Sollas, Professor of Geology at Oxford, who detested (in which he was not alone) Smith Woodward and would have wanted to discredit him; Arthur Conan Doyle, novelist, doctor, spiritualist and amateur palaeontologist, hatching perhaps a real-life Sherlock Holmes mystery; Father Teilhard de Chardin, bent, Stephen Jay Gould suggests, on playing a joke on the English; Sir Arthur Keith, who had the opportunity; and others. Then in 1996 came what is surely the denouement.

Brian Gardiner, Professor of Palaeontology in the University of London, had for some years been examining the contents of a trunk, discovered by workmen under the roof in one of the towers of the Natural History Museum. The trunk bore the initials of M. A. C. Hinton, Curator of Zoology at the time of the Piltdown find. It contained a myriad of rodent bones, for rodents were Hinton's speciality, but at the bottom lay the answer to the Piltdown mystery. There were bones and teeth, rich in chromium, deriving from the potassium dichromate that had been used to stain them and render them porous; an acid dichromate mixture had in fact been devised by Hinton for experiments on the origin of brown stains in early remains. The fragments were also enriched in iron in the form of brown ferric oxide. The orang-utan jawbone found at Piltdown was, by contrast, much more lightly stained, for the same treatment would have etched its two teeth and instantly revealed the skulduggery. Some teeth in the trunk were lightly stained and one had been painted brown.

Among the relics left for Hinton's executor were a series of tubes containing teeth stained in varying degrees: Hinton had clearly approached his task with painstaking professionalism. He had, it was also known, stained stone tools to simulate extreme antiquity and apparently passed them to Dawson. From him they had found their way into the collection of an expert on such implements, who had labelled them as fakes. Hinton, then, had evidently used the inept Dawson as a (presumably) unwitting projectile aimed at his real target, Smith Woodward. We cannot be certain that Dawson did not participate more intentionally in the deception, but the evidence points to Hinton as the sole forger. His motives remain a matter of conjecture, but he was known to be addicted to practical jokes (indeed his trunk also contained pieces of bone carved into fanciful shapes; these include a cricket bat, a fit possession for the First Englishman). Hinton would, moreover, have had no love for the pompous Smith Woodward, with whom he had had an altercation over pay for extramural work at the museum.

If it is a mark of a good joke that the punch-line should be delayed for decades, until all concerned were long dead, the Piltdown hoax stands supreme. It is reminiscent of a trick played by Ulysses S. Grant, when President of the United States, on the President of a liberal arts college at its inauguration. Grant handed this dignitary a cigar, which, rather than smoke, the recipient preserved as a kind of holy relic. Then, at the centenary celebration of the foundation, the cigar was brought out; the president's successor announced that it was now a fit moment to light up. The ceremony was enacted, there was a sharp report and President Grant's deferred joke—an exploding cigar—was at last consummated.

There have been innumerable books and articles about the Piltdown hoax. As good as any is Stephen Jay Gould's essay in his collection *The Panda's Thumb* (Norton, New York, 1980), in which he develops his conjecture that the hoaxer was Pierre Teilhard de Chardin. The standard work with full details of the find is *The Piltdown Forgery* by J. S. Weiner (Oxford University Press, Oxford, 1955); for an account of Brian Gardiner's detective work, which led to the identification of M. A. C. Hinton as the forger, see the article by Henry Gee in *Nature*, 381, 261 (1996).

~

123 Humphry Davy gives himself airs

Humphry Davy earned his seat in the scientific pantheon by a remarkable succession of discoveries, extending from sodium and potassium to the miners' safety lamp and Michael Faraday, whom he engaged as keeper of the laboratory records and inducted into the mysteries of scientific research. Davy got his start in natural philosophy, as science was then called, in 1798, when at the age of 19 he was appointed assistant in Dr Thomas Beddoes's Pneumatic Institute in Bristol. Beddoes, chemist, physician, and polymath, was something of a public figure, thanks to his well-advertised demonstrations of 'factitious airs'—the recently discovered gases, of which nitrous oxide, laughing gas, excited the greatest interest. Beddoes had high hopes of therapeutic uses for the gases. He even believed that gases emitted by cattle might cure tuberculosis and into his patients' sick room he piped effluvium from both ends of cows, which he kept on an adjoining lawn.

In 1799 Davy, at Dr Beddoes's instigation, breathed 16 quarts [18 litres] of nitrous oxide over a period of seven minutes. Here is how he described the ensuing sensation: the gas

absolutely intoxicated me. Pure oxygen gas produced no alteration in my pulse, nor any other material effect; whereas this gas raised my pulse upwards of twenty strokes, and made me dance about the laboratory as a madman, and has kept my spirits in a glow ever since.

Dr Beddoes's wife was acquainted through her sister, the fashionable novelist, Maria Edgeworth, with many of the literary lions of the day, including Samuel Taylor Coleridge and Robert Southey. Davy was introduced to this circle and the impression that he made, especially on Coleridge (who even expressed an interest in assisting Davy in the laboratory), is reflected in topical scientific allusions in Coleridge's verse. Joseph

Cottle, the Bristol publisher to both Davy and his poet friends, described in his *Early Recollections of Coleridge*, the effects of nitrous oxide on several literary men and on a game young member of what he calls 'the softer sex':

> Mr Southey, Mr Clayfield, Mr Tobin and others inhaled the new air. One it made dance, another laugh, while a third, in his state of excitement, being pugnaciously inclined, struck Mr Davy rather violently with his fist. It now became an object . . . to witness the effect this potent gas might produce on one of the softer sex, and he prevailed on a courageous young lady, (Miss ——), to breathe out of his pretty green bag, this delightful nitrous oxide. After a few inspirations, to the astonishment of everybody, the young lady dashed out of the room and house, when, racing down Hope-square, she leaped over a great dog in her way, but being hotly pursued by the fleetest of her friends, the fair fugitive, or rather the temporary maniac, was at length overtaken and secured, without further damage.

The famous Gillray cartoon of 1802 illustrates the public fascination with the physiological effects of nitrous oxide. It depicts a demonstration at the Royal Institution in London before a fashionable audience, among which can be identified Isaac d'Israeli and Benjamin Thomson, Count Rumford (whose famous experiments in a canon-boring workshop established the relation between mechanical work and heat). The lecturer is Dr Thomas Garnett and young Humphry Davy is seen administering the gas to Sir John Hippisley, with dramatic effect in the form of a violent flatulent eruption. It was some time before nitrous oxide came into use as an anaesthetic, in which guise it served especially dentists for a century and may still be in use in some quarters.

Maria Edgeworth, it might be mentioned, had acquired a keen interest in chemistry from her perusal of *Conversations on Chemistry* by Jane Marcet (1769–1858). The author belonged to a circle of London intellectuals, among whom were prominent scientists. The book is cast as a discussion between a knowledgeable older woman and two mettlesome young girls, whom she lectures on the attractions of chemistry. It was Mrs Marcet's book that first ignited the young Michael Faraday's [19] passion for science when, as an apprentice book-binder, he chanced upon its pages. The zealous Maria Edgeworth commended the study of chemistry to her readers. Chemistry, she wrote, is a science particularly suited to women, 'suited to their talents and to their situation. Chemistry is not a science of parade, it affords occupation and infinite variety, it demands no bodily strength, it can be pursued in retirement; there is no danger of its inflaming the imagination, because the mind is intent upon realities.'

It is related that *Conversations in Chemistry* brought succour to a younger Edgeworth sister, and may even have saved her life, after she had swallowed some unspecified acid. Maria had learned from the invaluable Mrs Marcet that acids could be neutralized with milk of magnesia (magnesium hydroxide), and was quick to administer the remedy.

Cottle's description of the effects of laughing gas is reproduced, together with much background on the subject, in *Humour and Humanism in Chemistry* (G. Bell, London, 1947) by John Read, who was Professor of Chemistry at the University of Aberdeen. The story about Maria Edgeworth is from *Letters for Literary Ladies* (Garland, New York, 1974; first published 1795).

~

124 Truth stranger than fiction

Aspirin is by a huge margin the most abundantly used of all drugs and new aspects of its multifarious beneficent activity are still coming to light. Its name derives from the willow tree, which was known from the earliest times to harbour in its bark an analgesic principle. Legend has it that this was first recognized when bears with broken or infected teeth were seen to strip and chew the bark. The active compound was identified in the nineteenth century as salicylic acid, but it soon transpired that the acid or its salt, sodium salicylate, a cheap and easily prepared chemical, while effective in suppressing pain, was intolerably bitter and also caused stomach disorders. Chemists at the German pharmaceutical concern, F. Bayer and Company, therefore set about synthesizing simple derivatives of salicylic acid. All accounts of the discovery of aspirin—acetylsalicylic acid—concur that it was prepared by a young chemist at the Bayer works by the name of Felix Hoffmann. His inspiration was the suffering of his father, crippled and in continuous pain from rheumatoid arthritis. Hoffmann made a pure preparation, which instantly relieved his father's worst discomfort, and after evaluation by the Bayer pharmacologist, Heinrich Dreser, aspirin reached the market in 1898.

The truth, it turns out, was quite other. Arthur Eichengrün joined the company in 1894 and at once took the salicylic acid problem in hand. His plan was to prepare an ester—a compound in which an acid group is blocked by coupling it to a another compound containing a hydroxyl group (that is to say an alcohol). Esters are in general resistant to decomposition by acid and so survive in the stomach, but in the alkaline conditions of the intestine they are broken down to regenerate the parent acid. In the case of aspirin, the stomach linings are thus spared the attention of the salicylic acid, but when this is liberated in the intestine it is absorbed and duly does its soothing work. The story of Hoffmann's discovery apparently originated in 1934 and, as Eichengrün bitterly recalled in old age, the Hall of Honour in the German Museum in Munich boasted in its chemical section a display of aspirin crystals, bearing the inscription, 'Aspirin, invented by Dreser and Hoffmann'. This display was erected in 1941, by which time the Jewish Eichengrün was languishing in the ghetto of Theresienstadt.

By good fortune Eichengrün survived the war and lived to tell his tale: Hoffmann was an assistant, whom he instructed to synthesize the ester, without troubling to explain the purpose, and Dreser had had no part at all in the work. As a Jew, Eichengrün was expunged from the record and the names of the two Aryans were written in. An examination of the laboratory notebooks in the Bayer archives confirmed Eichengrün's version. He had become director of the company's applied chemistry programme and had gone on to develop several more drugs, as well as cellulosic fibres, while Hoffmann had left research for pharmaceutical sales. In 1949, Eichengrün published his story in a German technical periodical, but myths die hard, and it took the researches of a scientist at the University of Strathclyde to confirm the truth and make it public.

The above account is taken from the articles by Walter Sneader, who uncovered the truth, in the *British Medical Journal*, **321**, 1591 (2000) and *The Biochemist* of August, 2001.

~

125 The mosquito bites back

DDT, or dichlorodiphenyltrichloroethane, which surfaced during the Second World War, heralded, so it appeared, man's final victory over malaria, typhoid, and other insect-borne diseases. A few years earlier its seemingly miraculous efficacy had been perceived by a chemist at the Swiss pharmaceutical concern, the J. R. Geigy company. His name was Paul

Müller, and for his discovery he was eventually rewarded with the Nobel Prize.

Müller, then 26, joined Geigy in 1925 and had worked on a variety of problems before settling on a search for the perfect insecticide. Geigy were especially interested in the prospect of an agent that would be more effective than mothballs (naphthalene) in discouraging the domestic clothes moth. Müller had been testing a range of synthetic chemicals by placing some of the substance into a glass tank, which he then filled with the insects. His fascination with this project earned him some mockery from his colleagues and the nickname *Fliegenmüller* (or fly-Müller).

DDT was one of a class of chemicals of which Müller had hopes, but it was at first sight a total failure, for the moths appeared unperturbed. But on this occasion, with no conscious motivation, Müller left the moths in the box overnight. The next morning all were dead. He repeated his experiment with more moths, with houseflies and other insects. All died overnight. Thus encouraged, Müller swabbed out his killing-tank with solvent and tried a range of related compounds, all seemingly with deadly effect. But that proved to be an illusion, for further study revealed that it was the residual traces of DDT, left behind even after a solvent wash, that was killing the insects. The Geigy company thereupon sent a canister of DDT powder to their American office, but it was some time before a chemist who could read the report in German of its properties, passed some of the material to the US Department of Agriculture, whence it came into the hands of the department's insect research station in Orlando, Florida. There it was tested and its prodigious and unparalleled toxicity against insects was confirmed, and in particular its activity against mosquitoes.

The discovery was timely, for the US military was by then engaged in the fight against the Japanese on the Pacific islands, and more of the troops were being struck down by malaria than by bullets and shells. There was also the constant fear of typhus, which had so ravaged the armies in the First World War. And so the entomologists of the Department of Agriculture went out to test the efficacy of DDT in the field. The results were startling: a dusting with DDT would protect a soldier against lice for a month. Soon aircraft were releasing DDT over landing beaches to protect the incoming marines. During the Allied invasion of Italy in 1944 a typhus epidemic in Naples was nipped in the bud by a dusting operation involving 1.3 million people.

When the war ended plans were made to eradicate the scourge of malaria from the planet once and for all. But objections soon arose. Could one be sure that DDT, deposited everywhere in such huge quantities, was not cumulatively toxic to humans? Animal experiments had shown no indications of toxicity and human subjects had breathed in air laden with DDT dust for hours. Two researchers swallowed some grams of the powder to convince the sceptics. But more serious was the emergence of resistant strains of mosquitoes. Only one in many thousand mosquitoes was naturally resistant, but it was these few that survived and bred to produce a highly resistant population. In 1962, DDT was denounced by Rachel Carson in her sensational and highly influential book, *Silent Spring*, because the substance had undoubtedly disturbed sections of the ecology. The destruction of insects reduced the populations of many bird species; some species of insect were unaffected by DDT and proliferated, thanks to the disappearance of other insect predators, such as parasitic wasps.

DDT is now little used and has been largely relegated to history, but it is thought to have saved lives running into millions by, while it lasted, eliminating mosquitoes from their breeding grounds, especially in Latin America and North Africa. It has been argued that more resolute and better organized action could have destroyed the mosquito population before resistant strains could establish themselves and that the failure to make best use of DDT was one of man's great missed opportunities.

After he retired from Geigy Müller continued to search for the perfect insecticide until his death in 1965. He had distributed the money for his Nobel Prize (awarded in 1948) to young researchers working on the control of insects.

A recent account of the war against the mosquito is *Mosquito: A Natural History of Our Most Persistent and Deadly Foe* by Andrew Spielman and Michael D'Antonio (Time Warner, New York, 2001).

~

126 Some are born great

Richard Feynman [89] was a young physicist, whose prodigious abilities had already been recognized, when he was summoned to a meeting at Princeton University, where he was still completing his doctoral thesis. It marked the beginning of the Manhattan Project to construct an atom bomb. Here is how, some 40 years later, he remembered one of the early discussions.

> One of the first interesting experiences I had in this project at Princeton was meeting great men. I had never met very many great men before. But there was an evaluation committee that had to try to help us along, and help us ultimately decide which way we were going to separate the uranium [to extract the very small proportion of the fissionable isotope [149]]. The committee had men like [A. H.] Compton and [R. C.] Tolman and [H. D.] Smyth and [H. C.] Urey and [I. I.] Rabi and [J. R.] Oppenheimer on it. I would sit in because I understood the theory of how our process of separating isotopes worked, and so they'd ask me questions and talk about it. In these discussions one man would make a point. Then Compton, for example, would explain a different point of view. He would say it should be *this* way, and he was perfectly right. Another guy would say, well, maybe, but there's this other possibility we have to consider against it.
>
> So everybody is disagreeing, all round the table. I am surprised and disturbed that Compton doesn't repeat and emphasize his point. Finally, at the end, Tolman, who's the chairman, would say, 'Well, having heard all the arguments, I guess it's true that Compton's argument is the best of all, and now we have to go ahead'.
>
> It was such a shock to me to see that a committee of men could present a whole lot of ideas, each one thinking of a new facet, while remembering what the other fella said, so that, at the end, the decision is made as to which idea was the best—summing it all up—without having to say it three times. These were very great men indeed.

From Richard Feynman, ed. Edward Hutchings, *Surely You're Joking Mr Feynman! Adventures of a Curious Character* (Norton, New York, 1985).

~

127 Victorian vitality

Victorian Britain was home to an army of naturalists. Some held professorial chairs of botany, zoology, geology, or palaeontology, but most were amateurs. This was pre-eminently the age of cataloguing of species and systematization. Learned societies arose and their meetings were crowded and sometimes occasions of uninhibited and passionate debate. Of the meetings of the Geological Society John Gibson Lockhart, the editor of an intellectual publication, commented: 'Though I don't care for geology, I do like to see the fellows fight'. The Victorians took their interests, not least in the wonders of nature, very seriously indeed and were willing, even eager, to suffer in their cause. William Buckland [13] and Adam Sedgwick, Professors of Geology at Oxford and Cambridge respectively, led their students on gruelling field trips and delivered lectures—in Sedgwick's case, five in one day—on horseback. Buckland on one occasion

> attracted an audience totalling several thousand for a lecture in the famous Dudley Caverns, specially illuminated for the purpose. Carried away by the general magnificence, he was tempted into rounding off with a shameless appeal to the audience's patriotism. The great mineral wealth lying around on every hand, he proclaimed, was no mere accident of nature; it showed, rather, the express intention of Providence that the inhabitants of Britain should become, by this gift, the richest and most powerful nation on earth. And with these words, the great crowd, with Buckland at its head, returned towards the light of day thundering out, with one accord, 'God Save the Queen!'

William MacGillivray (1796–1852), who was to become Professor of Natural History at the University of Aberdeen and author of a standard treatise on British birds, exemplified the Victorian virtues. Here is a splendid description of his journey to London, wither at the age of 23 he directed his steps, afire with impatience to examine the great bird collection at what was to become the Natural History Museum. MacGillivray was desperately poor, but

> What he did have in superabundance was energy. This led him to decide to make his way down to London entirely on foot—a distance of over 800 miles. He started off on 7 September [1819]—having risen, in fitting style, at approximately half past four in the morning and breakfasted around five. In his knapsack and his pockets he carried 'a penknife, a small inkpiece with pens, a small itinerary of Scotland, a glass for drinking by the way, and a trowel'. 'To my dress or clothing', he noted in his journal, 'I have added a great-coat and a pair of gloves. Of money I had just ten pounds sterling.' He subsisted on barley bread.
>
> Choosing initially a most circuitous route—west and then south, by Braemar, Strathspey, Fort William and Inveraray—he succeeded in covering some 500 miles in the first thirty days. At that point he had spent half his money. Undeterred, with the five pounds remaining he pressed on south: 'Bread and water will do very well for the greater part of my journey.'

But to his dismay, on entering Cumberland, he found Scottish banknotes were refused because of suspicion of forgeries and he was unable to purchase food or lodging before reaching Keswick. He slept under hedges, among heather, in barns, more often than in beds. By Manchester, he had to report, 'my trousers are ragged . . . plastered with mire . . . my shoes are nearly worn down, and my stockings are fairly finished'. By Northampton his total funds were down to one and three halfpence, so he decided from there onwards to dispense with breakfast. By the time he had struggled to St. Albans he was being obliged to sit for a time every two or three miles, to ease the appalling soreness of his feet.

He finally entered London on 20 October, six weeks after starting out—appropriately, in a torrential downpour. The very next day, refusing to admit his exhaustion, he duly inspected the British Museum. He stayed in the capital a week (presumably on borrowed money) and then returned to Aberdeen by steamboat.

Some twenty-five years later, when Professor of Natural History in his native city, he liked to take his students out on field excursions and would walk even the most active of them, it is recorded, 'into limp helplessness'. His death in the end was due to the effects of exposure.

The author of this passage, David Elliston Allen, gives more examples of the ways of these indomitable Victorian scholars. Here he quotes the biographer of the Reverend J. G. Wood:

His power of work was simply astonishing . . . He was always at his desk by half-past four or five o'clock in the morning, at all seasons of the year, lighting his own fire in the winter, and then writing steadily until eight. Then, in all weathers, he would start off for a sharp run of three miles over a stretch of particularly hilly country, winding up with a tolerably steep ascent of nearly a quarter of a mile, and priding himself on completing the distance from start to finish without stopping, or even slackening his pace. Then came a cold bath, followed by breakfast.

Allen resumes:

And so the day went on. Fully twelve hours, out of the twenty-four, we are told, were spent with pen in hand, 'recreation being reduced to a minimum, and indeed almost to vanishing point'. With all this, it is not perhaps surprising that throughout his life he suffered greatly from dyspepsia.

Small wonder that the enthusiasms of the day produced such an unimaginable mass of factual detail, in which was subsumed much of the foundation of modern anatomy. It also led, of course, to undirected activity of the sort that Darwin later likened to descending into a gravel pit and counting the pebbles, for, he remarked, all observations, to be useful, must surely be for or against something. It is striking that Buckland, MacGillivray, and especially Sedgwick, detested the theory of evolution when it appeared. Sedgwick called it mischievous and told his friends that Darwin's pronouncements had caused him to laugh until his sides ached.

All the passages reproduced above are taken from David Elliston Allen's masterly book, *The Naturalist in Britain* (Allen Lane, London, 1976; Penguin Books, London, 1978).

~

128 Charles Goodyear vulcanizes

It was early in the nineteenth century when Charles Macintosh and others first tried to produce waterproof materials by means of rubber coatings (whence, of course, the macintosh). These efforts had mixed success: in hot weather the rubber would become sticky and run, and in cold weather crack. Charles Goodyear, born in New Haven, Connecticut, in 1800, was reduced to penury by his long struggle to overcome this problem; he was forced, for instance, to write off a consignment of waterproofed mailbags, laboriously fabricated for the US mail service, and spent no little time in debtors' prisons.

Then in 1840 after he had unsuccessfully tried the effect, among many other materials, of sulphur on the rubber, he carelessly allowed a mixture of rubber and sulphur to come into contact with a hot stove. Instead of melting, the paste turned into a gummy mass, which only charred at the edges. Here is how his daughter described the moment:

> As I was passing in and out of the room, I casually observed the little piece of gum which he was holding near the fire, and I noticed also that he was unusually animated by some discovery which he had made. He nailed the piece of gum outside the kitchen door in the intense cold. In the morning, he brought it in, holding it up exultingly. He had found it perfectly flexible as it was when he put it out.

Goodyears's euphoria did not communicate itself to his inventor brother, Nelson, or to other interested parties. Here, in his own words is what happened (taken from his two-volume treatise on rubber, in which Goodyear always refers to himself in the third person):

He endeavoured to call the attention of his brother, as well as some other individuals who were present, and who were acquainted with the manufacture of gum-elastic, to this effect, as remarkable, and unlike any before known, since gum-elastic always melted when exposed to a high degree of heat. This occurrence did not at the time appear to them to be worthy of notice; it was considered as one of the frequent appeals that he was in the habit of making, in behalf of some new experiment.

He however directly inferred that if the process of charring could be stopped at the right point, it might divest the gum of its native adhesiveness throughout, which would make it better than the native gum. Upon further trial with heat, he was further convinced of the correctness of this inference by finding that India rubber could not be melted in boiling sulphur at any heat ever so great, but always charred.

Goodyear called his process, which he investigated further and improved, vulcanization, and it remains to this day the basis of the rubber industry and its products, from erasers to aircraft tyres. (The name, rubber, derives from the observation of the celebrated English chemist Joseph Priestley that a piece of caoutchouc would erase pencil markings from paper.) Goodyear received many honours, but consumed much of his capital in fighting patent infringements, which landed him again in debtors' prisons in England and in France. The Emperor Louis-Napoléon awarded him a medal and the Legion of Honour, which his son brought to him in the debtor's prison in Clichy.

See Robert Friedel in *American Heritage of Discovery and Invention*, 5(3), 44 (1990) and George B. Kauffman in *Education in Chemistry*, 20, 167 (1989).

~

129 Pasteur wields the tweezers

The young Louis Pasteur [172] came to fame with one of the most spectacular experiments in the history of science. While studying the chemistry of fermentation he fell to pondering the nature of optical activity [1]—an attribute of most compounds produced by living organisms.

One of the products of the fermentation of grape juice is tartaric acid. Its formula was already known and it had also been shown that, when purified from the ferment, solutions of the acid and its salts rotated the plane of polarized light to the left (or are said to be laevorotatory). Yet when a substance with the same formula was prepared in the laboratory by an organic chemist the solutions had no optical activity at all. This form of tartaric acid was known as racemic acid. Pasteur arrived at the inspired conjecture that racemic acid was a mixture (now called a racemic mixture) of equal parts of left- and right-rotating components. The salts of racemic acid, such as the sodium ammonium salt (for the acid possesses two acid groups), form handsome crystals and when Pasteur examined them under a lens he perceived that there were two forms, related to each other as a pair of mirror images. He picked out the two types of crystal with tweezers, dissolved them in water, and, indeed, the two solutions had opposite rotatory powers: they were dextro- and laevorotatory.

When Pasteur announced his discovery, it appeared to chemists so outlandish that many refused to believe it. The French Academy of Sciences accordingly demanded a demonstration before an independent expert, and a famous veteran physicist, Jean-Baptiste Biot [166], was appointed judge. Pasteur left his own account of the meeting with Biot. He called on the venerable savant in his laboratory at the Collège de France. Biot had himself procured the racemic acid and the sodium and ammonium hydroxides required to form the mixed salt, and he watched as Pasteur prepared the mixture. The solution was then left in Biot's laboratory to evaporate slowly and crystallize. When a good crop of crystals had formed Biot again summoned Pasteur. Handing him the dish of crystals he asked that the dextro-rotatory crystals should be placed by his right hand, the laevorotatory to the left. This accomplished, Biot said that he would do the rest. He took samples of the two heaps of crystals, weighed them, and dissolved them in water, and vanished into his laboratory, which housed the polarimeter, the instrument that measures the rotation of the plane of polarization of polarized light.

When all was ready Biot invited Pasteur to join him. He would start with the 'more interesting' sample—the left-handed form, which is generated by the fermentation process. Even before he had taken a reading it was obvious from the appearance in the instrument that the sample was optically active. 'The illustrious old man, who was visibly moved, seized me by the hand', Pasteur recalled, 'and exclaimed: "Mon cher enfant, j'ai tant aimé les sciences dans ma vie que cela me fait battre le coeur!"'

The story appears in biographies of Pasteur, of which there are many. See, for example, René J. Dubos, *Louis Pasteur, Free Lance of Science* (Da Capo, New York, 1986), and the more recent (revisionist) biography by Gerald I. Geison, *The Private Science of Louis Pasteur* (Princeton University Press, Princeton, 1995); see also Patrice Debré, *Louis Pasteur* (Johns Hopkins University Press, Baltimore, 1994).

~

130 The limits of logic

In 1931, an intellectual bombshell exploded in the sheltered world of mathematics. The perpetrator of the outrage was a young German, Kurt Gödel (1906–78), and the most illustrious of the casualties David Hilbert [11], doyen of German mathematicians. The Hilbert Project, as it was called, had the aim of establishing a complete system of axioms from which all of mathematics could eventually be rigorously developed. (Remote from everyday reality as such a preoccupation might appear, this, like other researches of Hilbert's and Gödel's, proved to have a profound bearing on topics in science and even technology.) Gödel proved by highly rarified mathematical reasoning, making use of paradoxes, that for some of the most important areas of mathematics no complete set of axioms can be formulated. This 'incompleteness theorem' essentially put an end to Hilbert's vision and made Gödel famous.

When the Nazis came to power Gödel seemed scarcely to notice, but when he was called up for military service he judged it a good moment to leave. He arrived by an indirect route in the United States and passed the rest of his life at the Institute for Advanced Study in Princeton. There Albert Einstein [161] became his closest friend, but after some years Gödel's behaviour grew increasingly odd. He continued to work, but withdrew from all human contact and would receive mathematical communications only through a crack in the door of his office. In due course he became persuaded that persons unknown were out to poison him; as time went on he stopped eating altogether and died of starvation (or, according to his death certificate, of 'innanition'). The following incident occurred when his paranoia was in its early stages.

One day, when taking lunch together, as was their custom, Gödel, Einstein, and Einstein's assistant, Ernst Straus, were discussing the news of the day. The year was 1951 and the main story was the return of General

McArthur from his campaign in Korea. Gödel was highly agitated, for he had been examining the photograph of an exultant McArthur on the front page of the *New York Times* and had formed the suspicion that the man riding in triumph down Madison Avenue was an impostor. He had now found his conjecture to be correct, for he had located an earlier photograph of the General, and had measured the ratio of two critical dimensions in the faces in the two pictures—the length of the nose and the distance between the tips of nose and chin. The ratios were different, and so, therefore, were the two uniformed men. What action he took and how his fears were calmed is not recorded.

The great physicist John Archibald Wheeler also testified to Gödel's idiosyncracies. He called on Gödel to ask whether there could be any relation between the Incompleteness Theorem and Heisenberg's Uncertainty Principle. Gödel, he recalled, was sitting in his overheated office with a blanket round his legs. Wheeler put his question and—'Gödel got angry and threw me out of his office'.

See Steven G. Krantz, *The Mathematical Intelligencer*, 12, 32 (1990).

~

131 As in a dream

Francis Crick [88] has argued that the function of dreaming is to efface from our consciousness the miasma of redundant memory, the detritus of daily experience. This, he suggests, serves to bring the memories that matter into sharper focus. Perhaps, then, that is why dimly perceived, inchoate ideas have so often crystallized in the minds of great scientists in dreams or dream-like states of reverie. This, at all events, was how the classification of the chemical elements presented itself to Dmitri Ivanovich Mendeleyev in his study in St Petersburg in 1869.

Mendeleyev was born in 1834 in a remote region of Siberia into a family of schoolteachers, the youngest of 17 children. He maintained in later life that he never fully mastered the Russian that was spoken in St Petersburg, where he was sent to study. There also, in the face of no little adversity, he became Professor of Chemistry at the Technological Institute. He appears in photographs as a patriarchal figure, wild of hair and heavily bearded. His command of detail and his capacity for work were legendary. It was his conviction that the chemical elements must conform to some all-embracing pattern. In this he was not alone, and indeed the concept for which he became famous had been prefigured—to general ridicule—by an English chemist by the name of John Newlands.

Newlands's Rule of Octaves implied that, when arranged in order of their atomic weights, the elements fell into related groups, repeating at intervals of eight, like the notes of the musical scale. He was told that he might just as well have arranged them in alphabetical order. Newlands's theory achieved no currency and was certainly not known to Mendeleyev when he dreamed his dream. He had made a determined effort over a period of three almost sleepless days and nights to impose some order on the collection of elements. Addicted to games of patience, he had written the names of the elements on cards and had arranged and rearranged them on his desk. Mendeleyev had an appointment on his small country estate at Tver with the Voluntary Economic Cooperative on the following day and was to leave in the morning, but, immersed in his 'game of chemical patience', he put off his departure until the evening. He had been trying in vain to capture a fugitive apprehension, fluttering somewhere in his brain. At length, exhausted with the mental effort, he fell asleep at his desk and dreamed. 'I saw in a dream a table', he later wrote, 'where all the elements fell into place as required. Awakening, I immediately wrote it down on a piece of paper.'

What Mendeleyev had realized was that when the elements were arranged in order of their atomic weights (as Newlands had foreshadowed) their chemical properties, all of which he knew by heart, recurred at regular intervals. He called his scheme the Periodic Table of the Elements. Thus, for instance, the elements known as the halogens—fluorine, chlorine, bromine, and iodine—which display a broad range of common characteristics appeared at equal intervals in his table.

Most striking were the gaps in the table, for to preserve the regularity of the repeats some elements had to be moved forward one position. The space thus opened up, Mendeleyev predicted, would be filled by an as yet undiscovered element. He specified what the properties of three of these elements would be. One such hypothetical element, designated 'eka-aluminium', would have properties similar to those of aluminium: it would, for instance, be metallic and have a characteristic valence [3] of three. Eka-aluminium duly turned up in France. Paul Lecoq de Bois-baudran announced: 'During the night before last, on 27 August 1875, I discovered a new element in a sample of zinc sulphide from Pierrefitte mine in the Pyrenees.' He dubbed the new element gallium, in homage to his country, Gallia (or perhaps to himself, for le coq is the *gallus*). In 1886, germanium followed, its properties exactly as predicted by Mendeleyev 17 years before. The discovery of the noble, unreactive gas, argon, in 1894 threatened to shake the edifice, for there was no space for a new element with such seemingly unique properties; but with the isolation of the whole series—a new group—of noble gases (helium, neon, argon, krypton, xenon, and radon) this eventually proved its triumphal coping-stone.

Mendeleyev received many honours for his discovery, though never the Nobel Prize. He may have been denied this ultimate accolade because the principle of the Periodic Table was also independently recognized by a German chemist, Lothar Meyer, who died before the Nobel Prizes were instituted. (Mendeleyev was in fact nominated in 1906, but lost the nomination by a single vote, and the palm went instead to a French inorganic chemist, Henri Moissan—the first man to isolate fluorine, calcium, chromium, and much else.) Mendeleyev made further contributions to chemistry, but, curiously, resisted the interpretation of his Periodic Table in terms of atomic structure, indeed, was never reconciled with the atomic theory of matter.

A few years before his death Mendeleyev came to the Royal Institution in London to receive the Chemical Society's Faraday award. The chemist Sir Edward Thorpe recalled the ceremony in his obituary notice.

> On the occasion of his delivering the Faraday lecture it fell to the writer's duty, as treasurer of the Chemical Society, to hand him the honorarium which the regulations of the society prescribe, in a small silken purse worked in the Russian national colours. He was pleased with the purse, especially when he learned that it was the handiwork of a lady among his audience, and declared that he would ever afterwards use it, but he tumbled the sovereigns out on the table, declaring that nothing would induce him to accept money from a society which had paid him the high compliment of inviting him to do honour to the memory of Faraday in a place made sacred by his labours.

Mendeleyev died in 1907. Fifty years later, his name was commemorated in the newly discovered transuranic element (one of the unstable, highly radioactive elements heavier than uranium, the heaviest natural element, and prepared by bombardment of lighter atoms). Element 101 is called mendelevium.

For a readable account of Mendeleyev's work and the history of the chemical elements generally, see *Mendeleyev's Dream—The Quest for the Elements* by Paul Strathern (Hamish Hamilton, London, 2000). Thorpe's obituary notice is in *Nature*, 75, 373 (1907).

~

132 Metal takes wing

The name of Ludwig Mond (1839–1909) is linked to Imperial Chemical Industries and to a range of commercial chemical processes. Mond was one of the greatest applied chemists. He regarded a background in academic chemistry as an indispensable preliminary to an industrial career—he himself had worked in Baeyer's laboratory in Munich [84] before emigrating to England, where he set up model commercial plants, run on benign and socially enlightened lines. Mond himself developed a series of ingenious industrial procedures. One of these generated ammonia for the Solvay soda works, which produced the important chemical, sodium carbonate. A by-product of the ammonia process was the so-called Mond gas, comprising mainly hydrogen and carbon monoxide, which was used as a smokeless fuel and for heating furnaces. Mond had also established a chlorine plant, but this was giving persistent trouble. In particular, the nickel stopcocks that controlled the passage of gas through the pipes became encrusted with a black deposit that soon blocked the flow. No such effect had been encountered in laboratory-scale experiments and Mond began to consider whether he should write off his losses and close the plant.

Mond had bought a family mansion, The Poplars, in St John's Wood, north London, and there, in the stables at the rear, had set up a small laboratory. Over it presided Mond's personal assistant, an Austrian chemist called Carl Langer, with his own young assistant, Friedrich Quincke. Mond brought back some of the noisome black deposit, which was quickly shown to be carbon. This was curious, but Mond and Langer soon tracked down the source: whereas in the laboratory experiments pure nitrogen had been used to flush ammonia out of the apparatus, the nitrogen supplied to the plant contained traces of carbon monoxide. But what could carbon monoxide be doing to metallic nickel, a chemically resistant material? Mond knew all about nickel: he used it in powdered form as a catalyst in the purification of his Mond gas. The gas was mixed with steam and passed over the hot nickel, the water and carbon monoxide reacted to form hydrogen and carbon dioxide, and the carbon dioxide was absorbed by bubbling through alkali. What, Mond now wanted to know, was the nickel doing in this reaction? What catalytic intermediate was it forming?

Carl Langer set up a simple apparatus for passing pure carbon monoxide over heated nickel powder. To prevent the escape of the toxic carbon monoxide into the atmosphere of the laboratory the gas issuing through a glass jet was burned off. At the end of the day, after Langer had gone home, it was left to Quincke to turn off the heat under the catalyst, stop the gas flow, wait for the flame to subside, and lock up. One evening in 1889 Quincke left early and it was Langer who turned down the heat and waited before closing the gas tap. Then a startling event occurred: as the apparatus cooled the pale-blue carbon monoxide flame turned luminous, grew brighter, and was suddenly suffused with green. Langer, excited by what he had witnessed, called Mond, who, according to current accounts, left his dinner table and his guests, and ran in his evening dress to the laboratory. In silent wonder the two men gazed at the green flame.

Mond's first thought was that the green colour betokened the presence of the gaseous arsenic compound, arsine, which he knew burned green. In an instant he had applied the classic forensic Marsh test for arsenic and arsine: a glass plate exposed to the gas acquires a black deposit. A deposit did indeed form, but it was shiny as a mirror and quite unlike arsenic. When analysed it proved to be pure nickel. But heavy metals are not volatile, nor were they supposed capable of forming gaseous compounds. For a long time Mond resisted the conclusion to which his own evidence pointed. He wondered whether a hitherto unknown element lurked in the nickel. But when the gas was condensed to a colourless liquid and froze to form needle-like crystals of the compound we now know as nickel carbonyl, Mond was finally persuaded. This was the first of many metal carbonyls revealed in the ensuing search. In the words of Lord Kelvin [10], Mond had given wings to heavy metals. Nickel carbonyl became the basis of a new and highly efficient process for preparing nickel from its ores, which Mond set up in a fine new factory in Swansea.

See J. M. Cohen, *The Life of Ludwig Mond* (Methuen, London, 1956).

133 A mathematical death

Abraham de Moivre was a French-born mathematician, famous for his work on probability theory—he was the first to grasp the principles of random distributions—on complex numbers (the device, inseparable from many areas of mathematics and physics, of representing a property in terms of a 'real' and an 'imaginary' part) and on trigonometry. He lived all his life in England and was an intimate of Isaac Newton. He died in 1754 at the age of 87 in curious circumstances, befitting his calling.

> The manner of de Moivre's death has a certain interest for psychologists. Shortly before it, he declared that it was necessary for him to sleep some ten minutes or a quarter hour longer each day than the preceding one: the day after he had thus reached a total of something over twenty-three hours he slept up to the limit of twenty-four hours, and then died in his sleep.

W. W. Rouse Ball, *History of Mathematics* (Macmillan, London, 1911).

~

134 Shocking experiment

Alexander von Humboldt (1769–1859) was born into the Prussian nobility, a milieu in which a military career was all but foreordained. But the young Alexander developed an unaccountable, and to his family wholly aberrant, interest in science. Persisting in the face of stern disapproval, he entered the Mining Academy in Freiburg and returned to Prussia as an Inspector of Mines. He developed modified versions of Humphry Davy's famous safety lamp and invented a breathing apparatus for miners. At the same time he was summoned to undertake diplomatic missions in an unsettled Europe. He somehow found time to establish his own Free Royal Mining School, where miners were taught about such subjects as geology. But Humboldt had set his heart all along on a life of science and exploration. He became possessed by a consuming fascination with geology, the Earth's magnetic field and, above all, 'animal magnetism', for Galvani's work on twitching frog muscles [58] had recently appeared. Humboldt developed a theory of his own: the metal electrodes, he decided, were not the cause of the twitches, they merely intensified an innate property of the muscle. He began a series of experiments on animal muscles and plants and then proceeded to test his ideas on himself.

> I raised two blisters on my back, each the size of a crown-piece and covering the trapezoid and deltoid muscles respectively. Meanwhile I lay flat on my stomach. When the blisters were cut and contact was made with the zinc and silver electrodes, I experienced a sharp pain, which was so severe that the trapezoid muscle swelled considerably, and the quivering was communicated upwards to the base of the skull and the spinous processes of the vertebrae. Contact with silver produced three or four single throbbings which I could clearly separate. Frogs placed upon my back were observed to hop.
>
> Hitherto my right shoulder was the one principally affected. It gave me considerable pain, and the large amount of lymphatic serum produced by the irritation was red in colour and so acrid that it caused excoriation in places where it ran down the back. The phenomenon was so extraordinary that I repeated it. This time I applied the electrodes to the wound on my left shoulder, which was still filled with a colourless watery discharge, and violently excited the nerves. Four minutes sufficed to produce a similar amount of pain and inflammation with the same redness and excoriation of the parts. After it had been washed my back looked for many hours like that of a man who had been running the gauntlet.

The effects were presumably engendered by electrode products, such as acid, released into and beneath his skin. Continued experimentation caused such alarming damage to his back that the doctor in attendance put a stop to it and bathed the lacerated skin in warm milk. A little later Humboldt caused himself convulsive agony when he stuck the electrodes into the cavity left by an extracted tooth, thinking apparently that such vigorous stimulation of the nerve might suppress the pain response.

Eventually the young experimenter collected together the results of numerous physiological experiments in a book, which he published with high expectations in 1797. But alas, Alessandro Volta [58], who had been properly sceptical about the notion of animal electricity, showed that it needed no animal tissues to create a battery. Humboldt was mortified, and his chagrin at his failure to discover the principle of the battery, which had brought Volta such renown, was never assuaged. He turned now to botany and published a well-received volume on German flora. But his real life's work still lay before him: he sailed to South America, which he explored for five years, cataloguing fauna, flora, and physical geography in a manner never before attempted. He discovered the Pacific 'Humboldt current', and suggested the idea of a Panama Canal. He climbed Mount Chimborazo in Ecuador, then the world's highest known mountain, an accomplishment that made him a hero throughout Europe. After 30 years of collating and writing up his observations in a series of books, Humboldt, then already 60 years old, set off on another journey of exploration, this time through Siberia.

During the last years of his life, passed in a state of some penury, Humboldt wrote, though never quite completed, his crowning work, *Cosmos*, in which he drew together his views of nature and the physical world. In the delirium of his last illness, at the age of 90, he was still attempting to dictate notes for this mighty opus.

Humboldt's life and work have been the subject of many studies. A very readable biography is *Humboldt and the Cosmos* by Douglas Botting (Sphere Books, London, 1973).

~

135 Galton surpassed

The erratic genius Francis Galton, cousin of Charles Darwin, conjectured that no two humans had the same fingerprint, and that fingerprints could be rigorously measured and compared. Fingerprints had in fact been used in the mid-nineteenth century by a scientifically informed administrator in British India for 'preventing personation and putting an end to disputes about the authenticity of deeds'. (They were introduced by a local governor, Sir William Herschel, grandson and great-grandson of eminent astronomers [156].) By 1905, *Nature* could report that the fingerprint department of Scotland Yard was in possession of 80,000–90,000 specimens. The fingerprint remained for another 80 years the principal instrument of forensic detection, until, in 1984, a discovery at the University of Leicester supplanted it.

Alec (now Sir Alec) Jeffreys was interested at the time in the evolution of genes and had chosen as his exemplar the gene for myoglobin, a protein that stores oxygen in muscle. He had begun with myoglobin from seals (in which, as in other diving mammals, the protein is especially abundant). The next experiment was to compare the gene with the myoglobin gene in humans. Jeffreys knew that the genome (the total assemblage of DNA, making up an organism's set of chromosomes) contains long tracts of seem-

ingly functionless repeating sequences of nucleotides [**88**]). These arise
from an erratic mechanism that from time to time, over the generations,
makes duplicates of certain favoured sequences and reintegrates these sur-
plus slivers of DNA into the chromosomes. Among the many repetitive
regions of the DNA are the 'hypervariable minisatellite sequences', in which
a pattern of about 20 nucleotides is repeated many times. But the repetitive
elements are not all exactly the same, although they are built around a
characteristic central pattern—namely GGGCAGGAXG, where X can be
any of the four nucleotides, A, C, G, or T. Because of their propensity to
replicate randomly over many generations, these sequences vary in number, as well as in exact identity, between different families of people or
animals.

One day in 1984 Jeffreys, while examining the DNA of the myoglobin
gene in a gel (a gelatinous matrix in which fragments of DNA migrate in an
electric field and separate according to their sizes), came upon a run of
minisatellites. This was odd, even though most genes do contain unused
segments of DNA, which are excised when the gene is expressed and copied
into the genetic messenger, RNA. Looking more closely, he perceived that
DNA samples from different individuals all had quite distinct minisatellite
patterns. Jeffreys's genius was to recognize immediately what this might
imply. No sooner had he published his inference than he was approached
by Home Office scientists, who saw in his discovery a sure way of establishing whether immigrants who claimed close kinship with a resident in
Britain were telling the truth. (In nearly all cases they were, much to the
discomfiture of the civil servants.)

Then one day in the summer of 1986, the body of a girl of 15 was discovered in a thicket outside the village of Narborough, not 10 miles from
Leicester. She had been raped and strangled. The trail of investigations led
the police to a hospital porter, Richard Buckland, who was arrested and
confessed to the crime. But Buckland would not confess to the precisely
similar rape and murder of a young girl in Narborough three years earlier.
Anxious to clear up that crime also, the police, who knew of Jeffreys's work
from stories in the press, called on him at the university. Could he help
them to identify Buckland as the murderer of the first victim? With semen
samples from the two bodies and blood from Buckland, Jeffreys went to
work. The DNA extracted and amplified by the polymerase chain reaction
[**108**] was the same in both semen samples, but when Jeffreys came to
examine Buckland's DNA from the white cells of his blood, he found that it

was unrelated. Buckland, then, was not after all the murderer. The police were incredulous and so samples were sent to the Home Office forensic laboratory, which had by then already set up Jeffreys's procedure. The outcome was the same, and the police reluctantly released Buckland. It was some months before they set about collecting blood from the citizenry of Narborough. Of the resulting 5,500 DNA samples not one matched that of the semen. And then one day an employee of a bakery in Leicester came forward with information: a colleague of his had been asked by another worker in the bakery to give a blood sample in his stead. The police closed in and arrested Colin Pitchfork, a Narborough man, who confessed to both murders, and this time the DNA samples all matched.

Since then the minisatellite test has been used to convict (and clear) countless criminal suspects, to establish paternity, and, in one prominent case, to identify the remains of the last Tsar of Russia and his family, taken from a shaft near Ekaterinburg, where they had faced their murderers. Jeffreys himself was asked in 1985 to examine the bones of Josef Mengele [164], the infamous Auschwitz doctor. The ageing fugitive had, it was said, drowned in Brazil in 1979, but despite matching dental records the Israeli authorities demanded better proof, for had this master of evasion not fooled them all those years? The bones, by the time of their exhumation, were in poor condition and Jeffreys was able to find only three intact cells from which to extract DNA, but thanks to the power of the polymerase chain reaction the sample proved sufficient for a genetic fingerprint. The only problem was that Mengele's son in Germany refused to cooperate, but was persuaded when he was told that if he remained obdurate all his family's graves would be opened. And so the Israelis and the world were reassured that Mengele was indeed dead (or at least that the man in the grave was Frau Mengele's son's father).

Good accounts of Alec Jeffreys's discovery can be found in *The Book of Man: The Quest to Discover Our Genetic Heritage* by Walter Bodmer and Robin McKie (Little, Brown, London, 1994, and in Matt Ridley's *Genome: The Autobiography of a Species in 23 Chapters* (Fourth Estate, London, 1999).

136 Beef and ale

'Possession's beef and ale' sang Oliver Goldsmith, and the great German chemist, Justus von Liebig (1803–73), had a hand in promoting both.

Liebig was the most celebrated organic chemist of his time. He was a combative man with a quick temper, and even his occasional spasms of generosity had a calculating edge. Liebig's school at the University of Giessen, described by a French rival as 'an evil hole', gave rise to many of the chemical luminaries of the next generation. Unusually for a German professor of the period, Liebig sometimes allowed his students to publish by themselves, for, he confided to a friend in a letter, 'if it is something good, a part of the credit is ascribed to me and I do not need to defend the mistakes. You understand?'

After he left Giessen, where his most productive years were spent (partly because, as he said, in this small and lustreless town there was nothing for his students to do except work), Liebig turned his attention to biological chemistry and especially to agriculture and nutrition. He established the nutritional value of fats, but wrongly insisted that soil nitrogen came only from ammonia in rain water; he also denied, against all the evidence, that yeast was a living organism, which brought him into conflict with Louis Pasteur (only one of his many vendettas against French chemists).

Liebig always kept an eye open for commercial opportunities. So hearing that in the mining region of Uruguay there was a superfluity of cattle, which were slaughtered for their hides, while most of the meat was thrown away, he devised a process for rendering the beef into a concentrated broth. This consisted in roasting the meat, pulverizing it, extracting the juices, and concentrating them in vacuum vessels. The extract, marketed by a company called Fray Bentos, from its location in Uruguay, was known as *extractum carnis Liebig* and was the precursor of the modern stock-cube. When Liebig tried the same trick with coffee the results were less satisfactory: essential oils were oxidized during drying and generated a foul taste, and a large part of the evaporated residue was insoluble. (Instant coffee had to wait for more advanced methods of extraction, first in a so-called Soxhlet extractor by a Swiss chemist, who was addicted to coffee but wanted to save time during his labours in the laboratory by simply pouring boiling water onto a concentrate, and later by the process of freeze-drying.)

Liebig's contribution to the marketing of beer came in 1852. A rumour had threatened to turn into a scandal, for it was bruited about that the pale ale produced by two of the leading brewers in Burton-on-Trent, Allsopp's and Bass, was being laced with strychnine to enhance its bitterness. The canard had apparently been started by a French analytical chemist, and to scotch it the brewers had approached the two most famous chemists in

England, Thomas Graham and August Wilhelm von Hofmann. Hofmann was a former pupil of Liebig's, who had been lured by Queen Victoria's consort, Prince Albert, to the newly established Royal College of Chemistry (which later became the Royal School of Mines) as its first professor. Graham and Hofmann both concluded from their analyses that the beer was harmless, but Hofmann suggested to Allsopps that their utterances would carry more weight if supported by the word of the greatest chemist of the day, Baron (as he by then was) Justus von Liebig.

For a fulsome open letter testifying to the excellence of English beer Liebig received the then not inconsiderable sum of £100. In his letter to Hofmann, he owned that 'the main test consisted in drinking a bottle with great enjoyment'. (He would, of course, have had full confidence in his pupil's analysis.) Liebig shamelessly complied with Allsopps' instructions as to how his commendation should be worded. Soon his testimonial appeared on billboards and in newspapers. Thereupon, as Hofmann and Liebig probably foresaw, Allsopps' competitor, Bass, asked for a similar favour, for which they paid the Baron an unknown sum.

Towards the end of his life Liebig's temperament evidently mellowed. He made his peace with his French adversaries, especially Jean Baptiste Dumas, with whom he had fought a running duel of words for decades, and in 1867 was a guest of honour at the Paris Exposition. In an after-dinner speech to an assembly of the Exposition's jurors Liebig recalled his days in the City of Light, where in 1823, as a young man, he studied with the great chemist Joseph-Louis Gay-Lussac. By then Gay-Lussac had become chemist to the government Commission des Poudres et Salpêtres (explosives), with a laboratory and residence in the Arsenal. Alexander von Humboldt [134] was in the audience at the Academy of Sciences when Gay-Lussac introduced a paper by Liebig on fulminates, accompanied by a demonstration (probably noisy) by Liebig. The valiant Gay-Lussac had earlier made a startling solo balloon ascent to a record height of 23,000 feet, thus eclipsing Humboldt's world altitude record on the summit of Mount Chimborazo and the two men had become friends. Humboldt had also befriended Liebig in Germany, and it was he who now persuaded Gay-Lussac to admit Liebig to his Arsenal laboratory. The two chemists had much in common and their interests at the time—Liebig's in the explosive fulminates and Gay-Lussac's in the related cyanogen compounds—complemented each other nicely. In his speech, four decades after the events, Liebig told his audience that that time had been the happiest of his life:

Never shall I forget the years passed in the laboratory of Gay-Lussac. When we had finished a successful analysis (you know without my telling you that the method and the apparatus described in our joint memoir were entirely his), he would say to me, 'Now you must dance with me just as Thénard [Louis Thénard had been Gay-Lussac's teacher] and I always danced together when we had discovered something. And then we would dance.

These events, and many others, are recorded in the definitive biography of Liebig *Justus von Liebig: The Chemical Gatekeeper* by William H. Brock (Cambridge University Press, Cambridge, 1997).

~

137 The wrath of fools

Alexandre Dumas opined that rogues are always preferable to fools, for rogues sometimes take a rest. In the Soviet Union, with science in thrall to Marxist ideology, the fools held sway. Here is a story related by George Gamow [81], famous as a physicist and cosmologist, who eventually abandoned his native country for a brilliant career in the United States.

Gamow was a forthright character, a large man with a loud voice and an anarchic sense of humour. He and his student, Ralph Alpher, performed a seminal calculation, concerning the formation of elements at the birth of the Universe. True to form, he co-opted Hans Bethe [62], the great mathematical physicist, to enlarge on aspects of this work, so that it could be published under the names of Alpher, Bethe, and Gamow. (Bethe's name was apparently initially added without his knowledge, to his considerable displeasure, but his eventual contribution was important.) Gamow drew another colleague, Robert Herman, into this endeavour and tried (unsuccessfully) to persuade him to change his name to Delter.

Gamow claimed in his memoirs that he acquired the habit of scepticism and aversion to authority at an early age: his father had bought him a small microscope and he used it to determine whether the bread soaked in red wine, administered to him during Communion, was indeed transubstantiated into the blood and body of Christ. He secreted a little of the mouthful in his cheek and hastened home to examine it with his microscope, having first prepared as a control a similar fragment of bread impregnated with wine. For a positive standard he shaved off a sliver of his skin and found that its appearance bore no resemblance to those of the two identical specimens of sodden bread. He could not, he confessed in his memoirs, have seen individual blood cells with his low-power instrument, so the proof was not iron-clad, but it sufficed to turn him away from religion and to science.

In 1925, Gamow was still in Moscow. One day he was deep in discussion with his close friend, the incomparable Lev Davidovich Landau (1908–68), when their colleague, Abatic Bronstein, entered the room, bearing the latest volume of the Soviet Encyclopedia. He pointed to an article about the luminiferous ether, an imponderable fluid, held by the classical physicists of the nineteenth century to pervade all space. It was in the ether that electromagnetic waves—of visible light for example—were propagated. Einstein's Theory of Relativity had eliminated the ether from physics, but relativity appeared to many of the older physicists counterintuitive and therefore philosophically unacceptable. But by 1925 physics had moved on; the subject was strongly represented in the USSR, relativity and other developments, such as quantum theory, had been assimilated into its fabric, and men of the calibre of Landau and Gamow had no patience with the old school of 'mechanists', as they were called, who adhered to Newtonian physics and rejected the new.

The article that Bronstein showed his colleagues was by the 'red director' of the physics department, Comrade Gessen, whose charge it was to ensure that the 'scientific director' and his staff did not deviate in their research from the path of Marxist orthodoxy. Gessen knew a little physics, having taught the subject in school. His account summarized the classical concepts of light, denounced Einstein, and asserted the material nature of the ether. It was, the article declared, the mission of Soviet physicists to study its properties. The three friends, with two of their students, resolved to send Gessen a letter, making fun of his vision of physics, together with a scurrilous cartoon. But Soviet officialdom was not to be mocked.

The text, which was, of course, originally in Russian, read as follows:

> Being inspired by your article on the light-ether, we are enthusiastically pushing forward to prove its material existence. Old Albert is an idealistic idiot!
>
> We call on your leadership in the search for caloric, phlogiston, and electric fluids [concepts from eighteenth-century science].
>
> G. Gamow   Z. Genazvali
> L. Landau   S. Grilokishnikov
> A. Bronstein

We expected that Gessen would blow up, but his explosion exceeded by far our expectations. He took our teleletter [letter transmitted by telegraph] to the Communist Academy in Moscow and accused us of being in open revolt against the principles of dialectical materialism and the Marxist ideology. As a result, by orders from Moscow, a 'condemnation session' was organized.

The consequences of the physicists' foolhardy act were serious, at least for the unfortunate students:

After the condemnation meeting, which lasted for hours, Dau [Landau] and Abatic came to my apartment and related what had happened. We were found guilty of anti-Revolutionary activity by a jury of machine-shop workers of the institute. The two graduate students who signed the telegram lost their stipends and had to leave town. Dau and Abatic were dismissed from their teaching jobs in the Polytechnic Institute (to prevent their infecting the minds of students with poisonous deviationist ideas) but were retained at their research positions at the Roentgen Institute. Nothing happened to me, since I was not connected with that outfit. But there were proposals to give us all 'minus five' punishment (a ban on living in the five largest cities of the USSR), which also never materialized.

For Landau it was not the first experience of the Party's displeasure, nor the last. He had already been in trouble and was to get into deeper hot water not long after. He was rescued, once when he was already in prison, by the intervention of the courageous Pyotr Kapitsa [170], who wrote to Stalin, guaranteeing his protégé's future decorum. As for Gamow, after a reckless and unsuccessful attempt to escape by row-boat across the Black Sea, he was finally permitted to attend a conference in Brussels in 1932, from which he did not return. Landau, a patriot and convinced communist, remained in the Soviet Union. His end was sad: he suffered severe head injuries when a car, driven by one of his students, crashed on an icy road. He recovered from his coma, but never regained his intellectual brilliance. He is supposed to have remarked, in characteristic style, 'I am no more Landau, but Zeldovich' (another first-rate Russian theoretician, if not Landau's equal, and disliked by Landau).

See George Gamow's *My World Line: An Informal Autobiography* (Viking Press, New York, 1970).

138 Domestic horror show

Claude Bernard (1813–78), the foremost physiologist of the nineteenth century, had a troubled domestic life. His wife was a strict Catholic, had no sympathy with, or understanding of science, and, worse, strongly disapproved of her husband's experiments on animals. She contributed money to an anti-vivisection movement and turned her three children against their father. There was some excuse for her distaste, especially because Bernard, a compulsive and passionate experimenter, would often bring work home with him. Here is a description of one such occasion by Bernard's biographer. It was at an early stage in their married life; they already had a two-year old child and Madame Bernard was again pregnant.

> Her husband brought to their small flat [in Paris] on an upper floor on Sunday morning a dog with an open wound in its side from which internal fluids were drawn off from time to time; the dog was in a state of extreme emaciation, yet with a voracious appetite, pus running from its nostrils, coughing as it was led up and down the stairs, suffering from diarrhoea, its faeces being of particular interest to the master of the house.

Bernard understood the revulsion that his physiological experiments excited in so many. He wrote: 'If an illustration were required to express my feelings in regard to the science of life, I should say that it is a superb *salon*, resplendent with light, which one can only enter by passing through a long and ghastly kitchen.' It is small surprise that the marriage eventually foundered. Later Bernard in his loneliness found solace in a probably platonic friendship with a vivacious and intelligent married woman, who took an interest in his work.

There are several biographies of Claude Bernard. The passage quoted above is from *Claude Bernard: Physiologist* by J. M. D. Olmsted (Cassel, London, 1939).

~

139 A ball on Mars

At the Jet Propulsion Laboratory in California preparations are in hand for the exploration of Mars. Instruments are to be installed on a roving vehicle, instructed by signals from the mission control centre. The surface of the red planet is covered by man-sized boulders and is scoured by strong winds. The vehicles used for exploration of the Moon had almost spherical wheels of large diameter, and the physicists of the Jet Propulsion Laboratory considered whether a single ball, with instruments at its centre, could serve at least as well, propelled by the afternoon winds of some 45 miles per hour. Models, the size of beach balls, were built and tried out in the Mojave Desert. They were a failure, for they regularly came to rest against small rocks or on slopes. Back then to the large-wheeled vehicles, and one of these, with inflatable spherical tyres, was being tested when a strange accident occurred: one of the tyres, 5 feet tall, broke away and, driven by a 20-mph breeze, accelerated rapidly and receded into the distance. It bounced over boulders and soared up steep sandy cliffs. The bounding sphere left the technician who had set off in pursuit far behind, and was caught only with the aid of a beach buggy. The effect was one of scale: a sphere smaller relative to the irregularities of the terrain had proved a poor model.

The plan at the Jet Propulsion Laboratory now is to build a giant inflatable sphere, 20 feet in diameter—the height, therefore, of a two-storey house—with radar equipment to detect subterranean water at its centre, along with other instruments. The afternoon winds are expected to drive it over all obstacles on the planet's surface, to be stopped at will by partial deflation and re-started by an inflation instruction from mission control. The Jet Propulsion Laboratory physicists refer to their device as a tumbleweed.

The story and a more complete description of the device can be seen and heard on the NASA web site, http://science.nasa.gov.

~

140 Boyle on the boil

The Hon. Robert Boyle (1627–91) once described, to Samuel Pepys's vast amusement, as 'son of the Earl of Cork and father of modern chymistry', played a large part in turning chemistry into a rational science. His influential book *The Sceptical Chymist* set out his quantitative credo. Boyle's Law, relating the pressure to the volume of a gas, is familiar to all schoolchildren and was first expounded in 1662 in a tract with the title *A Defence of the*

Doctrine touching the Spring and Weight of the Air. But Boyle never shook off his fascination with alchemy. He was one of the many attracted by the vision of the 'philosopher's stone', the substance that would transmute base metals into gold. In their search for the secret of transmutation the alchemists made many important discoveries, of which perhaps the most spectacular was the isolation of phosphorus. Boyle and others had been much engrossed by 'phosphoruses', a term applied to all substances that glowed in the dark. These included certain minerals, the *ignis fatuus*, or will-o'-the-wisp, which was supposed to lure unwary travellers into treacherous marshes, and many biological organisms, such as fireflies, luminescent planktons, and saprophytic bacteria, which feed on decaying plant and animal matter.

Boyle, a bachelor, lodged for the last 25 years of his life with his sister, Lady Ranelagh, in her mansion, Ranelagh House in Pall Mall in London. There, in the garden, he established his laboratory where he did much of his most important work. Here also he would entertain Fellows of the Royal Society, recently established by Charles II, to evening discussions. By the time of the famous meeting in 1677 word had reached England of a remarkable discovery in Germany. An alchemist, Daniel Kraft, had prepared a spontaneously inflammable substance, which glowed ceaselessly in the dark. Kraft, in fact, had obtained the secret from another alchemist, Hennig Brandt of Hamburg. Kraft's fame had spread and in 1677 King Charles, a dabbler in alchemy himself, invited him to visit London and demonstrate the miraculous new element (though, of course, it was not then recognized as such). In the evening of 15 September Kraft arrived with his paraphernalia at Ranelagh House, where Boyle and a group of Fellows had gathered. Boyle left his own account of what they witnessed.

'The windows were closed with wooden-shuts', he begins, 'and the Candles were removed to another Room by that we were in; being left in the dark we were entertained with the following Phaenomena.' Kraft first produced a glass globe, containing a suspension of some solid material in water, no more, Boyle thought, than two or three spoonfulls; 'yet the whole sphere was illuminated by it, so it seemed to be not unlike a Cannon bullet taken red hot out of the fire', and when he shook it the intensity increased and small flashes were seen. Another vessel, 'and the Liquore which lay in the bottom being shaken, I observed a kind of smoke to ascend and almost fill the cavity of the Vial, and near the same time there manifestly appeared as it were a flash of lightening that was considerably diffused and pleasingly surprised me'. But then Kraft brought out a solid lump of phosphorus, which, he said, had been glowing for two years. 'The Artist having taken a very little of his consistent matter, and broken it into parts so minute that I judged the fragments to be between twenty and thirty, he scattered them without any order about the Carpet, where it was very delightful to see how vividly they shined'—and, indeed, twinkled like stars, happily without damaging the valuable 'Turky Carpet'. Kraft next rubbed his finger over the surface of the phosphorus, drew luminous letters on a piece of paper and anointed his face and Boyle's hand, so that they glowed eerily in the darkness. From the paper arose a smell that Boyle found to resemble both sulphur and onions.

Some days later Kraft returned and demonstrated the combustibility of his phosphorus: a small piece, taken out of a bottle of water, when wrapped in paper caused a conflagration and another swiftly ignited a small mound of gunpowder. Boyle and his colleagues were mightily impressed. Boyle wanted immediately to perform his own experiments on the mysterious substance, but Kraft demurred when asked to leave a sample behind. When pressed as to its origin he would say only that it came from 'somewhat that belonged to the body of man'.

Boyle guessed that it must have been prepared from urine, for the yellow liquid had always been a provocation to the alchemists and iatrochemists, who wondered whether it perhaps harboured the essence of gold. Boyle worked on the problem for two years before he succeeded. He had instructed his assistant, Daniel Bilger, to collect and hoard huge quantities of urine from the privies of the great house and boil off the water, but to no avail, for we know now that the phosphorus in urine is in the form of phosphates, which are very stable salts. Boyle even considered that he might be on the wrong track altogether and that perhaps urine was not after all what Kraft had meant when he indicated that his phosphorus was derived from a human product; the wretched Bilger was accordingly sent to raid the cesspits. In the end Boyle hit on what had been the method of Kraft and of Brandt before him; or rather his later and more skilled German assistant, Ambrose Godfrey Hanckwitz, who had visited Brandt in Hamburg, alerted him to it. The key was to subject the solid residue recovered by evaporation to a very high temperature. When Hanckwitz tried it the retort cracked, but when Boyle went to survey the wreckage he perceived that the residue was now dimly glowing.

Boyle did many interesting experiments with phosphorus once quantities of the pure material could be prepared, but he published little and the preparative method was deposited in a sealed paper at the Royal Society, to be made public only after his death. The reasons for such coyness remains obscure. His posthumous account was published in 1694 and described the process in some detail, concluding with what he observed at the end of the heating step:

> By this means there came over [from the retort into the receiver] a good store of white Fumes, almost like those that appear in the Distillation of the Oil of *Vitriol* [sulphuric acid]; and when those Fumes were passed and the *Receiver* grew clear, they were after a while succeeded by another sort that seemed in the *Receiver* to give a faint bluish light, almost like that of little burning Matches dipt in Sulphur. And last of all, the Fire being very vehement, there passed over another Substance, that was judged more ponderous than the former because it fell through the Water to the bottom of the *Receiver*; whence being taken out, (and partly even whilst it stayed there), it appeared by several Effects and other *Phaenomena*, such a kind of Substance as we desired and expected.

Hanckwitz eventually became purveyor of phosphorus, much purer than that of Kraft, to the laboratories of Europe. Boyle ruminated on possible uses for the new element—for domestic lighting, lamps for underwater exploration, and even luminous clock dials. One of its first uses was for matches, the manufacture of which soon established its frightening toxicity, when the painful and disfiguring condition of 'phossy jaw' overtook workers in the industry. Hamburg was destroyed during the Second World War by incendiary bombs made out of the element first brought to light there.

See R. E. W. Maddison, *The Life of the Honourable Robert Boyle FRS* (Taylor and Francis, London, 1969). The story of Boyle's adventures with phosphorus is also divertingly told by John Emsley in *The Shocking History of Phosphorus: A Biography of the Devil's Element* (Macmillan, London, 2000).

141 The physicist as travelling salesman

With the advent of atom-smashing machines, colliders designed to accelerate particles to speeds approaching that of light and smash into each other, the practice of physics began to change. The cost of such experiments assumed significance on the scale of national budgets, and it was only men of boundless, even fanatical, self-confidence who could initiate the projects and direct the teams of hundreds needed to bring them to fruition. The leader became an advertising executive and salesman, much of whose work would be done on the road, travelling business class. With such astronomical stakes, intrigues against the competing laboratory became almost as great a preoccupation as the success of the experiment. 'This generation of high-energy physicists', in the view of Marty Perl, one of the leaders in the field of fundamental particles, 'could also have done very well in the retail garment trade.' One of the most predatory of this fearsome breed is an Italian, Carlo Rubbia, of Chicago and CERN, the pan-European laboratory in Geneva. Here is a tale that encapsulates the high intensity at which such laboratories operate.

A woman physicist had been waiting for a couple of weeks to steal just a few minutes of Rubbia's time and discuss what she considered a crucial and highly important piece of physics. Rubbia also thought it was important, but he had been flying around the world, coming and going, and the physicist had just about given up hope.

Finally, one morning she gets a call from Rubbia. She picks up the phone and Rubbia says, 'Okay, I have exactly twenty minutes to talk to you about your work.' This is great, she thinks. She slams down the phone, runs full speed to Rubbia's office, making it there in about ten seconds, only to find that his door is locked. She turns to Rubbia's secretary and says, 'Carlo's door is locked?'

'Yes,' the secretary replies, 'Carlo was calling from the airport in Zürich.'

Meanwhile Rubbia has called back and is saying to his secretary, 'What the hell is the matter with that woman? I tell her I can talk to her about her work, and she hangs up on me.'

For a fascinating account of the work and mores of the high-energy physics profession, see Gary Taubes's remarkable fly-on-the-wall reportage in *Nobel Dreams: Power, Deceit and the Ultimate Experiment* (Random House, New York, 1986), from which the above is taken.

142 Monsieur LeBlanc's solicitude

Sophie Germain (1776–1831) left her mark on both pure and applied mathematics. Among her achievements was a fundamental study of the theory of elasticity. She was born into a prosperous French bourgeois family with intellectual leanings. Her father's library was Sophie's university, and it was there that she read about Archimedes and his death at the hands of a Roman soldier [26]. From that moment Archimedes became her hero and mathematics her avocation. But it was not long before her questing intelligence transcended the material in the family library and she decided that she must learn from the best mathematicians of the day by correspondence. Her most loyal pen-friend was the French mathematician, Adrien-Marie Legendre, with whom she engaged in a voluminous exchange of letters on everything from number theory to topology. She also turned to the greatest mathematician of the day, Carl Friedrich Gauss (1777–1855).

Gauss's genius was recognized throughout Europe. The son of a Brunswick bricklayer, who wanted the boy to follow the same trade, he was able at the age of barely three to point out a mistake in his father's accounts. By the time he was 10 he was acquainted with such fundamental parts of algebra as the binomial theorem and infinite series. A perceptive schoolmaster overcame the father's objections and introduced the boy to the Duke of Brunswick, who arranged for his education. Long before he had finished his course at the Caroline University he was making the first of the abundant and diverse contributions to the advance of mathematics.

Gauss was a tetchy man and an unwilling correspondent, so when Sophie Germain, thinking it best to conceal her anomalous standing as a female mathematician, wrote to him under the name of M. LeBlanc, she received only a tardy and perfunctory reply. Gauss eventually showed enthusiasm, only when he discovered that M. LeBlanc was a woman in the following curious circumstances: in 1806, Napoleon's armies invaded Prussia and at the battle of Jena inflicted a crushing defeat on their enemies, so leaving much of the country at their mercy. Recalling the death of Archimedes, Sophie Germain became concerned lest the great Gauss should suffer a similar fate in Brunswick. The commander of Napoleon's artillery in Prussia was a family friend, General Pernety, and it was to him that she addressed her concerns. The General summoned a battalion commander by the name of Chantal to ride 200 miles to the already occupied city to find and protect the great savant. Chantal did as he was bidden, found Gauss, and reported that he was alive and had not been molested. From Pernety Gauss learned of the true identity of M. LeBlanc and he wrote her a warm letter.

It would be pleasant to report that a productive correspondence ensued, but Gauss, alas, seems after a brief interval to have resumed his curmudgeonly ways. Sophie Germain's contribution to mathematics and physics nevertheless endures.

See *Sophie Germain: An Essay in the History of the Theory of Elasticity* by Louis M. Bucciarelli and Nancy Dworsky (D. Reidel, London, 1980).

~

143 The emperor and the scientist

Napoleon I was a friend to science. He was patron to many of France's leading scholars, and several of the most distinguished, such as the mathematician and physicist Gaspard Monge, and the chemist Claude Louis Berthollet, were privileged members of his ill-starred Egyptian expedition. French scientists saw themselves and their brethren in other lands as citizens of the world. So it was that Humphry Davy [123], for instance, was allowed to travel unmolested through France at the height of the Napoleonic wars. Napoleon's regard for men of science can be gauged from an episode relating to Edward Jenner.

Jenner (1749–1823) is remembered now for introducing vaccination against smallpox, but he also did notable work in zoology, most famously on the life cycle of the cuckoo, and in palaeontology. Of his work on smallpox, Thomas Jefferson wrote to him: 'You have erased from the calendar of human affliction one of the greatest. Yours is the comfortable reflection that Mankind can never forget that you have lived'. Jenner, a country doctor in Berkeley in Gloucestershire, was struck by the apparent resistance of milkmaids to smallpox. It was, moreover, an old-wives' tale that a milkmaid exposed to cowpox, a mild affliction in humans, became resistant to smallpox ever after. When in 1796 a local herd of cows became infected with the cowpox, Jenner seized his opportunity for an experiment of highly dubious ethical propriety. He found a milkmaid, Sarah Nelmes, with a mild cowpox infection and into a pustule on her hand Jenner thrust his lancet. With it he then infected a youth by the name of James Phipps. After some weeks Jenner administered a smallpox extract. Phipps lived, and, not without vicissitudes, the practice was perfected and within a few years generally adopted.

The cowpox virus, as we now know, is related to the agent of smallpox. In honour of Jenner, Louis Pasteur [172] later coined the term, vaccination, from *vacca* or *vache*, a cow. It was a truth waiting to be uncovered, for forms of vaccination had been used in ancient times, and news of its application in Turkey had been brought back from her sojourn in Constantinople by Lady Mary Wortley Montagu (1689–1762), wife of the British ambassador there, who had her children 'variolated' after she herself had been ravaged by the disease (at the cost of her beauty and even of her eyebrows). Indeed, some 20 years before Jenner wielded his lancet to such effect, a farmer in Dorset, Benjamin Jesty, had 'with great fortitude tried the cow-pox on his wife and children'. Jenner's work largely eliminated the scourge of smallpox from Europe and brought him well-deserved fame:

[T]he world-wide eminence of Edward Jenner found no better illustration than his ability to secure the liberation of British prisoners from countries with which England was at war. One of the best known of these prisoners was the Earl of Yarmouth, the model of Thackeray's Marquess of Steyne and Disraeli's Marquess of Monmouth, on behalf of whom Jenner addressed in 1803 the following appeal to the National Institute of France: 'The Sciences are never at war . . . Permit me then as a public body with whom I am connected to solicit the exertion of your interest in the liberation of Lord Yarmouth.' In 1805 Jenner addressed himself directly to Napoleon requesting that two of his friends, Mr William Thomas Williams and Dr John Wickham, both men of science and literature, might return to England. According to Baron, the well-known biographer of Jenner, it was either on this or a similar occasion that Napoleon exclaimed: 'Jenner! Ah, we can refuse nothing to this man.' Jenner was also successful in obtaining the release of Sir George Sinclair, who had been arrested as a spy in Göttingen. Besides helping to liberate Englishmen detained on the Continent, Jenner issued certificates stating that travellers abroad were known to him and were undertaking a voyage in pursuit of science or health or other affairs entirely unconnected with the war, and were in his opinion entitled to protection and freedom.

This summary of a lecture in London is taken from *Nature*, **144**, 278 (1939). For the life of Jenner see, for instance, R. B. Fisher, *Edward Jenner 1749–1823* (Deutsch, London, 1991).

144 The man of principle

J. E. (John Edensor) Littlewood (1885–1977) was one of the foremost
mathematicians of the twentieth century. His absorption in his subject and
his productivity persisted almost until his death. At the age of 89 Littlewood
had a serious fall and entered a nursing home in Cambridge, where he
appeared to lose all interest in life. A young friend, Béla Bollobás, visited
him and tried to divert him with a new mathematical problem:

> In my desperation I suggested the problem of determining the best constant
> in Burkholder's weak L_1 inequality (an extension of an inequality Littlewood
> had worked on). To my immense relief (and amazement), Littlewood
> became interested in the problem. He had never heard of martingales [who
> has?] but he was keen to learn about them so he was happy to listen to my
> brief explanation and was willing to read some introductory chapters! All this
> at the age of 89 and in bad health.

The story bears witness to the restorative powers of intellectual stimulation.
Bollobás relates that Littlewood worked hard on the problem. It was
brought to a conclusion by Bollobás; the results were published, however,
only after Littlewood's death.

Bollobás's reminiscence can be found in his foreword to a collection of
Littlewood's *obiter dicta*, from which the following moral tale about the
Russian theorist, A. A. Markov:

> A Ph.D. this having admittedly failed, the other examiners were in favour
> of leaving it at that. Markov wished to read the man a severe lecture on the
> enormity of his performance, but allowed himself to be overruled. On his
> death-bed he said he had never forgiven himself for this weakness, and it sad-
> dened his end.

From *Littlewood's Miscellany*, ed. Béla Bollobás (Cambridge University Press, Cambridge,
1986).

~

145 The stolen invention

Dominique François Jean Arago [166] became, after his youthful adventures, a mandarin of the French scientific establishment and a force (for

good) on the national political scene. He was a formidable power-broker in the Academy of Sciences and sought to ensure that it was the deserving who were elected to its membership, not, as often in the past, those with the most influential patronage. Arago on three successive occasions thwarted the aspirations of a hydraulic engineer, Pierre-Simon Girard, who was successfully opposed each time by Arago's preferred candidate. All three of these had distinguished records and their names are commemorated in today's textbooks. The second of them was the noted applied mathematician Siméon-Denis Poisson (immortalized in the Poisson ratio of elasticity and the Poisson distribution in statistics). He was nominated by the venerated physicist and mathematician Pierre Simon, marquis de Laplace, but his election presented a tricky political problem.

The longevity of Poisson's mathematical confrères was prodigious and no seat (or *fauteuil*, as it is still called) in the section of geometry had fallen vacant for some years. Laplace therefore decreed that Poisson's candidature must come before the section of physics. Poisson had never approached any physical apparatus in his life, which was held to be a blessing, for he was a clumsy man and would assuredly have broken it. Laplace and Arago assembled a small caucus to consider how the academicians might best be persuaded to elect the misplaced candidate. How the objective was achieved is illustrated by the following conversation between Arago's friend, Jean-Baptiste Biot [166], a member of the inner circle, and the astronomer, Alexis Bouvard. Encountering him in the allée de l'Observatoire the day after the discussion, Biot asked who was to have his vote in the forthcoming election. 'Girard', Bouvard at once replied. 'You are mistaken,' Biot rejoined, 'you will be voting for Poisson. M. Laplace has charged me to tell you.' For Bouvard a wish of Laplace's was an order, and he voted for Poisson. (The loser, Girard, eventually made it as well.)

Biot, who had achieved eminence in physics, in fact occupied a *fauteuil* in the section of geometry, and in later years repeatedly asked that he and Poisson be allowed to change places, but such a logical move was held to violate the rituals by which the Academy governed itself.

As time went on, the two old friends, Biot and Arago, fell out. Biot was thought to have been resentful of the younger man's success, scientific, political, and social, for he had none of Arago's easy good nature and authority. When Arago was elevated to the position of *secrétaire perpétuel* at the Academy, he ran it as an autocrat, and yet retained the confidence of the members. The extent of the rancour that sprang up between him and Biot can be judged from a remarkable episode that did Biot no credit. He and Arago chanced to leave the Bureau des Longitudes (whence they had set out on their journey to the Balearic Islands many years earlier) at the same moment one Wednesday, and as they walked down the rue Saint-Jacques, Arago began to expound the principle of a photometer (an instrument for measuring light intensities) that he had just conceived. Biot was sceptical, and as they reached the church of Saint-Jacques-du-Haut-Pas, Arago took some keys from his pocket and proceeded to scratch a diagram on the near-est column, the better to explain his idea.

The following Monday there was a session of the Academy and Biot rose to speak. What he said stunned Arago, for Biot described as his own the principle of the photometer that Arago had confided to him five days before. Arago tried to interrupt, but Biot serenely continued. He concluded by drawing on the blackboard the very diagram that Arago had scratched on the pillar of the church. This was too much for Arago who leapt to his feet and shouted: 'This figure is precisely that which I drew for you to over-come your resistance to the principle that you now claim as your own.' Biot replied that he had no recollection of such a conversation. Arago thereupon demanded that the assembly instruct two *huissiers* to hasten to the church of Saint-Jacques-du-Haut-Pas, examine the said pillar and report back. Moreover, he added, a drawing with notes could also be found on his desk at the Observatory. The officers were summoned and despatched. Biot did not wait for their return; he left the Academy precincts and was not seen there again for another two years.

Many years later Biot vehemently opposed a plan of Arago's to admit the public to the Academy's sessions on the grounds that it might promote unseemly displays of vanity on the part of the savants, inhibit the candour of scientific exchanges, and cause embarrassment when aged academicians took the floor, no longer in a fit mental state to be exhibited before a youth-ful audience. Arago did at least secure entry for journalists, for whom a special bench was reserved.

Arago was assuredly no stranger to controversy. He managed even to stir up English anger when, having written a biography of James Watt, whom he much admired, he spoke in Glasgow and Edinburgh. He was invested as an honorary citizen of Glasgow and was introduced at a public meeting by a politician, Lord Brougham. Arago credited Watt not only with inventing the steam engine but also with discovering the composition of water. This was, of course, a false claim, which engendered wrath in the Royal Society, for the Fellows knew that the achievement belonged not to the Scot but to an Englishman, the Hon. Henry Cavendish [156]. Arago was accused of allowing his political convictions to outweigh his objectivity, of elevating the genius of Watt, the practical working man, above the lucubrations of an effete scion of the nobility. The Royal Society demanded and received a retraction.

The details of most of the above events are given in *Arago: La Jeunesse de la Science* by Maurice Daumas, 2nd edn (Belin, Paris, 1987, first published in 1943).

~

146 The Jesuits and the bomb

The Jesuits have always taken an interest in science. Among their ranks have been noted astronomers and astrophysicists, and indeed the astrophysical laboratory in the Vatican appears to have been established with the aim of seeking revelations from celestial meteorites. But in 1896 Father Frederick Odenbach of the Jesuit College of St Ignatius in Cleveland, Ohio, developed an interest in the science of meteorology. This evidently did not long satisfy his scientific appetites, for by 1900 he had turned his attention to seismology. That year he constructed a seismograph and began to make observations. After working for some years the Father had a brainwave: the Jesuit order was represented in almost all countries and maintained regular contacts between its outposts. Why not, then, organize a chain of seismological observation stations around the world and plot the movements of the Earth's crust globally?

In 1909, Father Odenbach wrote to all the Jesuit colleges in North America, soliciting their help. 'With a small outlay at a number of our colleges', he told them, 'we would be in a position to do *the great thing in seismology*.' Soon 18 seismographs were in action in Jesuit colleges across the United States and Canada. But the Great War and the preoccupations that came with it intervened and the project lapsed, until, in 1925, another Jesuit, Father John Macelwane, Professor of Geophysics at Washington University in St Louis, Missouri, reanimated the scheme. There were by then seismological stations in Jesuit colleges in Australia, Bolivia, China, Colombia, Cuba, England, Grenada, Hungary, Lebanon, Madagascar, the Philippines, and Spain, but they mostly opted for an informal arrangement, fearing that any officially sanctioned organization might lead to 'misunderstanding'—the threat, that is to say, that the order was hatching a global conspiracy. But there was evidently great enthusiasm for the project, which met the inescapable condition of cheapness, and the participants set about it with professional zeal.

The Jesuits' most famous success came in 1954, when one of their number, Father Rheinberger in Sydney, Australia observed a small seismographic signal, which appeared to coincide with the hydrogen-bomb explosion at Bikini Atoll in the Pacific. The Jesuit stations around the world were enjoined to consult their records, and it emerged that all four recent thermonuclear bomb tests had been detected by their instruments. So began the worldwide monitoring of nuclear tests. There was also a bonus for geophysicists, for the records showed that the explosions were always set off at five minutes past the hour. The observers could thus prepare for the event and follow the course and attenuation of seismic shock waves through the Earth's crust. But the request for hydrogen bomb explosions to be arranged for the exclusive benefit of the seismologists was predictably rejected.

See *The Dark Side of the Earth* by Robert Muir Wood (Allen and Unwin, London, 1985).

~

147 Ferreting out a virus

Research on the viruses of the common cold and flu became possible only when the first such virus was cultured in the laboratory. For many years these rhinoviruses, as they are called, resisted all such endeavours. Success finally came at the National Institute for Medical Research in London. A group of researchers there had been trying everything they knew, inoculating many species of animals—guinea-pigs, mice, rabbits, hamsters, hedgehogs, and monkeys—with throat garglings from human flu patients and lung tissue taken at autopsy from victims of the disease. These preparations were injected into abdomens, brains, and testes, all to no avail.

Then in 1933, Wilson Smith decided to try his preparations on ferrets, which were being used in the same laboratory for studies on dog distemper. Moreover, he thought of what in retrospect seems the obvious course of administering the virus through the nose. Smith inoculated two ferrets with washings from the throat of one of his colleagues, who had gone down with flu. A few days later Smith recorded in his notebook: 'Ferret I looks somewhat seedy—crusts round the nose and slight discharge with suggestion of pus—eyes also watery, sneezing.' Ferret II was soon similarly afflicted. Not long thereafter, Smith himself caught flu, it was thought because one of the ferrets had sneezed in his face. The virus recovered from Smith's throat turned out indeed to be the strain that was thriving in the ferret and not the one spreading that winter through the human population. And so transmission of the virus from man to ferret and back to man was established, and the so-called WS strain became the classic vehicle for research on flu.

The events, as recalled by different participants, are to be found in *The Lancet*, ii, 66 (1933), *Nature*, 207, 1130 (1966), and *Biographical Memoirs of Fellows of the Royal Society*, 12, 479 (1965).

~

148 A man's world

Cecilia Payne-Gaposchkin (1900–79) was an astronomer of great lustre, who would undoubtedly have achieved even more had she not had to struggle against the prejudice of a hidebound profession. A Cambridge undergraduate just after the First World War, she intended at first to become a biologist, but physics formed part of her Natural Science Tripos course and so she found herself in the Cavendish Laboratory, terrorized by the predominantly misogynistic lecturers and demonstrators, especially Ernest Rutherford [16], at whose lectures she was forced, as the only woman, to sit by herself in the front row, the recipient of the great man's Olympian ironies.

The laboratory work was the province of Dr Searle, an explosive, bearded Nemesis who struck terror into my heart. If one made a blunder one was sent to 'stand in the corner' like a naughty child. He had no patience with the women students. He said they disturbed the magnetic equipment, and more than once I heard him shout 'Go and take off your corsets!' for most girls wore these garments then, and steel was beginning to replace whalebone as a stiffening agent. For all his eccentricities, he gave us excellent training in all types of precise measurement and in the correct handling of data.

Cecilia Payne's epiphany came one evening, when, as she put it, the door into a new world swung dramatically open for her:

There was to be a lecture in the Great Hall of Trinity College. Professor Eddington [76] was to announce the results of the eclipse expedition that he had led to Brazil [sic] in 1918. Four tickets for the lecture had been assigned to students at Newnham College and (almost by accident, for one of my friends was unable to go) a ticket fell to me.

The Great Hall was crowded. The speaker was a slender, dark young man with a trick of looking away from his audience and a manner of complete detachment. He gave an outline of the Theory of Relativity in popular language, as none could do better than he. He described the Lorenz–Fitzgerald contraction [a manifestation of relativity], the Michelson–Morley experiment [measurement of the velocity of light] and its consequences [the elimination of the ether from physics, in accordance with Einstein's theory]. He led up to the shift of the stellar images near the Sun as predicted by Einstein and described his verification of the prediction.

The result was a complete transformation of my world picture. I knew again the thunderclap that had come from the realization that all motion is relative. When I returned to my room I found that I could write down the lecture word for word . . . For three nights, I think, I did not sleep. My world had been so shaken that I experienced something very like a nervous breakdown.

Cecilia Payne was henceforth wholly wedded to astronomy. She read every book on the subject that she could find in the library. Henri Poincaré's formidable *Hypothèses cosmogoniques* became, she recorded, a perennial source of inspiration.

Presently I learned that there was to be a public night at the Observatory. I bicycled up the Madingley Road and found the visitors assembled in the Sheepshanks Telescope, that curious instrument which, in the words of William Marshall Smart [a resident astronomer], 'combined all the disadvantages of a refractor and a reflector'. . . The gruff, kindly Second Assistant, Henry Green, was adjusting the telescope, and presently I had a view of a double star whose components (as he pointed out) differed in colour. 'How can that be,' I asked him, 'if they are of the same age?' He was at a loss for an answer and when I persisted in my questions he gave up in despair. 'I will leave you in charge', he said, and fled down the stairs. By that time he had turned the instrument to the Andromeda Spiral. I began to expatiate on it (Heaven forgive my presumptuousness!) and was standing with a small girl in my arms, telling her what to look for. I heard a soft chuckle behind me and found Eddington standing there.

As I heard him tell it later, when I had come to know him, Henry Green had gone to 'The Professor's' study and told him: 'There's a woman out there asking questions', and asked for help. The moment had come and I wasted no time. I blurted out that I should like to be an astronomer. Was it then or later that he made the reply that was to sustain me throughout many rebuffs? 'I can see no *insuperable* objection.' I asked him what I should read. He mentioned several books, and I found that I had read them all. So he referred me to the *Monthly Notices* and the *Astrophysical Journal*. They were available in the library of the Observatory which he said I was welcome to use. To paraphrase Herschel's epitaph [William Herschel, the eighteenth-century astronomer [156]], he had opened the doors of the heavens to me.

Cecilia Payne's enthusiasm and determination won her the esteem of the younger and brighter Cambridge astronomers. Here is how she made the acquaintance of one of the most celebrated:

One afternoon I bicycled up to the Solar Physics Observatory with a question in my mind. I found a young man, his fair hair tumbling over his eyes, sitting astride the roof of one of the buildings, repairing it. 'I have come to ask,' I shouted up at him, 'why the Stark effect [of an electric field on the position of lines in a spectrum] is not observed in stellar spectra.' He climbed down and introduced himself as E. A. Milne [64], second in command at the Observatory. Later he became a good friend and a great inspiration to me. He did not know the answer to my question, which continued to exercise me.

Despite the support of Milne and Eddington, Cecilia Payne could make no headway in the stuffy world of British astronomy, and so she betook herself to Harvard, where she pursued a notable career. Her most famous work concerned the composition of the Sun. She showed that the accepted interpretation of lines in the spectrum of sunlight—that they reflected the presence in the Sun's interior of iron in high abundance—was wrong. The Sun, she found, was made up predominantly of hydrogen, and the rest was helium. This result, elaborated in her doctoral thesis, was too revolutionary for the Harvard establishment and attracted only scorn, especially from the doyen of the American astronomers, the pompous and powerful Henry Norris Russell.

It took some years for Cecilia Payne's work to be confirmed and accepted, and it gave, of course, the explanation—nuclear fusion—for the Sun's seemingly inexhaustible supply of energy. She was vindicated by a theoretical analysis by none other than Russell himself, who belatedly gave her full credit, without, however, confessing to his earlier repudiation of her work. Harvard still did nothing to promote her career and, despite the magnitude of her achievements, she was saddled with so large a teaching load that her research was almost extinguished. She was much admired as a teacher and managed late in her career to collaborate in a research project with her daughter, who followed her into astronomy in a more enlightened era. By then she had herself become a professor and head of the Harvard astronomy department. She had married a boisterous Russian astronomer, Sergei Gaposchkin, whom she had met in Europe when he was down on his luck and had managed to infiltrate into the Harvard faculty. He was never in truth much more than his wife's assistant and was once heard to say with apparently unconscious hyperbole, 'Cecilia is an even greater scientist than I am'.

In her memoirs Cecilia Payne advised aspiring scientists:

> Young people, especially young women, often ask me for advice. Here it is, *valeat quantum*. Do not undertake a scientific career in quest of fame or money. There are easier and better ways to reach them. Undertake it only if nothing else will satisfy you; for nothing else is probably what you will receive.

The passages quoted here are from *An Autobiography and Other Recollections* by Cecilia Payne-Gaposchkin, ed. Katherine Haramundanis (Cambridge University Press, Cambridge, 1984). See also *Portraits of Discovery* (Wiley, New York, 1998) by George Greenstein, himself an astronomer.

~

149 The shock of recognition

Frederick Soddy, born in Eastbourne in 1877, was a chemist, recruited by Ernest Rutherford [16], then occupying his first professorial chair at McGill University in Canada, to assist in the analysis of radioactive elements. Together, in 1901, they made a staggering discovery: the radioactive metal, thorium, gave rise spontaneously to a radioactive gas, a different and new element. Soddy managed to prepare enough of this gas to liquefy it and show that it resembled the inert gas, argon. This 'thorium emanation' was later named radon.

I was overwhelmed by something greater than joy—I cannot very well express it—a kind of exaltation, intermingled with a certain feeling of pride that I had been chosen from all chemists of all ages to discover natural transmutation.

I remember quite well standing there transfixed as though stunned by the colossal import of the thing and blurting out—or so it seemed at the time: 'Rutherford, this is transmutation: the thorium is disintegrating and transmuting itself into an argon gas.'

The words seemed to flash through me as if from some outside source.

Rutherford shouted to me in his breezy manner, 'For Mike's sake, Soddy, don't call it *transmutation*. They'll have our heads off as alchemists. You know what they are'.

After which he went waltzing round the laboratory, his huge voice booming 'Onward Christian so-ho-hojers' which, as H. R. Robinson declared, was more recognizable by the words than the tune.

The warning was wise: public announcements caused a sensation and, according to another of Rutherford's collaborators, A. S. Russell, a company was floated in Glasgow, on the promise of turning lead into mercury and gold. Soddy later wrote:

Nature can be a sardonic jester at times, when you come to think of the hundreds of thousands of alchemists in the past few thousand years toiling and broiling over their furnaces, spending laborious days and sleepless nights trying to transmute one element into another, a base into a noble metal, and dying unrewarded in the quest, whilst we at McGill, by my first experiment, were privileged to see, in thorium, the process of transmutation going on spontaneously, irresistibly, incessantly, unalterably! There's nothing you can do about it. Man cannot influence in this respect the atomic forces of Nature.

This was as much a hostage to fortune as Rutherford's famous declaration, some 25 years later, that harnessing of atomic energy was 'moonshine'.

For the discovery of radioactive transformation Rutherford was awarded the 1908 Nobel Prize for chemistry—much to his amusement, for, as he liked to quip, his own transformation into a chemist had been instantaneous. Soddy, the real chemist in the partnership, never overcame his resentment that his contribution had not been similarly recognized.

Later more radioactive emanations were discovered, all with similar properties, but differing slightly in atomic weight. They were, indeed, all the same element, differing only in the number of neutrons in the nucleus, and therefore in weight. The revelation that elements could exist in such different forms with no chemical difference between them cleared up several mysteries that had exercised chemists for generations. Soddy called these forms isotopes, and for their discovery he was rewarded with the Nobel Prize for chemistry in 1921. It did surprisingly little to assuage his bitterness at the earlier slight (as he conceived it). By then he had been appointed to the chair of physical chemistry in Oxford, but he did not thrive there. His plans for the reform of research and teaching met with obstruction by the college tutors and he withdrew into a prolonged sulk. He did no more research and his department atrophied, while he devoted himself to developing a universal monetary theory and other equally fruitless concerns. Finally, at the age of 59, he resigned his chair and ended his life in embittered and paranoid obscurity. Frederick Soddy died in 1956.

Soddy related his reaction to the discovery of transmutation to his first biographer and friend, Muriel Howorth, in *Pioneer Research on the Atom: The Life of Frederick Soddy* (New World Publications, London, 1958). See also the more recent and less adulatory biography by Linda Merricks, *The World Made New: Frederick Soddy—Science, Politics and the Environment* (Oxford University Press, Oxford, 1996).

~

150 Fruits of the sea

The discovery of anaphylaxis was a turning-point in immunology. This phenomenon—the often lethal reaction of a sensitized subject to a minuscule amount of an agent, such as a bee sting, penicillin, a morsel of shellfish, or a grain of hazelnut—is linked to the name of Charles Richet (1850–1935), whose researches into the subject began on the ocean-going yacht of Prince Albert I of Monaco. The Prince, a reforming monarch, who transformed an impoverished Mediterranean fishing port into a prosperous democracy, was a passionate marine biologist. On his yacht he had installed a lavishly appointed biological laboratory, and he would invite biologist friends to accompany him on his cruises. One such was Richet, then Professor of Physiology at the Sorbonne, who had been occupied for some years with problems of immunization. He had noted that dogs injected repeatedly with foreign blood serum sickened and sometimes died. Also on board the yacht that summer of 1902 was another of the Prince's scientific friends, Paul Portier, a physiologist and later professor at the

Oceanographic Institute in Paris. Here, in Richet's words, is how the project came about:

> During a cruise on Prince Albert of Monaco's yacht the Prince and G. Rickard suggested to P. Portier and myself a study of the toxic properties of Physalia [the Portuguese man-o'-war] found in the South Sea. On board the Prince's yacht experiments were carried out, proving that an aqueous glycerine extract of the filaments of Physalia is extremely toxic to ducks and rabbits. On returning to France I could not obtain any Physalia, and decided to study comparatively the tentacles of Actinaria which resemble Physalia in certain respects, and are easily procurable.

Richet was evidently fascinated by the extreme toxicity of the venom preparations that he and Portier had extracted, and he may have wondered whether they related to the sensitization effect, which had engaged his attention before. He decided, at all events, to immunize animals with the toxin to see what happened. From the tentacles of *Actinaria*, the sea anemone, he and Portier again prepared a toxin extract and injected it into dogs.

> While endeavouring to determine the toxic dose we soon discovered that some days must elapse before fixing it; for several dogs did not die until the fourth or fifth day after administration, or even later. We kept those that had been given a dose insufficient to kill, in order to carry out a second investigation upon them when they had completely recovered. At this point an unforeseen event occurred. The dogs which had recovered were intensely sensitive and died a few moments after administration of small doses.

Richet and Portier described the death throes of an unfortunate dog called Neptune, which expired 25 minutes after the decisive injection, 26 days after inoculation. The experimenters were astonished by the result, which they realized was at the root of allergic reactions in humans. Richet alone, who continued with this work, was rewarded with the Nobel Prize in 1913. In later life he became increasingly embroiled in parapsychology and was all too easily duped by the fraudulent mediums who abounded in Paris at the time.

For more background, see, for instance, W. D. Foster's *A History of Medical Bacteriology and Immunology* (Heinemann, London, 1970).

151 A slice of pi

The nature of π, the ratio of the circumference of a circle to its diameter, has been a source of frustration and fascination to mathematicians and philosophers for millennia. It is obvious from the dimensions of the squares drawn inside and outside the circle and touching its circumference that π must be greater than 2 and less than 4, but there is nothing to indicate that it is an irrational number—one that cannot be expressed as a ratio of two integers. The value of π has been calculated on high-speed computers to billions of places of decimals, but no pattern or recurrence has emerged [30].

A Canadian mathematician, Simon Plouffe, chose to memorize π to 4,096 digits (2^{12}, and therefore to a mathematician a nice round number). 'I was young', he explained when questioned, 'and had not much else to do.' The strangest attempt to rationalise π dates from 1894, when Edward Johnston Goodwin, a doctor and amateur mathematician of conspicuous self-esteem, who lived in a small town in Indiana, published in the *American Mathematical Monthly* an article with the title 'Quadrature of the circle'. In a succession of steps he derives a value for π of 3.2 (in place of 3.14159...), although from a close analysis of the arguments that he erects, no less than eight other values, ranging from 2.56 to 4, can be extracted. At all events, Goodwin notes in his article, that he has copyrighted his value of 3.2 in the United States, Britain, Germany, France, Spain, Belgium, and Austria. In 1896, he approached his representative in the Indiana state legislature, Mr Taylor I. Record, and called on him to bring a bill before the lower chamber, the Indiana House of Representatives, 'for an act introducing a new mathematical truth and offered as a contribution to education to be used only by the State of Indiana free of cost', while everywhere else a royalty would be exacted. In January 1897, the House Bill 246 to this effect duly reached the chamber, and after passage through two committees was approved by 67 votes to nil. In February, despite some derision in the local press, the bill was sent by the responsible committee to the upper house of the legislature, the Senate, 'with the recommendation that said bill do pass'.

At this stage a fortunate stroke of fate intervened in the shape of C. A. Waldo, Professor of Mathematics at Purdue University, who chanced to be in the House on university business. Waldo was astonished to find that a bill on a mathematical topic was up for debate in the Senate that very day. In an article written 19 years later he recalled:

An ex-teacher from the eastern part of the state was saying: 'The case is perfectly simple. If we pass this bill which establishes a new and correct value of π, the author offers to our state without cost the use of his discovery and its free publication in our school text books, while everyone else must pay him a royalty.' . . . A member then showed the writer a copy of the bill just passed and asked him if he would like an introduction to the learned doctor, its author. He declined the courtesy with thanks remarking that he was acquainted with as many crazy people as he cared to know.

With the urging of Professor Waldo, the Senators decided that the subject of the Bill was not after all a matter for legislation and it was postponed *sine die*. It may well, therefore, be still on the statute book of the State of Indiana.

The story has been several times told. An excellent account and analysis of Dr Goodwin's geometrical reasoning is by David Singmaster, 'The legal values of pi', *The Mathematical Intelligencer*, 7 (no. 2), 69 (1985); see also an internet article by Mark Brader (http://www.urbanlegends.com/legal/pi_indiana.html).

~

152 The last of the true amateurs

Science, and especially physics, has become too specialized and too expensive to shelter amateurs. The de Broglie brothers, Louis (who first gave expression to the particle-wave duality [113]), and especially his elder brother, the duc Maurice de Broglie, were among the last.

Louis-César-Victor-Maurice de Broglie (1875–1960) was born into an ancient and illustrious French family; the diplomatic corps and the army were the only careers thought fitting for the scion of such a noble line, and it was only after delicate negotiations with his grandfather, the head of the clan, that he was permitted to enter the navy. Posted to the Mediterranean fleet, his scientific bent soon asserted itself: it was Maurice de Broglie who installed the first shipboard wireless on a French warship. But he became restless at the limited opportunities for indulging his interests and asked leave to resign from the navy and devote himself to science. The grandfather was scandalized: science was an amusement for old men and not an occupation for a de Broglie. But, after pleading his case, it was agreed that the young Maurice might convert a room in the family *hôtel* in Paris into a laboratory in which to amuse himself when not at sea. It was only after the patriarch's death that Maurice, then 33, finally felt able to resign his commission (although he returned to work on submarine communications for the duration of the First World War). He then studied spectroscopy at the Collège de France and completed a thesis under a distinguished physicist, Paul Langevin, with whom he had worked on the submarine project. Thereafter Maurice de Broglie withdrew to his superbly equipped private laboratory, where he was joined by a succession of assistants, including his brother, Louis, who later won a Nobel prize. The Swiss crystallographer, P. P. Ewald, told in a lecture in 1953 that:

> Progress in x-ray diffraction came from many European countries in those early years [from just before the Great War to some years after]. Maurice de Broglie in Paris was quick in developing his own spectroscopic methods and in training co-workers like Trillat and Thibaut. Some members of this audience may remember the unique setup of his laboratory in his private *hôtel* in the rue Lord Byron where cables for the current came in by holes cut in the Gobelins adorning the walls.

The very last of the true amateurs, who never sought paid employment but made important contributions to science, was probably Alfred Loomis, born into a prosperous New York family in 1887. His liberal education at Yale did not incline him towards science, but he loved and built mechanical gadgets. He had a peculiar fascination for ballistics and, joining the army when the United States entered the First World War, he developed new methods of measuring the velocities of artillery shells. Several of the country's leading physicists came to the Aberdeen Proving Grounds to bring their skills to bear on matters of concern to the military, and Loomis got to know several of them. He developed a special rapport with R. W. Wood [45], who became a lifelong friend. Loomis had already entered into a career as a lawyer and banker, and his wealth grew until he was able to indulge his burgeoning interest in science by establishing a private laboratory in his mansion on Long Island. Wood told how this came about:

> Loomis was visiting his aunts at East Hampton and called on me one afternoon, while I was at work with something or other in my barn laboratory. We had a long talk and swapped stories of what we had seen or heard of science in warfare. Then we got onto the subject of postwar research, and after that he was in the habit of dropping in for a talk almost every afternoon, evidently finding the atmosphere of the old barn more interesting if less refreshing than that of the beach and the country club.

One day he suggested that if I contemplated any research we might do together which required more money than the budget of the physics department could supply, he would like to underwrite it. I told him about Langevin's experiments with supersonics [now called ultrasound] during the war and the killing of fish at Toulon Arsenal. It offered a wide field for research in physics, chemistry, and biology, as Langevin had studied only the high-frequency waves as a means of submarine detection. Loomis was enthusiastic, and we made a trip to the research laboratory of General Electric to discuss it with Whitney and Hull.

The resulting apparatus was built at Schenectady and installed at first in a large room in Loomis' garage at Tuxedo Park, New York, where we worked together, killing fish and mice, and trying to find out whether the waves destroyed tissue or acted on the nerves or what.

As the scope of the work expanded we were pressed for room in the garage and Mr Loomis purchased the Spencer Trask house, a huge stone mansion with a tower, like an English country house, perched on the summit of one of the foothills of the Ramapo Mountains in Tuxedo Park. This he transformed into a private laboratory deluxe, with rooms for guests or collaborators, a complete machine shop with mechanic and a dozen or more research rooms large and small. I moved my forty-foot spectrograph from East Hampton and installed it in the basement of the laboratory so that I could continue my spectroscopic work in a better environment.

The work on ultrasonics by Loomis and Wood was widely recognized as laying the foundations of this new subject. But Loomis, with and without Wood and a series of visitors, experimented in many areas of physics, primarily in the design of precision apparatus. He continued for many years to divide his time between Wall Street and his laboratory until he tired of the financial scene. He gave anonymous support to the American Physical Society and to indigent physicists, and, as the Second World War approached, he began to devote more and more of his time to military projects, especially the development of radar. His contributions to the work of the radar laboratory at the Massachusetts Institute of Technology encompassed both the technical and managerial aspects; he sat on or chaired several important committees and his close friendship with his cousin, Henry Stimson, the Secretary for War, gave him quick access to the centres of power and funding. His interest in science continued into old age, when he turned increasingly to the study of *Hydra*, a tiny freshwater creature, sometimes in association with his biologist son. He continued to invent gadgets, such as a rolling cart for delivering food to diners seated at his long table. Alfred Loomis died, laden with honours, in 1975.

For an account of Loomis's life and work see the article by Luis W. Alvarez, 'Alfred Lee Loomis—last great amateur of science' in *History of Physics*, ed. Spencer R. Weart and Melba Phillips (American Physical Society, New York, 1985). For the career of Maurice de Broglie see the entry by Adrienne Weill-Brunschwicg and John L. Heilbron in *Dictionary of Scientific Biography*, Vol. 1, ed. C. C. Gillespie (Scribner, New York, 1970).

~

153 Dr Pincus's pill

The birth-control pill is associated with the names of Gregory Pincus, John Rock, and Carl Djerassi—the physiologist, the doctor, and the organic chemist—and quite a number of others, for many man-years of patient research went into its development. Much of the initial impetus came from Margaret Sanger, who had sought to liberate women from the tyranny of unwanted pregnancies, and the philanthropist Katherine McCormick. The first success came in 1955, when Rock, a professor at Harvard Medical Scool and an expert on fertility, began cautiously testing the effect of the hormone analogue, progestin, on a plucky 'cagefull' of ovulating women, willing guinea-pigs all.

> Rock, as a practicing physician, instinctively emphasized the safety of the volunteers over the laboratory researcher's concern for the outcome of the experiment. As a result, Rock's caution reached beyond normal practices of the time. A young Yale-trained obstetrician assisting Rock in the trials, Dr Luigi Mastroianni, has recalled: 'I don't really think I sensed the true significance of what was being done. The concept of informed consent that is so talked about now, and is a legal requirement of any research project involving human volunteers, didn't exist then. But Rock practiced it before it was ever defined.' ...
>
> The work became known within the Rock team as the PPP—the Pincus Progesterone Project. But that was soon translated irreverently as 'the pee-pee-pee' project to honor Mastroianni's endless task of checking daily urine samples from each of the fifty subjects.
>
> The results were perfect. Not one of the fifty women ovulated. While requiring more proof than a trial with only fifty women, Pincus and Rock knew that they had identified an oral birth control pill.

Especially with as cautious a partner as Rock, Pincus was not about to run around shouting the triumphant news. Not yet. But he had to pour it out somewhere, to someone.

His wife, Elizabeth, who had a talent for distilling complex episodes into succinct summations, would never forget the moment her husband brought the news home. Using his intimate name for her, he said, 'Lizuska, I've got it'.

'What have you got?'

'I think we have a contraceptive pill.'

'My God, why didn't you *tell* me?'

He replied that he was telling her now.

'Did you think you could *ever* get the pill?' she asked in awe.

Pincus replied grandly, 'In science, Lizuska, everything is possible.'

The account is taken from Bernard Asbell, *The Pill* (Random House, New York, 1995).

~

154 The frivolous *philosophe*

The following story of Denis Diderot's embarrassing encounter with Leonhard Euler (1707–83), the celebrated Swiss mathematician, was attributed to Augustus De Morgan (1806–71), English mathematician and writer. It is repeated in many popular works of mathematics, and goes like this: the renowned *encyclopédiste* was visiting the Russian Imperial court, expounding his doctrine of atheism. His dissertations aroused much interest, mingled with outrage. One day he was informed that Euler, who had also been received at the Empress Elisabeth's court, was in possession of an algebraical proof of the existence of God. He would demonstrate it to Diderot before the court if Diderot were willing to attend. The *philosophe* readily consented. Euler was summoned, advanced towards Diderot, and in a tone of solemn conviction addressed him thus: '*Monsieur, $a + b^n/n = x$, donc Dieu existe: répondez!*' Diderot, who was innocent of all mathematical knowledge, was struck dumb. Mortified, and a prey to general mirth, he begged leave to withdraw and departed immediately for France.

Despite its wide currency the tale is, at best, inaccurate and, indeed, carries no conviction, for Diderot, a man of monumental learning and intelligence, would scarcely have allowed himself to be so foolishly duped. Moreover, he had taught mathematics and even made some modest advances in the subject. The origins are to be found in a book by De

Morgan, in which he makes reference to the memoirs of one D. Thiébault, entitled *Souvenirs de vingt ans de séjour à Berlin,* and published in 1804. Thiébault does not vouch for the veracity of the story, for he was not himself present at the encounter, but it was, he tells his readers, believed 'by the inhabitants of the North'. The year was 1774 and the place St Petersburg. Diderot's adversary in Thiébault's account was an anonymous 'Russian *philosophe savant mathématicien,* a distinguished member of the academy'. (Euler's name seems to have been introduced gratuitously by De Morgan.) The Russian savant spoke as De Morgan relates (though there are variants on the equation with which Diderot was taxed). Diderot, annoyed and affronted by the ridiculous prank, and fearing that he would be subjected to more in the same vein, left and soon thereafter returned to France.

The clarification was in an article, 'A story concerning Euler and Diderot', by Dirk J. Struik in *Isis* **31**, 431 (1939).

155 Koch on cooking

Paul Ehrlich (1854–1915), one of the founders of medicinal chemistry, is famous for inventing the first effective anti-microbial agent and the first drug to combat syphilis without the side-effects that were as destructive as the disease. One of his great discoveries was the so-called 'acid-fast' staining procedure for tubercle bacilli. This came about quite accidentally. The bacilli, which have a fatty coating, had resisted all attempts at detection by staining, until one day Ehrlich left a preparation on the laboratory stove. It must have been an unexpectedly cold day, for a member of the laboratory lit the stove. Ehrlich's dish was heated, the bacteria absorbed the hot staining compound, and came up in bold colour under Ehrlich's microscope. Robert Koch, the discoverer of the tubercle bacillus, wrote: 'We owe it to this circumstance alone that it has become the general custom to search for the bacillus in sputum.' This quickly established itself as the universal diagnostic method.

See M. Marquardt's *Paul Ehrlich* (Heinemann, London, 1949).

156 The heat of the light

Wilhelm Friedrich Herschel, eventually Sir William, was born in 1738 in Hanover, the son of a musician, in whose profession he was trained. At the age of 19 he travelled to England and soon established himself as a composer, conductor, teacher, and church organist in Bath. In 1766, he began to take a deep interest in astronomy and before long had built his own reflecting telescope. For this purpose he ground his own mirrors, fashioned from speculum, an alloy of copper and tin. Every spare moment, even during the intervals of concerts, Herschel would hasten to his workshop to put in a little time on his mirrors. The Astronomer Royal of the day, Nevil Maskelyne, invited to examine Herschel's telescope in Bath, pronounced it superior to any in London. Before long Herschel had made a series of discoveries, above all that of a new planet, Uranus. He wanted to call it Georgium in honour of George III, but this was vetoed by the Royal Society.

Herschel's fame spread and soon the King summoned him to Windsor as his house astronomer. The two Hanoverians hit it off at once, and the King remained a faithful patron. During his career Herschel constructed at least 400 telescopes with his own hands. He searched the heavens and came upon many nebulae, which he rightly conjectured to consist of clusters of stars. He discovered two moons orbiting his planet, Uranus, and he was the first to observe the existence of double-stars—two stars conjoined in orbit about their common centre of mass.

An observation of Herschel's that caused a great stir was the outline of the fixed stars: thanks to the high quality of his mirrors, which reduced optical distortion (aberration), the images of the stars were round and sharply defined, devoid of radiating rays. This excited the Hon. Henry Cavendish (1731–1810), arguably the greatest experimental scientist of the time, an inordinately shy and eccentric bachelor, who avoided human contact and even had a second staircase built in his mansion to avert the danger of encountering a servant face to face.

Sir John Herschel was fond of recalling this anecdote which he had heard from his father in this connection. At a dinner given by Mr Aubert [a respected amateur astronomer, who had built a private observatory at Deptford, just outside London, and installed there a reflecting telescope of a type known as Short's Dumpy and considered to be as good as any in England] in the year 1786. William Herschel was seated next to Mr Cavendish, who was reputed to be the most taciturn of men. Some time passed without his uttering a word, then he suddenly turned to his neighbour and said: 'I am told that you see the stars round, Dr Herschel.' 'Round as a button', was the reply. A long silence ensued till, towards the end of the dinner, Cavendish again opened his lips to say in a doubtful voice: 'Round as a button?' 'Exactly, round as a button', repeated Herschel, and so the conversation ended.

During much of his life Herschel's spinster sister, Caroline, kept house for him and assisted in his observations, becoming eventually a skilled astronomer herself. The King appointed her assistant astronomer with an emolument of £50 per annum. For her several discoveries, and especially for editing and extending John Flamsteed's famous catalogue of heavenly bodies, first published 60 years earlier, she was awarded the Royal Astronomical Society's gold medal. (Much acrimony, it might be added, had attended the publication in 1712 of Flamsteed's great compilation, which he called *Historia Coelestis*. Flamsteed was a compulsive checker of data and would let nothing out of his Greenwich observatory without first mulling over it for years. This exasperated Isaac Newton, who, urgently wanting the data, accused Flamsteed of withholding information that he was bound by the terms of his public office to make available. Newton finally obtained what he needed by a subterfuge and secured the publication of all Flamsteed's data. The enraged and chagrined Flamsteed accused Newton of theft. He managed to retrieve 300 of the 400 printed copies and burned them.)

It was on 11 September 1800 that William Herschel made his most remarkable and serendipitous discovery. He set out to determine whether light engendered heat, and how this effect might vary with colour. He caused a beam of sunlight from a slit in the blackout of a darkened room to fall on a prism and projected the emerging rainbow colours onto a screen. Before each component of the light in turn he placed his thermometer and waited for it to record an increase in temperature. Presently he went to lunch and found on his return that the Sun had moved on and the thermometer was now beyond the red end of the spectrum of colours. But to his astonishment the temperature had risen. Herschel quickly realized that the heat came from a radiation which he could not see. He had discovered the infrared, the source of radiant heat.

William Herschel was knighted in 1816 and in 1821, the year before his death, was elected President of the Royal Astronomical Society. His son, John, also later knighted, followed him in his vocation, and achieved fame especially for his studies on photometry, the measurement of the intensity of light from stars.

See the biography by C. A. Lubbock, William Herschel's grand-daughter, *The Herschel Chronicle: The Life-Story of William Herschel and his Sister Caroline Herschel* (Cambridge University Press, Cambridge, 1933).

~

157 A world of science in a teacup

R. A. (later Sir Ronald) Fisher was one of the founders of applied statistics. He devised methods, standard to this day, of analysis of biological data and of design of such undertakings as clinical trials of drugs. Fisher was born in north London in 1890 and in 1910 he began his research career at the Rothamsted Experimental Station near London, then and still a centre of agricultural research. Four years earlier the life of the laboratory had been enriched by the arrival of its first female staff member, Muriel Bristol, an authority on algae. It was out of consideration for her that the custom of afternoon tea in the common-room was initiated.

One day, soon after his arrival at Rothamsted, Fisher courteously handed Dr Bristol a cup of tea, which, however, she declined, protesting that she preferred the milk to be added to the tea, rather than the tea to the milk. (This fine distinction was long seen as a shibboleth of English social class.) Fisher was astonished: surely she, as a scientist, could not believe that it made any difference to the taste. But Dr Bristol insisted: she could certainly tell the difference. Fisher decided to put the matter to the test with a blind tasting, and he and William Roach, a chemist at the laboratory, devised a trial. The question was resolved in Dr Bristol's favour: she could indeed tell the difference (though with what statistical certainty is not recorded). It was this episode that caused Fisher to ponder the general principles of statistical evaluation, which culminated in his magisterial treatise, *Statistical Methods for Research Workers*, published in 1925.

Much later Fisher fell into bad odour for rejecting the evidence of a causal link between smoking and lung cancer. His explanation for the correlation was that there was a common genetic disposition towards smoking and cancer. Fisher was then, it became known, in the pay of a tobacco company.

The episode of the milky tea appears in an article by George V. Mann, 'Chance encounters', in *Perspectives in Biology and Medicine*, **25**, 316 (1982).

~

158 A copper or two

Rudolf Schoenheimer was a German biochemist of exceptional accomplishment. As a Jew, he lost his position in Germany before the Second World War and found refuge, along with many others in the same predicament, in the medical school of Columbia University in New York. There the Head of the Biochemistry Department, Hans Thatcher Clarke, assembled a brilliant, polyglot, and captious *galère* of Europeans.

The years immediately after the war saw a succession of remarkable advances in the chemistry of physiological processes, thanks in large part to the advent of radioactive isotopes [149]; it was now possible to render radioactive, and thus label, substances involved in metabolism and follow their chemical transformations in a cell or in an animal. But radioactive isotopes were still in short supply and precious. Schoenheimer wanted to experiment with radioactively labelled urea, the metabolic end-product excreted by animals and man. A leader in the field of isotope purification was Harold Urey, who agreed to make available to Schoenheimer a minute quantity of ammonium nitrate, greatly enriched in the isotope of nitrogen with an atomic weight of 15, which makes up only a vanishingly small proportion of the Earth's nitrogen, with its atomic weight of 14. Urey had prepared the material from bulk ammonium nitrate, a substance that can explode if provoked, which he had illicitly driven into New York from a plant in New Jersey through the Holland Tunnel in the rumble-seat of his coupé. The glass ampoule, which he presented to Schoenheimer, contained the bulk of the world's supply of the purified isotope, ^{15}N.

To turn the ammonium nitrate into urea was the task of Schoenheimer's young research assistant, DeWitt Stetten. Schoenheimer and Stetten had first to decide on a method of synthesizing urea; there were many such, for urea had been the very first organic molecule ever prepared in the laboratory from an inorganic starting material (ammonium cyanate). They began, as organic chemists still do, by consulting the German bible of synthetic organic chemistry, Beilstein's handbook, and settled on a seemingly simple procedure: ammonia, generated from a solution of the ammonium nitrate, was to be bubbled through molten diphenyl carbonate. The reaction would yield urea with 100 per cent efficiency and none of the precious ^{15}N would be lost. Stetten set up the experiment and proceeded to try out the method with some ordinary ammonium nitrate. Here is how things fell out:

> To my deep chagrin, apparently no reaction whatsoever occurred. Ammonia that was introduced into the system passed out through the diphenyl carbonate without chemical alteration. I naturally tried all likely variations of the procedure, but to no avail, and was finally forced to the conclusion that the German author who had described the method had simply lied. Deeply aggrieved, I so reported to Rudi. 'When', he asked me, 'was the synthesis reported?' 'In 1880', I replied. With a dash of chauvinism, Rudi answered, 'In 1880 there was no need for German organic chemists to lie.' He next explored with me the details of the original description of the synthesis, and we noted that, whereas I had been working with small amounts of diphenyl carbonate, the original description called for kilogram quantities of reagents. Suddenly Rudi's face broke into a grin. 'I can remember when I was a graduate student in Thomas's laboratory in Leipzig, there were hanging high on the walls and covered with dust large copper retorts and other reaction vessels. Glass in those days was not as strong as it is today. Reactions involving kilogram quantities of reagents, particularly when run at elevated temperatures, were usually done in copper vessels. Could it be that copper might catalyze this reaction?' I therefore undertook to do the synthesis once more with the now additional feature that a small amount of copper would be added to the diphenyl carbonate. I found, in the chemical stockroom, a small quantity of finely divided metallic copper which had apparently been prepared to be added to paint to impart a metallic sheen. A pinch of this copper dust was all that was required. The ammonia now generated was indeed quantitatively taken up by the diphenyl carbonate, and the urea was formed in the expected yield.

The synthesis of the labelled urea was the first use of the ^{15}N nitrogen isotope and opened a new chapter in biochemistry. As time went on Rudolf Schoenheimer began to show signs of irrational behaviour and suffered intermittent bouts of deep depression, and one night in 1941, at the summit of a brilliant career, took his own life.

DeWitt Stetten tells the story of the urea synthesis in *Perspectives in Biology and Medicine*, 25, 354 (1982).

159 The savants and the chauvinists

The planet Uranus, discovered by William Herschel in 1781 [156], began in the nineteenth century to cause concern to astronomers. Alexis Bouvard [145] discovered that its observed course was deviating from the predictions based on Newton's laws, which took into account the gravitational influences of the Sun and of the other planets. Surely Newton's laws could not be in error? Was there perhaps, then, another unidentified planet affecting its motion? In 1843, acting on this supposition, the youthful John Couch Adams, recently graduated from Cambridge, began to calculate what the orbit of such a planet would be. The problem had been considered intractable, but after some three years of labour Adams came up with a rough answer. When he took the results to the Reverend Professor John Challis, director of the Cambridge observatory and guardian of its telescope, he was fobbed off with the advice that he should apply to the Astronomer Royal, Sir George Airy. Sir George was also less than helpful: more detailed calculations, he told Adams, would be required before he was prepared to initiate a search for a planet. These took the exasperated Adams another year.

Meanwhile, at the École Polytechnique in Paris, Urbain Jean Joseph Le Verrier (1811–77) had set out with the same design and in June of 1846 he published his predictions for the position and likely size of the mysterious planet. Le Verrier, too, had difficulty in animating the star-gazers of France, and so addressed himself to the Director of the Berlin Observatory. Le Verrier's letter arrived in Berlin on 23 September and that same night an assistant, Johann Gottfried Galle, began the search. By a happy circumstance the observatory had commissioned and just then received a new and superior map of the heavens. With its aid Galle took only a matter of hours to locate the elusive body. Excitedly he wrote to Le Verrier, 'The planet whose position you indicated *really exists*. The same day I received your letter I found a star of the eighth magnitude that was not in the excellent Carta Horta XXI.'

When Le Verrier's publication appeared Airy must have felt some remorse and he quickly asked Challis in Cambridge to initiate a search. But Challis's star map was not the equal of the Carta Horta XXI, and only after Galle announced his success did he realize that he had also already seen the new planet. Adams, a modest and reticent man, bore no grudges and when he and Le Verrier met some time later in Cambridge they became friends.

The discovery of Neptune, as it was designated by the French Bureau of Longitude (despite Airy's insistence, on grounds of consistency with Greek mythology, on Oceanus, and Galle's preference for Janus), caused a sensation. In the first place it was a triumphant vindication of the power of Newton's laws and the correctness of Kepler's model of the planetary system. But more especially the dramatic vindication of the theoretical prediction caught the imagination of the public and indeed of the politicians, who were made aware for the first time of the value of scientific inquiry. It has been argued that this event marked the beginning of government interest in science.

Sir John Herschel [156], son of William, wrote a popular account of the discovery in the journal, *The Athenaeum*, making flattering mention of Adams's work, and this engendered a deplorable effusion of chauvinistic passion in the press. Adams's calculations had still not been published and in France suspicions of plagiarism by the perfidious English were easily aroused. The English press hurled counter-accusations of distortions of the truth, and theft by the French of a British discovery. Le Verrier complained bitterly to Airy that Herschel was trying to rob him of the credit for what was in truth a prodigious achievement: Le Verrier's calculations were more accurate than Adams's and had taken half the time. Airy replied in mollifying tones:

> Dear Sir—I have received your letter of the 16th [October 1846], and I am very sorry to find that a letter published by Sir John Herschel has given you so much pain. I am certain that Sir John Herschel would be equally sorry, for he is the kindest man, and the most scrupulous in his endeavours to do justice to all persons without giving offence to any that I ever saw . . .

More recriminations and exchanges in the same vein followed, though never between Adams and Le Verrier, for they acknowledged that their work had been done independently. Both received many honours; Adams in due course succeeded Airy as Astronomer Royal, and Le Verrier followed Arago [166] as director of the Paris Observatory. His brusque, irascible nature did not endear him to all his colleagues. Arthur Schuster (1851–1934), in later life Professor of Physics in Manchester, had occasion, as a young man, to visit the observatory. On arriving in Paris he first called on Marie Alfred Cornu, an optics specialist who, hearing that Schuster was proposing to seek advice from Le Verrier, shook his head mournfully. 'Je ne sais pas' he told Schuster, 'si M Le Verrier est l'homme le plus détestable à

Paris, mais je sais qu'il est le plus détesté.' Arriving at the observatory, Schuster was warned that the *patron* had just attended a funeral and was in a bad mood. 'Who are you and what do you want?', Le Verrier demanded when Schuster was ushered into his presence. Schuster gave his name and stated his business—to solicit help with the design of a mirror. Le Verrier indicated that he had already given instructions for one of his staff to give the visitor every assistance and dismissed him. When Schuster went at the end of his visit to make his farewells, the director was interrogating a terrified assistant over a failed calculation. The interview over, Le Verrier turned to Schuster and chuckled: he had known all along what the assistant's mistake was, but he wanted to see whether he could work it out for himself. Schuster was then taken for a stroll in the gardens and retained ever after a warm recollection of Le Verrier's gruff good humour.

There are many accounts of the discovery of Neptune; see, for example, *Astronomy* by Fred Hoyle (Macdonald, London, 1962); and (despite its title) *A History of Astronomy from 1890 to the Present* by David Leverington (Springer, London, 1995).

∼

160 The case of the flaccid ears

Lewis Thomas was a distinguished medical scientist, who became the head of the Sloan-Kettering Institute for Cancer Research in New York, and achieved wider fame for his collections of lucid and elegant essays on science, medicine, and related subjects. One of his first adventures in research occurred in 1936 when he was an ambitious medical student. Thomas had become intrigued by something called the Shwartzman phenomenon. The eponymous researcher had observed that when a rabbit's skin was injected with bacterial endotoxin (a rather mildly toxic secretion from certain bacteria), the only result was a little local inflammation; yet when the process was repeated, though only within the very narrow time interval of 18–24 hours, extensive skin lesions and haemorrhages ensued. If the second dose of toxin was injected into a vein nothing happened, but when both injections were intravenous the result was catastrophic, culminating in destruction of the kidneys.

A few days after reading a paper about the Shwartzman reaction, which showed a picture of a rabbit's necrotic kidneys, Thomas was sitting at a conference table in his professor's office during a weekly pathology seminar.

I forget what was being talked about, but I remember leaning back in my chair, bumping my head against a heavy glass jar on the shelf of tissue specimens behind me, and knocking it over. I picked it up to replace it and saw that it contained a pair of human kidneys with precisely the same lesion as the one in the photograph. The label said that the organs were from a woman who had died of eclampsia [elevated blood pressure due to toxaemia during pregnancy], with severe bacterial infection as well.

Thomas resolved to get to the bottom of the strange anomaly, and he and colleagues at four universities studied it for 10 years. They never did resolve all aspects, but they established how the blood supply was cut off to cause tissue death, and they showed that white blood cells were the primary agents of the mayhem. By eliminating these cells from the circulation (until a new batch could be formed a day later), or by inhibiting clotting of the blood, Thomas and his friends could cure the rabbits of the Shwartzman response. But in the course of their researches they ran into another interesting manifestation:

It occurred to us that the release of a proteolytic enzyme [one that attacks and digests proteins] by damaged tissue cells might be one way of rupturing small vessels, and we guessed that such an enzyme might be the sort most active in the acid environment which we knew existed in the rabbit's prepared skin. So, without much hard thinking, we injected small amounts of a plant enzyme of this type, papain (from papaya latex), into rabbit skin, and within an hour had a fair copy of the hemorrhagic necrosis of the local Shwartzman phenomenon.

This, we thought, was surely the way to go. The next step was to inject papain intravenously, in order to reproduce the generalized reaction, kidney necrosis and all. We did this, and nothing happened. The animals remained in good shape, active and hungry, and their kidneys were unblemished. We repeated it, using various doses of papain, always with negative results. But then we noted that the rabbits, for all their display of good health, *looked* different and funny. Their ears, instead of standing upright at either side, rabbit-style, gradually softened and within a few hours collapsed altogether, finally hanging down like the ears of spaniels. A day later, they were up again.

I am embarrassed to say how long it took to figure out what had happened. I first observed the papain effect in 1947, and examined microscopic sections of the affected ears without finding anything wrong in the cells, fibrous tissue, cartilage, or other structures in the ear, and dropped the matter. Every few months I returned to it, sometimes to demonstrate the extraordinary change to friends and colleagues, but never with any sort of explanation. It was not until six years later that it dawned on me that since a rabbit's ears are held upright by cartilage there simply had to be something wrong with the cartilage plates in those ears. I went back to it, comparing the quantity of cartilage matrix—the solid material between the cartilage cells— in the ears of papain-treated rabbits with normal rabbits, and found the trouble immediately: although the cartilage cells themselves appeared perfectly healthy, almost all the supporting matrix had vanished after papain. Moreover, the same changes occurred in all cartilaginous tissue, including the trachea, bronchial tubes, and even the intervertebral discs. Parenthetically, several years after my published report on this business, some orthopedic surgeons introduced the use of papain as a method of getting rid of ruptured intervertebral discs without surgery.

Thomas owns that there were no other clinical consequences of his observation, but recalls being interviewed by two sociologists, who, on discovering that another researcher had come upon the phenomenon but had not published on it or pursued it further, wanted to know why Thomas (and not the other) had followed it up. They later wrote a learned paper on the results of their inquiries, but Thomas, while doing his best to justify cherishing for so long such an essentially frivolous preoccupation, confessed in the end to no worthier motive than that he was amused by it.

The story comes from Lewis Thomas, *The Youngest Science: Notes of a Medicine Watcher* (Oxford University Press, Oxford, 1985).

∼

161 The good-natured philosopher

Albert Einstein (1879–1955) was one of the first Jewish scientists to leave Germany. He had been the target of savage vilification by the champions of what was known as the *Deutsche Physik*, notably two Nobel laureates, sympathizers both of Hitler's, who held that the purity of German science had been polluted by alien—that is to say Jewish—influences. They especially abhorred the new counterintuitive developments in theoretical physics, relativity, and quantum mechanics. Aryans who embraced these concepts were denounced as 'white Jews'. (Chief among these was Werner Heisenberg [180], who propounded, among other revolutionary new concepts, the Uncertainty Principle, anathema to the traditionalists.) A Jewish friend of Einstein's, the foreign minister Walter Rathenau, was assassinated by fascist thugs and Einstein himself had been threatened with the same fate. He now felt he had endured enough and left his country, never to return.

In the course of his peregrinations, before finally settling in Princeton, Einstein spent a brief period in Oxford. It was a troubled and anxious time, but Einstein could always find tranquillity in his thoughts. Once, in the United States, a young physicist begged an audience, and Einstein, as he always did, acceded. They arranged to meet at a street corner in New York, but the young man was held up by traffic and arrived an hour late. He was mortified and offered profuse apologies. Einstein smiled; it mattered not at all, he said, he could work at the side of the street with cars surging by as well as anywhere else. Here is a description of an encounter with Gilbert Murray, the classical scholar, in the Oxford college, Christ Church, where Einstein was lodging:

> Entering Tom Quad one day, Murray caught sight of Einstein sitting there with a far-away look on his face. The far-away thought behind that far-away look was evidently a happy one, for, at that moment, the exile's countenance was serene and smiling. 'Dr Einstein, do tell me what you are thinking', Murray said. 'I am thinking', Einstein answered, 'that after all, this is a very small star.'

Murray's colleague Arnold Toynbee, who gave the story to posterity in his memoirs, interprets Einstein's meaning like this:

> All the universe's eggs were not in this basket that was now infested by the Nazis; and, for a cosmogoner [sic!], the thought was convincingly consoling.

Einstein evidently also found solace in what his General Theory of Relativity told him about the nature of eternity. 'For a believing physicist', he wrote towards the end of his life, 'the distinction between the past, the present and the future is only an illusion.'

Einstein was amiable and unassuming and had a puckish sense of humour. His penchant in later life for the occasional Delphic utterance, which could engender mystified debate among his admirers, is nicely caught in the following reminiscence by the science journalist J. G. Crowther. The year was 1948.

> [Oswald] Veblen [the mathematician and director of the Institute of Advanced Study in Princeton] [64] took me across the Princeton campus to

Fine Hall. On the way he suddenly remarked, 'There's Einstein, you must meet him'. We sprinted gently up to him, and Veblen introduced me as 'The Scientific Correspondent of the *Manchester Guardian*'. Einstein bowed and said, 'The *Manchester Guardian* is the greatest newspaper in the world'. That was all he said, and he then walked on. After I had returned to London, I was rung up by one of the stalwarts of the *Manchester Guardian* in its London office, who asked me whether I had heard rumours about Einstein. 'What rumours?' I asked. 'Well, there is a rumour in Fleet Street that he has gone mad.' 'If that is so,' I said, 'it is unfortunate for the *Manchester Guardian*.' 'What do you mean?' my interlocutor asked. 'I met him a short time ago, and he told me "the *Manchester Guardian* is the greatest newspaper in the world".' 'Oh,' said the *Manchester Guardian* man solemnly, 'it is evident that our information is incorrect.'

The first two extracts are from Arnold Toynbee's *Acquaintances* (Oxford University Press, Oxford, 1967); J. G. Crowther's reminiscence is in his book *Fifty Years with Science* (Barrie and Jenkins, London, 1970).

162 A myth and its genesis

An anecdote in circulation for two decades, told anew and embellished wherever biologists meet, is variously attributed but seems to have appeared in print only once:

A worker wrote to another laboratory asking for a λ-phage which had recently been described from there. A letter came back refusing the phage, making it clear the worker making the request was not a proper member of the club. Despite this obvious lack of qualification to receive the gift, the miscreant had the effrontery to realise that such phages 'get around in laboratories', and he succeeded most elegantly in culturing the λ-phage by incubating the refusal letter itself. The end of the story is veiled in obscurity, but one somehow hopes the director of the refusing laboratory has now added fumigation of all outgoing letters to his other censorship duties.

The lambda-phage referred to is a DNA-containing bacteriophage [55], which has had wide use in genetic engineering. The story, which has achieved the status of an urban myth, probably owes its enduring appeal to the pre-eminence in the field of a small and brilliant coterie of insiders, who developed much of molecular genetics from the beginning. It has its genesis in a prank by Sydney Brenner, a leader of that group and a member of the famous Laboratory of Molecular Biology, still one of the dominant intellects of molecular biology. Brenner has divulged all in an article.

The bacteriophage in question was called f2 (and was, unlike lambda, an RNA-, not DNA-containing type), discovered in New York sewage by a geneticist, Norton Zinder. Hearing about it, Brenner thought of asking for a sample, but held back, suspecting that Zinder would not believe him if told the truth—that Brenner wanted it for research on bacterial 'sex factors' (genetic elements found in some bacteria, which can be transferred to a recipient bacterium); Zinder would assume that it was into *his* interest— the replication of RNA, the phage's genetic material—that Brenner would in reality want to intrude. Others, too, were interested in the f2 bacteriophage, and it was then that Brenner started the canard about culturing a letter from Zinder, hinting possibly that he had already done it. Brenner had, in fact, devised the inverse thought-experiment of salting letters, whether to rivals or to nuisances making tiresome requests, with another bacteriophage, called T1: this is a virile and rugged invader, which withstands even drying, so that if infiltrated into a laboratory it would establish itself in all cultures, and all research involving bacteriophages would come to an abrupt and catastrophic halt.

Brenner, at all events, did not culture any letter, but decided he could find his own RNA bacteriophages in the local sewage works. His illustrious friend François Jacob, he adds, found his bacteriophages in local Parisian pharmacies, where preparations, derived from sewage, were sold as remedies for gastrointestinal complaints. Now that so many pharmaceutical materials are produced in genetically modified bacteria, biotechnology companies are naturally very careful to sterilize laboratory effluent; indeed Brenner recollects looking in a sample sent to him of the commercially valuable protein, interferon, prepared in genetically engineered bacteria. He wanted to see whether some live bacteria remained that he could culture, but, alas, there were none.

The first passage, which does not identify the supposed protagonists, is from an article by A. C. Fabergé, 'Open information and secrecy in research', in *Perspectives in Biology and Medicine*, 25, 263 (1982). For Sydney Brenner's article, see 'Bacteriophage tales', *Current Biology*, 7, R736 (1997).

163 'Where the hormones there moan I'

Casimir Funk (1884–1967), the Polish biochemist, is now remembered (if at all) for giving us the word vitamin, but he deserves better of history. Funk began his career in Warsaw and then emigrated to France, where he worked in a pharmaceutical concern. Later he founded his own industrially supported laboratory in a suburb of Paris, the Casa Biochemica, and there from 1928 to 1939 he turned his attention to hormones. He isolated a male sex hormone preparation from the gonads of cocks, and soon thereafter he reported finding traces of sex hormones in the animals' blood. He also sought and found hormones in the blood and urine of pregnant women, and was able to distinguish chemically between male and female hormones. When he reported these findings at conferences in the United States, he encountered considerable scepticism.

A man of some initiative, Funk caused a minor diplomatic incident during the Italian Abyssinian (Ethiopian) war in 1936. He read in a newspaper report that the Ethiopian irregulars had the custom of castrating their Italian prisoners: here, then, was the ideal source of material that he was after, and he attempted to negotiate with the Ethiopian regime for a consignment of severed testicles. The Italian government came to hear of it and Funk's plan was construed as an insult to the flag. Moreover, the Ethiopian tribesmen evidently had other uses for the material, for Funk appears never to have received any of the precious offal.

When the Second World War approached Funk decamped to the United States, and there continued his work on sex hormones—preparing milligram quantities from hundreds of gallons of urine—and on other biological compounds until his death.

The life and career of Casimir (Kazimierz) Funk are described in *Casimir Funk und der Vitaminbegriff* by Bernhard Schulz (thesis, University of Düsseldorf, 1997), but his skirmishes with the Italian government are omitted.

∼

164 The disagreeable man

One of the villains of twentieth-century science lived and thrived in Germany, though born in 1877 in Switzerland. His name was Emil Abderhalden and he was a student of Emil Fischer, the great organic chemist. Fischer was at the time preoccupied with the structure of proteins and had developed methods of synthesizing peptides—amino acids strung together in a chain of the kind found in natural proteins; but whereas proteins consist of chains of hundreds, even thousands of amino acids (though of only 20 kinds) in defined order, the chemistry of the day could assemble only a very few. When Abderhalden achieved independence as a professor in a veterinary school in Halle, he continued in the same line of work and his assistants synthesized peptides in vast numbers. To do something useful with these materials Abderhalden turned to the study of proteolytic enzymes, the digestive enzymes that break down proteins into small fragments.

In 1909, Abderhalden announced his most egregiously spurious discovery: when foreign substances enter the body, he claimed, new enzymes are generated that destroy the alien molecules. It was already established that the body does, indeed, have defence mechanisms in the form of antibodies, but what was then known about the immune system bore no relation to Abderhalden's 'protective' or 'defence' enzymes, as he called them. The scope of the discovery quickly widened, and in due course impinged on medicine. For Abderhalden now announced that fetal proteins enter the bloodstream of pregnant women, inducing the formation of the defence enzymes. Here, then, was an early test for pregnancy. Abderhalden's procedures were taken up by clinical laboratories and apparently confirmed, but the enzymes were a mirage, as was recognized by the more respectable biochemists who had taken an interest in the matter. Their objections were met with abuse from Abderhalden, who was by now a public figure with a powerful and consistently malign influence in the German academic world.

The advance of the illusion could not be halted: protective enzymes were induced by tumours and by other medical conditions, including neurological diseases, and innumerable papers from many clinical centres were published on the subject. Worse, the Nazi anthropologist, the Freiherr Otmar von Verschuer and his favourite disciple, Josef Mengele of infamous memory, embarked on a study of the defence enzymes of different races, and when the time came procured samples from Mengele's fiefdom in Auschwitz. It was only in 1947 that, at a conference in Germany devoted to the subject of defence enzymes, it was resolved that their existence was at best unproven. Abderhalden died in 1950, but in some corners of the German academic system work on his enzymes continued for some years, championed by his son.

Abderhalden's personality is encapsulated in the following anecdote, related by Professor John Edsall of Harvard University. As a young man in the 1920s Edsall had spent some years in the Cambridge laboratory of Sir Frederick Gowland Hopkins, one of the leading biochemists of the day. There he met a young English biochemist, recently returned from a stay in Germany, who related to him his experience in Abderhalden's laboratory. Arriving in Halle to begin a year of postdoctoral work, he told the Herr Professor about the research project he had recently completed in Cambridge. '*Wann publizieren Sie, Herr Doktor?*' Abderhalden asked, obviously intrigued. Soon, was the response, a manuscript had already been drafted. The Englishman then left for a climbing holiday in the mountains in the south. Forewarned, he had locked his papers into a desk drawer, reinforced by a stout padlock. On his return he found the lock forced and his manuscript missing. On pursuing the matter he discovered that his paper was already in press, unchanged, except that Abderhalden's name now stood at the head of the list of authors. It scarcely counts as mitigation that in many laboratories at the time, in Germany and elsewhere, it was the custom to attach the professor's name to all publications emerging from his department; but this would hardly have applied to work done in another laboratory in another country.

Abderhalden's infamous career, and in particular the sorry tale of his imaginary enzymes, is portrayed by Ute Deichmann and Benno Müller-Hill in their article, 'The fraud of Abderhalden's enymes', *Nature* 393, 109 (1998).

165 The crackle that made history

In January of 1891 Sir William Preece, Chief Engineer of the British Post Office, opined in a newspaper interview that 'we have done as much with wireless telegraphy as is likely to be done'. Ten years later, on a windswept eminence in Newfoundland, Guglielmo Marconi clasped a telephone receiver to his ear and heard above the crackle of static a signal cast into the void at Poldhu in Cornwall 1,800 miles away. Preece had asserted—and many experts agreed with him—that 'bridging the Atlantic' was a pipe-dream, for 'the curvature of the earth will send the waves out into space'. Preece, it must be said, seems to have enjoyed a remarkable record where prediction was concerned. When Alexander Graham Bell exhibited his first telephone Preece gave evidence before a committee of the House of Commons. His confident evaluation was: 'Americans have need of this invention, but we do not. We have plenty of messenger boys.' (Americans, by contrast, were on the whole cautiously optimistic. 'One day', said the mayor of Chicago after witnessing a demonstration of the instrument, 'there will be one in every city.' A Senator, on the other hand, when told that Maine would soon be able to speak to Texas, riposted, 'What should Maine have to say to Texas?')

The young Marconi (he was 27 and entirely self-educated), whose Anglo-American Telegraph Company was barely solvent, had installed himself in an old hospital overlooking the harbour of St John. His purpose was to detect the transatlantic signal, but to maintain secrecy he had let it be known that he wished merely to communicate with a ship, the SS *Lucinda*: he was testing a system for preventing shipwrecks. This would have caused little surprise, as signals were already commonly transmitted over a distance of a hundred miles or so. On Thursday, 12 December a fierce gale was blowing, but Marconi resolved to press on and hoisted his aerial attached to a kite to a height of 400 feet. He had chosen to use a telephone receiver as a detector on the grounds that the ear would better discriminate than any apparatus a signal of faint clicks against a strong background of static. Marconi recalled later:

> Suddenly, about half past twelve there sounded three sharp little clicks of the 'tapper', showing me that something was coming, and I listened intently.
>
> Unmistakably, the three sharp little clicks corresponding to three dots [the letter S in Morse code] sounded several times in my ear; but I would not be satisfied without corroboration.

Marconi passed the receiver to his assistant, George Kemp, who confirmed what he had heard.

> I knew then that I had been absolutely right in my calculations. The electric waves which were being sent out from Poldhu had traversed the Atlantic, serenely ignoring the curvature of the earth [refracted in fact by a dense layer of the atmosphere], which so many doubters had considered would be a fatal obstacle.

When the result was announced in the press the local authorities immediately and indignantly expelled Marconi and his little team, for the

Anglo-American Telegraph Company, they stated, had no business transmitting or receiving signals on their territory. The enthusiasm created by Marconi's triumph, moreover, was less than unanimous. How could he be so certain that he had heard clicks and not atmospheric disturbances? Preece and Sir Oliver Lodge (the physicist, who had discovered radio waves independently of Hertz [**94**]) were among the sceptics; and the churlish Thomas Alva Edison called the report 'a newspaper fake'. Marconi prevailed and in time was created *marchese* and joined the Fascist party.

See *Marconi and the Discovery of Wireless* by Leslie Reade (Faber, London, 1963) and *Marconi: A Biography* by W. P. Jolly (Constable, London, 1962).

~

166 The physicist's peregrinations

If the French revolutionaries cut off Antoine Lavoisier's head, declaring that the Revolution had no need for scholars [**90**], the Republic nevertheless upheld the primacy of reason and encouraged the growth of science and technology. One of its monuments is the metric system of measurement, which succeeded in all spheres except that of time. The standard of length laid down, the metre, was to be one ten-millionth of the separation of the North Pole from the Equator along the Paris meridian. The Bureau des Longitudes was charged in 1806 with determining this length with the greatest possible precision. Some preliminary measurements based on the distance from Dunkirk to Barcelona were made and a provisional metre bar already reposed in Paris, but greater certainty was demanded: the measurements were to be extended to the Balearic Islands, through which the meridian also ran, passing south from Barcelona. Dominique François Jean Arago (1786–1853) and Jean-Baptiste Biot (1774–1862) were chosen for the task; Arago was then 20, Biot 12 years older.

The distance of the islands from Barcelona precluded direct observation of flares. Therefore

Biot and Arago took triangulations from mountain tops down the Spanish coast to Denia, then across the shorter stretch of sea from there to Ibiza and Formentera—still a difficult task with visual observations—and then finally to Majorca. On Ibiza they used the mountain Camp Vey, and on to Formentera, the highest point on the island, La Mola. At the end of 1807, Biot returned to Paris with the observations from Ibiza and Formentera, and left Arago alone to complete the final readings from Majorca. On Majorca, Arago chose S'Eslop, a peak on the northwest coast, as his viewpoint of Ibiza and Formentera. He had a hut built on the summit and settled in with his instruments for the final series of measurements. But events didn't go according to plan.

War broke out between France and Spain in June 1808, while Arago was on the summit of S'Eslop. Soon Majorcans were commenting that the nightly bonfires were signals and that Arago must be a French spy, and a detachment of soldiers was sent up the mountain to capture him. Arago got wind of this; in his memoirs he recounts what happened next: 'We set off for Palma and we encountered the troops that had come to look for me. Nobody recognised me because I spoke Mallorquin perfectly. I urged the platoon to continue on the path, and we continued on our route towards the city.' (Arago spoke Mallorquin, a version of Catalan, as he was born in the French Pyrenees, a Catalan-speaking region of France.) His escape was only temporary, though—he eventually ended up in Bellver Castle, overlooking Palma de Majorca. Now a tourist attraction, it was then a prison.

In the end, Arago managed to persuade the authorities that he wasn't a spy, and left the island for Algiers. From there he took a ship headed for Marseilles, but his bad luck continued—the ship was intercepted by Spanish pirates and escorted to Catalonia, where he was again imprisoned. Once more, he got himself released, and again set sail for Marseilles. This time it was not pirates but bad weather that intervened—it was now December 1808. The ship was forced by a storm to land at a small port in Algeria and was unable to make the winter crossing to Marseilles, so Arago headed back overland to Algiers. Here he was held captive again, this time as a hostage by the Algerians until the French paid for goods sent to France.

This was resolved in July 1809, and after a year-long odyssey, Arago finally arrived back in France to file his scientific reports, and received a triumphal reception in Paris. Arago and Biot's labours confirmed the accuracy of the original measurements; in the end, the prototype metre differed from the original meridian definition by just 0.02%.

Arago and Biot's meridian is marked in Paris by a series of plaques in the pavements. Both Arago and Biot went on to distinguished careers in physics (see also [145]). Biot's name is commemorated in Biot's Law and the Biot–Savart Law, describing properties of light propagation. It was Biot who was invited to look into the veracity of Pasteur's famous experiment on the separation of optical isomers [129]. Arago was also responsible for many advances in optical physics, among his innovations being the Arago prism and the Arago disc. He made his mark in politics as a minister of the Republic. He was a friend of Jules Verne, who used his adventures in the Balearic Islands in a novel.

On Formentera, this fiction is remembered more clearly than the reality, because on La Mola stands a monument—not to Arago and Biot, but to Verne, who probably never visited the islands.

A notable biography of Arago remains untranslated: *Arago: La Jeunesse de la Science* by Maurice Daumas, 2nd edn (Belin, Paris, 1987, first published in 1943); the passage reproduced above is from an excellent and entertaining article by Julyan Cartwright, *Nature*, 412, 683 (2001).

~

167 From mania to miracle

Of all the drugs discovered in the past 50 years or so, lithium has probably done the most good. Lithium chloride, closely similar to sodium chloride, common salt, is taken in large quantities by those with clinical depression and related conditions. It is cheap, essentially devoid of long-term side-effects, and has made life tolerable for many desperate people. There is no logical route by which its efficacy could have been divined and it came into being by a bizarre train of faulty reasoning.

Dr John Cade was a psychiatrist in a small medical centre in Australia. He conceived the idea that manic disease is caused by an endogenous toxin. Were this so, one might then reasonably expect that, like many known toxins, it would be continually eliminated from the body and would show up in the urine. It was not an unreasonable conjecture, especially since there had been reports (albeit later disproved) of a characteristic component in the urine of schizophrenics. Cade decided to test for an excreted toxin by injecting the urine of his patients into guinea-pigs. The animals became sick, but equally so when normal control urine was injected. Cade did not repine, but took the curious step of trying pure urea, the primary excreted metabolic end-product that makes up the bulk of dissolved matter in urine. The guinea-pigs fared even worse, in fact died when quite low concentrations of urea were injected, presumably of kidney failure. In any case it turned out, not surprisingly, that the concentration of urea was no higher in the urine of patients than in that of normal subjects.

The logic now becomes completely opaque, for Cade chose next to challenge the guinea-pigs with uric acid. This substance is chemically related to urea, and is the product of excretion of some animals, especially birds. But uric acid is insoluble. On the other hand, as Cade discovered from a visit to the library, its lithium salt does dissolve in water. No sooner thought of than tried, the lithium salt [65] proved innocuous, even mitigating the toxic effects of urea, and exerted a 'calming effect' on the justifiably agitated guinea-pigs. Now Dr Cade came down to earth: he asked himself whether the beneficial effect might even have been due to the lithium and not the uric acid at all. He reached for a bottle of lithium carbonate, and indeed it soothed the rodents. Thus encouraged, Dr Cade tried lithium carbonate on a patient. The results were miraculous and the seriously deranged patient returned to something approximating a normal mental state. This, to be sure, was not a clinical trial, for which the means could not have been found, but a paper was written, published in an insignificant journal in 1949 and at once sank from view, to be rediscovered five years later in the course of a literature search by a Danish research worker, Mogens Schou. Schou thought the report might be worth following up, with the consequence that lithium is now celebrated as a triumph of clinical science. But patients with manic disease have no toxin in their urine and Dr Cade's guinea-pigs became sluggish only because the lithium carbonate made them sick.

See Alexander Kohn, *Fortune of Failure: Missed Opporunities and Chance Discoveries in Science* (Blackwell, Oxford, 1989).

~

168 Madame la savante

Gabrielle-Émilie le Tonnelier de Breteuil, marquise du Châtelet, born in 1706, was the first to render the published work of Isaac Newton [15] into French. Her translation, with explanations, of Newton's most important work, the *Principia*, established her reputation as a savante. It burst on the French intellectual scene like a bombshell, and very soon Newton's explanation of planetary motion as governed by gravitational attraction had supplanted Descartes's theory of 'elementary vortices' and had changed the direction of mathematical thought in France.

Mme du Châtelet had befriended and enchanted Voltaire, then at the height of his fame, who took up residence at her husband's Château of Cirey. She came to public notice when, in 1736, she and Voltaire entered a prize competition established by the Academy of Sciences. She wrote a *Dissertation sur la nature et la propagation du feu*. She and Voltaire set up a laboratory at Cirey for the purpose of this study and burned and weighed a variety of materials, including metals, wood, and vegetables. The results were less than conclusive, for some substances lost and others gained weight, and little could be said about the 'weight of fire', but Mme du Châtelet's contribution must have been judged to have merit, because although she did not win the prize (which was shared by Leonhard Euler [154] and two lesser mortals) the Academy made favourable mention of her contribution in its report: 'submission number 6', it stated, 'is by a lady of high rank, Mme la Marquise du Chastellet'. This announcement was enough to establish her as a public figure. She went on to conquer new heights. 'She read Virgil, Pope and algebra', it was said, 'as others read novels.' Her mathematical facility was clearly exceptional. It was whispered with awe that she could 'multiply nine figures by nine others in her head', and no less an authority than Ampère [17] called her a genius in geometry. Cirey became a place of pilgrimage for many leading European scholars, and its habitués were dubbed 'the *Émiliens*'. Besides her translation with explanations of the *Principia*, Mme du Châtelet published an influential work entitled *Institutions de Physique*, a dissertation on space, motion, and energy.

Mme du Châtelet was not, of course, elected to membership of the Academy, which remained for more than another century a male preserve [9], but she received much acclaim. Frederick the Great of Prussia, the patron of Voltaire, called her Venus-Newton, and many verses were written in her honour. She became pregnant at the age of 42 and died, as she had dreaded, of childbed fever. During her life, and especially after her death, she attracted the malice of the supercilious *salonnières*, Mme du Deffand and Mme de Staël, who let drop spiteful aspersions on her character. An outraged Voltaire, who had already composed a moving epitaph of his friend (*L'Univers à perdu la sublime Émilie . . .*), responded with his *Epistle on Calumny*, which Tobias Smollett rendered into English verse; it concludes:

> You ne'er of virtue made parade,
> To hypocrites no court you've paid,

Therefore, of Calumny beware,
Foe to the virtuous and the fair.

Mme du Châtelet was not the only notable savante of the eighteenth century, and as a mathematician she was surpassed by her Italian contemporary, Maria Gaetana Agnesi, born in Milan in 1718. A prodigy, she had by the age of nine mastered several languages. Her *magnum opus* was the two-volume treatise on calculus, *Le Instituzione Analitiche*. It was said that often, after pondering a refractory problem, she would rise at night in a somnambulistic state, go to her study, write out the solution, and return to bed; the next day she would have no recollection of her nocturnal excursion. It took 50 years for Maria Agnesi's work to reach England in translation, when it was extolled by John Colton, the occupant of the Lucasian chair of mathematics (once graced by Newton himself) in Cambridge. She received the homage of many leading scholars and numerous honours came her way, including an invitation from the Pontiff of the day to the chair of mathematics at the University of Bologna, the most illustrious in Italy (but she did not want to leave Milan and declined). The French Academy was so impressed by her work that a member of its committee was delegated to write her a fulsome letter of esteem. It concluded with the assurance that she would have been elected an Academician, but that women were, alas, excluded from that distinction. To general astonishment and dismay, Maria Gaetana Agnesi forsook her mathematical and scientific avocation while still in her twenties and dedicated herself to good works among the poor. She died in Milan at the age of 81.

Mathematics seems always to have exerted a special attraction for intellectually gifted women, perhaps because it could be pursued without access to the competitive and, in the past at least, exclusive world of experimental science. Emmy Noether and the formidable Mary Cartwright were outstanding examples in the twentieth century. (The diffident, cricket-loving G. H. Hardy [10] would plead not to be seated next to Mary Cartwright at the dinners of the Cambridge Mathematical Society, because 'her fast ball', he complained, 'is so very devastating'.)

The first woman mathematician to achieve distinction was probably the famous Hypatia, born in Alexandria in about AD 370 and murdered there in 415. She was the daughter of a mathematician, Theon of Alexandria, in whose work she is thought to have assisted in her youth. She became head of the school of philosophy in her native city, and her teaching in mathematics, science, and philosophy attracted intellectuals from afar. One of her pupils, Synesius, Bishop of Ptolemais, wrote her letters, some of which survive, asking her advice on such matters as the construction of scientific instruments. Hypatia's tolerant views alienated the more pious elements of the Alexandrian citizenry and she was eventually set on by a Christian mob and done to death. (The Christians then further distinguished themselves by destroying the library of the Serapium, which probably housed most of Hypatia's written work, none of which survives.)

For a brief life of Mme du Châtelet, see Esther Ehrman, *Mme Du Châtelet: Scientist, Philosopher and Feminist of the Enlightenment* (Ber, Leamington Spa, 1986); for further details of her life and that of Maria Gaetana Agnesi, see H. J. Mozans, *Women in Science* (Appleton, New York, 1913, reprinted by MIT Press, Cambridge, Mass., 1974). See also the article by G. J. Tee, 'Pioneering women in mathematics', *The Mathematical Intelligencer*, 5, 27 (1983).

~

169 The gold standard

The origin of the Earth's magnetism was a source of fierce debate from the sixteenth century onwards. During the Second World War two distinguished British physicists were drawn into a project to combat the danger to Allied shipping of magnetic mines. Edward Bullard and Patrick Blackett fell to arguing as they worked, about terrestrial magnetism. At the war's end both returned to Cambridge and continued to ponder the problem.

Blackett was much admired both as a theoretician and an ingenious experimenter, and later, in 1948, won a Nobel Prize. He had an unusual background: born into a seafaring family, he entered the Royal Navy as a cadet at the age of 13 and served through many of the engagements of the First World War. When peace returned he was chosen as one of a small group of junior career officers sent to Cambridge on a six-month course. Blackett had already by then demonstrated an exceptional technical facility and had devised a gunnery aid used throughout the Royal Navy. At Cambridge he went to visit the Cavendish Laboratory out of curiosity to see what a physics laboratory was like. Instantly smitten, he resigned his commission and enrolled in the university as a mature undergraduate. There he developed political interests and his sympathy with the Soviet Union was to keep him out of the British nuclear-bomb programme after the Second World War, although he had participated in the Manhattan Project in the

United States. By the end of his life he was sitting on the Labour benches in the House of Lords as Lord Blackett of Chelsea; he died in 1974.

Blackett's idea, nurtured by his conversations with Bullard, was that the Earth's magnetism was the result of rotation, that in fact any massive rotating body would generate a magnetic field. An attractive feature of this hypothesis was its possible bearing on Einstein's vision of a nexus between gravity and electromagnetism. Blackett set about detecting such an effect, to be generated by a non-magnetic body in rotation. This demanded the measurement of magnetic field strengths far lower than any that could be discerned by existing techniques. Blackett accordingly devised and constructed a magnetometer of unparalleled sensitivity. The environment of a university laboratory was unsuitable because of background sources of magnetism and Blackett put up a shed made of wood, held together with copper nails, in a field at Jodrell Bank in Cheshire, where his friend, Bernard Lovell, had sited his radio telescope. In the hut he erected a concrete block with a central cavity, resting on a cushion of soft rubber and in the cavity he hung his rotating body. Thanks to his wartime connections, for his country had recognized his services, he managed to borrow from the Bank of England sufficient gold to cast into a cylinder 10 centimetres across, weighing more than 15 kilograms.

Blackett made his measurements, but the rotating cylinder engendered no perceptible magnetic field. The theory, it had to be conceded, was wrong. But the technical accomplishment was prodigious and afforded a means of measuring the magnetic contents of minerals. This in turn inaugurated a new chapter in geophysics, for measurements of the strength and direction of residual magnetism in rocks led to new insights into movements of the Earth's crust over the ages. The plasticity of the crust was mooted in the nineteenth century by George Darwin, son of Charles, who in consequence came into conflict with the panjandrum of Victorian physics, William Thomson, Lord Kelvin [10]. Darwin's father egged him on: 'Hurrah for the bowels of the earth', he wrote to George, 'and their viscosity and for the moon and for the Heavenly bodies and for my son George.' He would have been delighted by all that flowed from Blackett's experiment in the hut at Jodrell Bank.

See *P. M. S. Blackett: A Biographical Memoir* by Sir Bernard Lovell (Royal Society, London, 1976); also *The Dark Side of the Earth* by Robert Muir Wood (Allen and Unwin, London, 1985).

170 Margin of error

Pyotr (or Peter) Leonidovich Kapitsa (or Kapitza) was a Russian physicist, whose formative years were spent at the Cavendish Laboratory in Cambridge during Rutherford's reign [16]. Kapitsa arrived in Cambridge as a young man, fresh from his studies in Moscow and sought an interview with Rutherford, for he had set his heart on working with the great man.

Rutherford refused to consider Kapitsa because the laboratory was already seriously overcrowded. Impetuously the young Russian asked him, 'How many research students have you?' 'About thirty', was the reply. 'What is the customary accuracy of your experiments?' was the next question, to which Rutherford replied, 'About two or three percent'. 'Well then', Kapitsa beamed, 'one more student would not even be noticed within that accuracy!'

Rutherford, so the story has it, could not resist such an ingenious appeal and Kapitsa soon became his favourite protégé. Although Kapitsa had a strong streak of traditional Russian authoritarianism, which he exerted over those who later worked with him, he adored Rutherford. As a permanent member of the Cavendish, he did notable work on low-temperature physics. In 1934, on his annual visit to his family in Russia, he was detained on Stalin's orders. Appeals to the Soviet government from colleagues and statesmen in the West were without avail. Kapitsa was installed in a laboratory in Moscow and informed that his duty was to the Soviet Union and not England or the international community. Rutherford eventually conceded defeat and had Kapitsa's apparatus shipped to Moscow.

Kapitsa distinguished himself by resolutely defending Russian colleagues who had fallen foul of Stalin's regime and probably saved many of them from death or the Gulag. Stalin clearly had a soft spot for this brave and resolute man and kept him out of the clutches of the mephistophelean head of the NKVD, Beria, who had demanded his head. Kapitsa nevertheless spent five years under house arrest, pursuing science as best he could in a laboratory that he himself constructed in a shed, with his son as assistant. It was only in old age that Kapitsa was allowed out of the country, to receive a belated Nobel Prize (in 1978) and make a sentimental visit to Cambridge.

The story of Kapitsa's first encounter has been often told. The version here is from Lawrence Badash, *Kapitza, Rutherford and the Kremlin* (Yale University Press, New Haven and London, 1985).

171 Pomposity and circumstance

Samuel Pierpont Langley (1831–1906) was an American physicist of high standing, though not as lofty as he believed. He was Professor of Physics in Pittsburgh and director of the Allegheny Observatory. Langley was notoriously pompous and self-important, with an unshakeable faith in his infallibility. Here is a recollection by Sir Arthur Schuster [159], Professor of Physics at Manchester University, who himself made many important contributions, especially in the field of spectroscopy:

Langley's invention of the bolometer [an instrument for the measurement of radiant heat], and his pioneer work in the construction of the flying machine, are achievements sufficiently great to ensure a reputation which will outweigh the recollection of defects due to an exaggerated consciousness of dignity, accompanied by a marked inability to see the humorous side of things. I first met Langley on the occasion of the total solar eclipse in August 1878, when he established an observing station on the top of Pike's Peak in order to obtain, if possible, a measure of the thermal radiation of the solar corona. Unfortunately he suffered severely from mountain sickness, and had to be carried down before the day of the eclipse.

In the following year, Langley visited England and expressed to me the desire to become acquainted with Clerk Maxwell [44]. I was working at the Cavendish Laboratory at the time, and was able to assure him that Maxwell would be interested to meet him as he had, in my presence, referred in very eulogistic terms to a method proposed by Langley to eliminate the personal equation [that is, subjectivity] in transit observations. Clerk Maxwell was just then editing Cavendish's scientific manuscripts, and conscientiously repeating every experiment that was described in them. He was specially interested in the method which Cavendish [156] had devised for estimating the relative intensities of two electric currents, by sending the currents through his body and comparing the muscular contraction felt on interrupting the currents: 'Every man his own galvanometer', as Maxwell expressed it. When Langley arrived, I took him to the room where Maxwell stood in his shirt sleeves with each hand in a basin filled with water through which the current was laid. Enthusiastic about the unexpected accuracy of the experiment, and assuming that every scientific man was equally interested, he tried to persuade Langley to take off his coat and have a try. This was too much for Langley's dignity; he did not even make an effort to conceal his anger, and on leaving the laboratory he turned round and said to me: 'When an English man of science comes to the United States we do not treat him like that.' I explained that, had he only had a little patience and entered into the spirit of Maxwell's experiment, the outcome of his visit would have been more satisfactory.

As an experimenter Langley takes a high rank, though the numerical results he derived were sometimes based on calculations that were not entirely free from defects. This led him occasionally to an optimistic judgment of their accuracy. In sending out an assistant to repeat his measurement of the so-called solar constant, which expresses the total solar radiation in certain units, his final words to him were: 'Remember that the nearer your result approaches the number 3, the higher will be my opinion of the accuracy of your observations.' The assistant, who since then has himself attained a high position among American men of science, fortunately was a man of independent judgment and skilful both in taking and reducing his observations, with the result that the number 3 is now altogether discredited.

From Sir Arthur Schuster, *Nature*, 115, 199 (1925).

~

172 The use of vacations

Here is how Louis Pasteur (1822–95) came upon one of the main tenets of vaccination. It illustrates perfectly his famous maxim that fortune favours the prepared mind: he had been studying fowl cholera in chickens and interrupted his work by a vacation. When he returned he tested the cultures of the cholera bacteria and found that they had become inactive, in fact had died: subcultures (growth media seeded with bacteria from the original cultures) did not grow and the birds into which they were injected developed no sign of disease. Pasteur prepared to start over with new cultures, but instead of merely abandoning the abortive experiment, he decided, for no reason that he could articulate, to re-inject the same birds with a new, active culture. One of his colleagues reported the outcome:

> To the surprise of all, and perhaps even of Pasteur, who was not expecting such success, nearly all these fowls withstood the inoculation, although fresh fowls succumbed after the usual incubation period.

This inspirational experiment established the principle of immunization with attenuated bacteria, which was also later crucial in protecting against other pathogens, including viruses.

It is only fair to add that there is some doubt about the exact veracity of the original account, which probably came from Pasteur's most devoted apostle, Émile Duclaux. There is some evidence that Pasteur's younger colleague, Émile Roux, continued the experiments after Pasteur had left for his vacation and introduced his own methods for generating an attenuated vaccine preparation. Both Duclaux and Roux later became directors of the Pasteur Institute.

For the original story see, for example, W. I. B. Beveridge, *The Art of Scientific Investigation*, 3rd edn (Heinemann, London, 1960).

~

173 The lecturer's craft

Many of the world's great scientists have been egregiously bad lecturers. The opacity of Niels Bohr's public performances was legendary [31]. His friend Ernest Rutherford was an ebullient speaker [16], but would dissolve into incoherence when forced to manipulate algebraic equations. He would on occasion round angrily on his audience: 'You all sit there like numskulls and not one of you will tell me where I've gone wrong!' For others, more theoretically inclined, mathematical derivations were all too easy and the students would be left in a dazed state while the lecturer vaulted over the intermediate steps of an argument. It was related that the mathematician, G. H. Hardy [10], began his peroration to a lecture by the declaration: 'It is now obvious that . . .' Thereupon he stopped and turned to contemplate in silence the equations that he had written on the blackboard. After an interminable interval, he broke into a smile and reassured his listeners that it was indeed obvious.

Norbert Wiener was a visionary, famous for his pioneering work in cybernetics, a neologism that he coined. He was a professor at the Massachusetts Institute of Technology, and his mathematical and analytical brilliance, his vanity, and his absence of mind gave rise to many legends. On one attested occasion he developed the proof of a mathematical proposition before a class of students at the blackboard, leaping from one logical crag to another without offering any explanation of the reasoning. When asked by a member of his bewildered audience whether he might not repeat the exercise more slowly, he affably consented, then stood, silent and motionless, before the blackboard for a few minutes and, smiling with satisfaction, added a triumphant tick to the last line.

Sir Joseph (J. J.) Thomson [73] recalled in his memoirs the lectures of his teacher in Manchester, Osborne Reynolds (1842–1912), a noted physicist and engineer, who gave his name to the Reynolds number, which describes the nature of fluid flow.

> Occasionally in the higher classes he would forget all about having to lecture and, after waiting for ten minutes or so, we sent the janitor to tell him that the class was waiting. He would come rushing into the room pulling on his gown as he came through the door, take a volume of Rankine [a standard textbook of the time] from the table, open it apparently at random, see some formula or other and say it was wrong. He then went up to the blackboard to prove this. He wrote on the board with his back to us, talking to himself, and every now and then rubbed it all out and said it was wrong. He would then start afresh on a new line, and so on. Generally, towards the end of the lecture, he would finish one which he did not rub out, and say this proved Rankine was right after all. This, though it did not increase our knowledge of facts, was interesting, for it showed the workings of a very acute mind grappling with a new problem.

Sir Arthur Schuster [159], another Manchester alumnus, remembered Reynolds's lectures to an elementary class:

> In his lectures Osborne Reynolds was often carried away by his subject and got into difficulties. Some humorous incidents are related with regard to the manner in which he got out of them. He was once explaining the slide-rule to his class; holding one in his hand, he expounded in details on the steps necessary to perform a multiplication. 'We take as a simple example three times four', he said, and after appropriate explanation he continued, 'Now we arrive at the result; three times four is 11.8.' The class smiled. 'That is near enough for our purpose', said Reynolds.

See *John von Neumann and Norbert Wiener: From Mathematics to the Technology of Life and Death* by Steve J. Heims (MIT Press, New York, 1980); *Recollections and Reflections* by J. J. Thomson (G. Bell, London, 1936); Sir Arthur Schuster, *Nature*, 115, 232 (1925).

~

174 The purple cloud

With the destruction of the French fleet during Napoleon's Egyptian expedition, the British naval blockade in the Mediterranean took effect. An early consequence was that the supplies of potassium nitrate, saltpetre, the main ingredient of gunpowder, which had been mainly imported through the southern French ports, began to dry up. The material was derived from the contents of the cesspits of North Africa by bacterial fermentation. (For the massive demands of the First World War, Germany initially imported its saltpetre from the mines of Chile, and the supply again was interrupted by a naval blockade; this time it was left to Fritz Haber [96], the celebrated chemist, to solve the problem by devising a chemical, rather than biological, 'nitrogen fixation' process.) The French sought to ferment their own cesspits and the products of the farms, slaughterhouses, and beaches, and chemists were employed to improve the saltpetre yields from these sources. One of them was Bernard Courtois (1777–1838), but he took a different route and sought instead to collect potassium compounds from seaweed.

Courtois extracted the ash from burnt seaweed with boiling water, evaporated the resulting solution, and experimented on the products. Adding sulphuric acid to the residue one day in 1811, he observed a remarkable phenomenon: clouds of purple smoke rose from the hot mixture and condensed to lustrous black crystals. Courtois had discovered iodine. The full exploration of its properties was left to others, but Courtois did uncover the reaction of iodine with ammonia, which yielded the highly explosive nitrogen triiodide.

Courtois's discovery proved also to be of prime medical interest, for it had been said since ancient times, dating back to the Chinese two millennia ago, that burnt seaweed or sponges had the capacity to relieve the symptoms of goitre. In 1820, a Swiss physician, Jean-François Coinder, tried iodine solutions on patients with goitre with some success (but there were disagreeable side-effects). An effective way to administer iodine (in the form of a mixture of sodium chloride and iodide) was later developed by a doctor in Cleveland, Ohio. By then the presence of iodine compounds in the thyroid gland had supposedly been discovered when an experimenter spilled some concentrated acid on an excised thyroid gland and saw the purple vapour rise.

See John Emsley, *Nature's Building Blocks: An A–Z Guide to the Elements* (Oxford University Press, Oxford, 2001).

175 The fist in the fistula

It was a strange accident that gave an alert amateur physiologist access for the first time to the digestive processes in the human stomach. William Beaumont was an American military surgeon in the first half of the nineteenth century. He was the product of the prevailing system of training, when aspiring doctors qualified not by going to medical school but by serving an apprenticeship in the private practice of a physician or surgeon.

While in post as surgeon at a fort in a remote fastness in Michigan, Beaumont was summoned one evening to tend the victim of an accident in a nearby fur trading post. Alexis St Martin, a young Canadian from the far north, had been hit by a shot-gun blast at close range and was lying unconscious in a pool of blood when Beaumont arrived. Wadding, shot, and fragments of clothing had penetrated his ribcage and stomach, creating a hole through which a man's fist could pass. To general astonishment the victim did not die, but because he was not strong enough to work, the authorities at the trading post, unwilling to support an invalid, resolved to send him home to Canada. Beaumont doubted whether St Martin would survive the journey of 2,000 miles, and so took him in, 'nursed him, fed him, clothed him, lodged him and furnished him with every comfort, and dressed his wounds daily and for the most part twice a day'. St Martin recovered fully and returned to his trade as a woodsman. But on the left side of his torso remained an aperture into his stomach. Beaumont could dose him with medicines 'as never medicine was before administered to man since the creation of the world—to wit, by pouring it through the ribs at the puncture of the stomach'.

As time passed, Beaumont came to realize that the gastric fistula presented him with a unique means of observing what went on within. 'I can pour in water with a funnel', he wrote, 'or put in food with a spoon, and draw them out again with a syphon. I have frequently suspended flesh, raw and wasted, and other substances into the perforation to ascertain the length of time required to digest each; and at one time used a tent of raw beef, instead of lint, to stop the orifice, and found that in less than five hours it was completely digested off, as smooth and even as if it had been cut with a knife.'

Beaumont tried the effects of the gastric juices, both inside and outside St Martin's stomach, on many kinds of food, which he would withdraw and inspect at intervals. He examined the action of bile on the digestive processes and measured temperature and acidity.

Eventually St Martin wearied of his role as an experimental subject, a mere ambulant stomach, and vanished into the northern wilderness. In time he found employment with the Hudson Bay company, married, and fathered two children. The distraught Beaumont went to great lengths to find him again, and, when he succeeded, paid for St Martin and his family to return to the United States. Defections and retrievals of the errant St Martin continued for some years until Beaumont had accumulated results of 238 experiments. By then St Martin had had enough. Beaumont published his book, *Experiments and Observations on the Gastric Juice and the Physiology of Digestion* in 1833 and a second, enlarged edition followed after an interval of 14 years. The results served as a basis for a mass of later work, including that of Claude Bernard [138] and of Ivan Petrovich Pavlov, who both opened fistulas into the stomachs of dogs. Alexis St Martin outlived his rescuer by 28 years, dying in Canada at the age of 83.

A good account (one of many) of this story is to be found in Victor Robinson's *The Story of Medicine* (Tudor, New York, 1931).

~

176 Tripping over an answer

Isidor Rabi [21] grew to scientific maturity just at the time that the new physics of wave and quantum mechanics was emerging from Germany. In 1926, he was in the late stages of his Ph.D. at Columbia University in New York, hugely excited by the new developments. Roman candles, as he later put it, were going off all around. Erwin Schrödinger [116] had just succeeded in reconciling his wave mechanics with Werner Heisenberg's quantum mechanics, at first sight entirely different, but as Schrödinger ultimately showed, different mathematical formulations of the same principles. Heisenberg's method called for a new mathematics [180], while Schrödinger's, though difficult, was of a form that a well-educated physicist could recognize.

Rabi, and his somewhat older and more experienced friend Ralph Kronig, found Schrödinger's approach more congenial and resolved to see where it led. Where Schrödinger had calculated the permitted energy states only of atoms, Rabi and Kronig wanted to try his method on molecules. They chose to investigate molecules of a form known as a symmetrical top. They set up their version of Schrödinger's formulation and found themselves faced with an equation of a kind they had not seen before and did not know how to solve. They tried it on three colleagues, all of whom conceded defeat.

> Rabi's penchant for eschewing the onerous demands of daily work by escaping to the peaceful serenity of the library broke the impasse. This was a time when he should have been busy. Not only was he teaching twenty-five hours a week at CCNY [City College of New York], but he was also under pressure to finish his dissertation, to keep on top of developments in the new quantum mechanics, and to work with Kronig on their quantum mechanical problem. Nonetheless he was sitting in the library reading for pleasure the original works of Carl Gustav Jakob Jacobi, eminent mathematician of nineteenth-century Germany. As he browsed through the pages of Jacobi, an equation seemed to leap from a page. 'My God!' he thought. 'That's our equation!' The equation had the form of a hypergeometric equation [pertaining to a mathematics that transcends normal geometry], which Jacobi had already solved. The solution was expressed in terms of a hypergeometric series; and, in terms of this series, the intractable equation could now be solved.

The outcome was the demonstration that molecules of the symmetric-top family could only exist in certain defined energy states. This result brought about a change of thinking in the field of molecular spectroscopy [70].

From John S. Rigden's biography, *Rabi: Scientist and Citizen* (Basic Books, New York, 1987).

~

177 Throwing the hounds off the scent

Superconductivity was discovered in The Netherlands in 1911 by Heike Kamerlingh Onnes—*le gentleman du zéro absolu,* as he was known. Kamerlingh Onnes had dedicated his life to the attainment of low temperatures and had succeeded in liquefying helium, which he found to have a boiling point 4.2 degrees above absolute zero. Absolute zero, the temperature at which molecular motion (approximately) ceases, is −273 degrees Celsius, designated 0 K (degrees Kelvin) on the absolute scale. Kamerlingh

Onnes and his students at Leiden proceeded to measure electrical properties of solids at temperatures down to that of liquid helium. They expected that the resistance of metals would fall to a very low level, but the results were startling: somewhere near the boiling point of the liquid helium the resistance fell abruptly to below the level of detectability. The electrical resistance of the metals became, in effect, zero, so that a current in a circuit at this temperature would go on circulating for ever. The phenomenon plagued physicists for most of the twentieth century and it took many decades of effort before a theory of the process finally emerged [69]. The search also began for a material that might become superconductive at higher temperatures, for the technological possibilities this would afford seemed limitless.

The theory of superconductivity made possible a more rational basis for such a search, and in 1985 two scientists in Switzerland prepared a complex mixed metal-oxide ceramic material that became superconductive at a critical temperature of 35 K. The publication of this work (which led to a Nobel Prize in 1987) precipitated a frenzied rush in university and industrial laboratories around the world for materials with a still higher critical temperature. The prospect of the rewards in terms of fame, patents, and riches that success would bring were intoxicating. One of the most determined of the searchers was Paul Chu, Professor of Physics at the University of Houston. By 1987 he and his students had prepared and examined a vast range of complex mixtures and one day early that year their efforts bore fruit in the shape of a material that became superconductive at 90 K. This was a dramatic improvement on what had gone before.

But success presented Chu with a dilemma: how to get his result into print (and patented) without letting his competitors in on the secret of the composition. The journal of choice for rapid publication in physics is *Physical Review Letters*. Like all other respectable journals, it is 'peer reviewed'; that is, papers submitted for publications must be critically scrutinized by the editor and generally two referees, who are, of necessity, experts in the field and, therefore, in such a hyperactive area as superconductivity then was, probably competitors. It is a gross impropriety for referees to make any use of the information in a manuscript, but here the frailty of human nature and paranoia can converge, and Chu, at any rate, was taking no chances. He called the editor of the journal and asked leave to submit his report without actually identifying his superconductive material. The editor predictably demurred, so the paper was submitted with what purported to be full details, and duly accepted for publication. In the interim Chu gave a press conference, announcing his discovery without divulging the composition of the material, and the University of Houston prepared a patent application.

The announcement set the world of physics abuzz with excitement, and in laboratories around the world researchers tried to divine what the material might be. A picture in *Time* magazine showed Chu holding a sliver of a greenish substance. Green might imply a nickel compound, but this proved a false trail. Then rumours began to circulate that the magic ingredient was ytterbium, one of a group of closely related elements, the rare earths, also known as lanthanides; this was found no more effective than nickel. Chu's submitted manuscript gave the composition of the ceramic superconductor, expressing it only in chemical symbols for the elements, Yb, Ba, Cu—not their names, ytterbium, barium, and copper. Repeating the result of Chu and his associates should have been simple, but in the laboratories in which it was tried there was only failure. And presently a shameful story emerged.

Ytterbium takes its name from 'the village of the four elements', Ytterby in Sweden, where in the late eighteenth century a new ore was discovered. It was named yttria, and contained, as was later established, four new elements, all very similar and belonging to the rare-earth family. They were given the names, ytterbium, terbium, erbium, and yttrium. The symbol for yttrium is Y, that for ytterbium, quite logically, Yb. Chu's superconductor contained yttrium, not ytterbium, as his paper implied. When charged by his indignant brother-physicists with deliberate deception, he denied any such intent. No, his secretary had merely typed Yb for Y every time it occurred in the manuscript—a pure accident. Moreover, for she was clearly prone to lapses of concentration and Chu to careless proofreading, she had also made a mistake in the proportions of the components in the complex. On the last day before the journal went to press Chu called the editorial office to correct the typos. Some physicists, when questioned, owned that in Chu's place they would have resorted to deception to protect their priority, others were less forgiving. Worse, however, was the leaked information of Chu's incorrect recipe, and, indeed, a rumour had also begun to circulate before the paper was published of the yttrium-ytterbium substitution.

Whether confidentiality was breached by a referee or in the editorial offices of *Physical Review Letters* was never disclosed, but the story points a moral: consciences become elastic when the stakes are high. For some of the protagonists in the race to high-temperature (though still 183 °Celsius below freezing) superconductivity there were some painful regrets: ytterbium will after all, it later emerged, form a high-temperature superconductor if the complexes are prepared in the right way; and another research team found that they had prepared the same material as Chu's but had decided against testing it because an examination of its structure revealed heterogeneity, always previously taken as a mark of a worthless preparation.

See an article with the title 'Yb or not Yb? That is the question' by Gina Kolata, *Science*, **236**, 663 (1987), and Bruce Schechter's book, *The Path of No Resistance: The Story of the Revolution in Superconductivity* (Simon and Schuster, New York, 1989).

~

178 The body's bounty

Beliefs about the curative virtues of human bodily products were once common, but the value of oil secreted from a philosopher's head probably has no precedent. The philosopher in question was Jeremy Bentham, who founded University College London, as a free-thinking institution: its members were liberated from the tyranny of the Anglican Church under which the ancient universities of England were labouring. When Bentham died in 1832, his body was anatomized, as he had laid down in his will, and his mummified remains repose still in a mahogany case in the college foyer, to be exposed to view on ceremonial occasions. Thomas Love Peacock, the novelist, had been a friend of Bentham's, of whom 'he talked at much length'.

He mentioned among other things that when experiments were being made with Mr Bentham's body after his death, Mr James Mill [philosopher and father of the utilitarian philosopher, John Stuart Mill] had one day come into Mr Peacock's room at the India House and told him that there exuded from Mr Bentham's head a kind of oil, which was almost unfreezable, and which he conceived might be used for the oiling of chronometers which were going into high latitudes. 'The less you say about that, Mill,' said Peacock, 'the better it will be for *you*, because if the fact once became known, just as we see now in the newspapers that a fine bear is to be killed for his grease, we shall be having advertisements to the effect that a fine philosopher is to be killed for his oil.'

From Sir Mountstuart Elphinstone Grant Duff, *Notes from a Diary*, Vols 1 and 2 (John Murray, London, 1897).

~

179 The flat-earther's revenge

Alfred Russel Wallace was a passionate naturalist, whose long life stretched from 1823 to the eve of the Great War in 1913. He spent his early years exploring the wildlife in remote corners of the world from Sarawak to the Amazon. He pondered the nature of speciation, and the concept of natural selection crystallized in his mind as he lay shivering, too weak to move, in the grip of a malarial fever.

Darwin, who had been toiling for many years at his mighty work, *On the Origin of Species*, was shocked to find the essential principles of his grand conception prefigured in a paper, which Wallace dispatched in 1858 from his base on a Malayan island. Darwin sought reassurance from his friends: could he, having read Wallace's manuscript, now honourably publish an abstract of his own, which he would otherwise not have done? He wrote to his friend and supporter, the geologist Charles Lyell, of his mortification at his own 'trumpery feelings'. 'I would far rather', his letter concluded, 'burn my whole book than that he or any man should think I had behaved in a paltry spirit.' But Wallace was a humble man, who recognized Darwin's genius, and was happy with the accommodation that was quickly reached: an abstract, or 'sketch' of Darwin's theory would be read, together with Wallace's offering, at a meeting of the Linnean Society in London. Wallace was content henceforth to appear as 'the Moon to Darwin's Sun'.

After his return to England Wallace wrote incisive articles in defence of Darwin and natural selection, and interested himself in many subjects, including unfortunately spiritualism and phrenology (the study of cranial bumps as indicators of moral and intellectual proclivities). Then in 1870, Wallace, who was perpetually hard-up, entered into what was to prove a disastrous wager. In January of that year there appeared in a popular journal, *Scientific Opinion*, an advertisement, placed by one John Hampden, who offers to 'deposit from £50 to £500, on reciprocal terms, and defies all the philosophers, divines and scientific professors in the United Kingdom to prove the rotundity and revolution of the world from Scripture, from reason or from fact. He will acknowledge that he has forfeited his deposit if his opponent can exhibit, to the satisfaction of any intelligent referee, a convex railway, river, canal or lake'. For Hampden, of course, was one of the vociferous company of flat-earthers, which survives to this day, unconverted by the evidence of circumnavigation or satellite pictures; these, they hold, reveal only the edge of the disc or shallow bowl on which we reside. (The headquarters of the present-day Flat Earth Society is in California.)

With the encouragement of Charles Lyell, who was eager to see these pests finally squashed, Wallace took up the challenge, and the money was duly deposited with an independent referee, John Walsh, editor of *The Field* magazine. Wallace set up his demonstration on the Bedford Canal, where a straight six-mile stretch separated two bridges. The iron parapet of Welney bridge, Wallace recorded, stood 13 feet and 3 inches above the surface of the water, while the Old Bedford Bridge was a little higher. To this bridge, in the presence of Hampden, the referee, and two other witnesses, Wallace, who in his youth had trained as a surveyor, affixed a length of white cloth on which was painted a black line at the same height above the canal as the parapet of Welney Bridge. Half-way between the bridges he erected a pole bearing two red discs, one centred at the height above the water of the black line and the iron parapet, the second four feet below.

Wallace mounted his telescope on the iron parapet. He had calculated that the Earth's curvature would offset the upper disc from the line joining the two reference points by five feet and six inches, which the effect of refraction would reduce by about one foot; it would appear, therefore, four feet and five inches above the position that Hampden's world-view demanded. Mr Walsh took a sighting and pronounced the demonstration conclusive. But Hampden refused to put his eye to the telescope, asserting that the very thought of curved water was an affront to common sense. Walsh attempted to reason with him, was rebuffed, published an account of what had happened in his magazine, and presented the £500 to Wallace.

The infuriated Hampden now invoked a clause in the law on wagers that the bargain had to be implemented by the stakeholder without delay. Walsh had sought to avoid alienating Hampden and making him see reason before handing over the money, so after prolonged legal wrangling was forced to recover the entire sum from Wallace. Not content with this, Hampden proceeded to abuse and publicly vilify the unhappy Wallace, to whose wife he addressed a threatening and abusive letter. Wallace at this point took action and had Hampden brought before a magistrate. Hampden, by now totally unhinged, was unrepentant and continued his vendetta for 15 years, in the course of which he was three times imprisoned. Wallace suffered much vexation and estimated that the affair had cost him far more than the £500, which he eventually recovered, in legal expenses. It was only in his last years that income from his books brought him and his family a measure of security.

There are several accounts of Wallace's life and the flat-earth story in particular. A good one is Amabel Williams Ellis's *Darwin's Moon—A Biography of Alfred Russel Wallace* (Blackie, London, 1966).

$\sim$

180 Potholes on the path to fame

Werner Heisenberg was one of the small band of theoreticians who wrought the revolution in physics during the first half of the old century. Born in 1901, he was in his early twenties when he constructed a mathematical foundation for quantum mechanics. His patron at the University of Munich, Arnold Sommerfeld, recognized Heisenberg's genius and did all he could to promote his career. By the time he came to present his doctoral dissertation Heisenberg had already solved some formidably difficult theoretical problems, but his exploits in the laboratory had been less spectacular.

A certain tension prevailed in the Munich physics department between Sommerfeld and the professor of experimental physics, Wilhelm Wien. Wien disapproved of his colleague's laissez-faire attitude to the training of research aspirants. Although no mean theoretician himself, Wien had won his Nobel Prize in 1911 for his experimental studies on radiation from hot objects, and he observed with disapproval the conceptual convulsions that were eroding the foundations of classical physics. He had set Heisenberg a problem, as part of his doctoral course of training, which involved the measurement of the wavelengths of lines in the spectrum of mercury vapour, split into two components under the action of a magnetic field (the so-called Zeeman effect). Heisenberg was given apparatus for the purpose, especially a Fabry–Pérot interferometer, an instrument for the accurate determination of wavelengths of light. He later claimed that he had not been told he could make use of the departmental workshops and had therefore tried to rig up his experiment with pieces of wood from cigar boxes. This had incurred the professor's wrath, and Heisenberg had probably not concealed his strong preference for theoretical work.

Nemesis overtook Heisenberg in the oral examination for his doctoral degree. He told the story towards the end of his life (he died in 1976) to the historian and philosopher of science, Thomas Kuhn, in an interview. At first the examination had gone well, but then it was Wien's turn:

> I had not, as I ought to have done, occupied myself with the principal questions relating to my experimental exercise. In the examination Wien asked me about the resolving power of the Fabry-Perrot [sic] interferometer [the smallest wavelength difference that the instrument could discriminate] . . . And this was something I had never studied. During the examination I tried, of course, to derive it, but in this short time I could not manage it. Thereupon he grew angry and asked about the resolving power of a microscope. As I did not know this, he asked about the resolving power of a telescope, and I did not know that either.
>
> So he asked me about the operation of a lead storage battery, and that I likewise did not know . . . I am not clear whether he wanted to fail me. Probably there was afterwards a heated discussion with Sommerfeld.

Heisenberg's performance had indeed been monstrous, for Wien's questions were such as a schoolchild studying physics might have been able to answer. Doctoral candidates in Germany were awarded a single mark for their ability in theoretical and experimental physics combined, and Wien and Sommerfeld had, therefore, to reach a compromise. Wien's report contained the phrase 'bottomless ignorance', whereas Sommerfeld referred to his protégé's 'unique genius'. The highest possible mark was one and the most abject a five, so Heisenberg was awarded the average of those—a three, the narrowest of passes.

Wien's animosity towards Heisenberg did not entirely abate with the passage of time. In 1925, two years after Heisenberg's ordeal, Erwin Schrödinger [116] gave a lecture in Munich in which he outlined his newly conceived wave mechanics; this, he claimed, would supplant Heisenberg's quantum mechanics. In the discussion that followed Heisenberg could make little impression, Wien applauded Schrödinger's achievement and brutally denounced Heisenberg's 'atomic mysticism', and even Sommerfeld did not come to his favourite pupil's defence.

Heisenberg, of course, prevailed in the end, not without vicissitudes. Sommerfeld had urged the university to appoint Heisenberg as his successor in the chair of theoretical physics, but by then the Nazis were in power and Heisenberg, like Sommerfeld, was denounced by the champions of German nationalistic physics as 'a white Jew'—an Aryan who had espoused the counterintuitive new physics, associated with the names of Einstein [161], Pauli [25], and Born [73], all of them Jews. Hans Bethe [62], another brilliant pupil of Sommerfeld's, and a Jew, was present when Sommerfeld began a lecture by winding up the blackboard on which he had, the previous day, written equations. There was an appalled silence from the audience and Sommerfeld turned to see the words, 'damned Jews', scrawled across the board. In the event, Sommerfeld's chair was filled by a party nonentity and physics at the university atrophied until Heisenberg's return years later.

Heisenberg's fate in his doctoral examination occasioned much mirth among physicists around the world, but it was perhaps small wonder that the German atom-bomb project, with its vast experimental demands, did not thrive under his leadership. Heisenberg's part in that episode is, however, a more complex and controversial chapter of history. In 1944, the OSS (forerunner of the CIA) despatched an agent, Moe Berg, to attend a lecture by Heisenberg in Zurich in neutral Switzerland. Berg was a sporting hero, a baseball star, who spoke German (and several other languages) and knew some physics. He was to divine from Heisenberg's discourse whether the German atomic bomb project was making progress. If he were able to form such a judgement his instructions were to shoot the lecturer. Berg sat through the lecture, his hand on his gun, but, perhaps not surprisingly, Heisenberg did not allude to the matter, and the thwarted assassin returned quietly to base.

The outstanding biography of Heisenberg is David C. Cassidy's *Uncertainty: The Life and Science of Werner Heisenberg* (Freeman, New York, 1991); but see also *Heisenberg's War: The Secret History of the German Bomb* (Jonathan Cape, London, 1993) by Thomas Powers. The passage quoted above is translated from a brief biography by Armin Hermann: *Heisenberg in Selbstzeugnissen und Bilddokumenten* (Rohwolt, Hamburg, 1976).

181 Humboldt's stratagem

Joseph-Louis Gay-Lussac [136] was an illustrious French chemist, remembered for, amongst many other achievements, his law relating the combining volumes of gases—an important advance in atomic theory. He was assisted in this work by the young Alexander von Humboldt [134]. Their experiments necessitated special thin-walled reaction vessels, which had to be procured from Germany. Humboldt applied his native ingenuity to the problem of evading customs duty on the imports, which at that time was exceptionally steep. He instructed the German glass-blowers to seal the long necks of the vessels and label the containers: *Handle with care—German air*. The French *douaniers* had no instructions concerning duty on 'German air' and so let the consignment through. Humboldt and Gay-Lussac cut the ends off the sealed vessels and proceeded with their experiments.

The story is told in *Was nicht in den Annalen steht* by Josef Hausen (Verlag Chemie, Weinheim, 1958).

Credits

The author and the publisher would like to thank the following for granting permission to quote extracts from their books (the numbers in parentheses refer to the anecdotes in which the quoted passages appear):

Adam Hilger (27.1); Allen Lane (127); the American Center for Physics (69); the American Chemical Society (3); the American Institute of Physics (89); the American Society for Biochemistry and Molecular Biology (50); Annual Reviews (98); the British Medical Journal (80); Califonia Monthly (72); Cambridge University Press (15, 20, 28, 37, 54, 65, 116, 144, 148); Cassell & Co. (138); Christian Ejlers' Forlag (75); Cold Spring Harbor Laboratory Press (88); Columbia University Press (86); Constable & Robinson Ltd (71); Elek Science (107); Elsevier Science Ltd. (34, 104); Fourth Estate Ltd. (62); George Bell (1, 173); Victor Gollancz Ltd. (42); Harcourt Ltd. (45); HarperCollins (25, 52, 53, 55); Harvard University Press (12); Heinemann Publishers, Ltd. (59, 74, 97); Hodder & Stoughton (85, 115); Houghton Mifflin Company (25, 52); the Institute of Physics (60); the Johns Hopkins University Press (158, 162); Little, Brown (41, 71); Macmillan Publishing (10, 57, 61); Marcel Dekker, Inc. (5); MIR Publishing, Moscow (31); MIT Press (20); The Nature Publishing Group (2, 73, 84, 106, 119, 143, 166); Neal-Schuman Publishers, Inc. (66); W.W. Norton & Company, Inc. (65, 89, 101, 126); Oxford University Press (17, 18, 21, 31, 40, 79, 83, 88, 160, 161); Penguin Putnam, Inc. and Penguin UK (10, 12, 18, 22, 60, 67, 87, 137); Pergamon Press, Inc. (112); Perseus Books Group (2, 62, 25, 95, 176); Princeton University Press (83); Proceedings of the National Academy of Sciences (105); Random House (64, 67, 72, 85, 141, 153, 161); The Scientist (92); Simon & Schuster, Inc., and Simon & Schuster UK Ltd. (29, 51, 56); Springer Verlag New York Inc. (11, 27, 68, 86, 103); Taylor & Francis Group (90, 109); the University of California Press (48, 56, 76); Weidenfeld and Nicholson (88); Wiley-VCH Verlag GmbH (96); the Wisconsin State Journal (42). The first extract in anecdote 72 (page 117) is reproduced from *Lawrence and Oppenheimer* by Nuel Pharr Davis (Copyright © Nuel Pharr Davis 1969) by permission of PFD on behalf of the author. The extracts in anecdote 85 (pages 137–138) concerning J.B.S. Haldane are reproduced from *J.B.S.: The Life and Work of J.B.S.*

Haldane by Ronald Clark (Copyright © Ronald Clark 1968) by permission of PFD on behalf of the Estate of Ronald Clark.

Name index

Subject index

Popular Science from Oxford

Galileo's Finger: The Ten Great Ideas of Science
Peter Atkins

Ten great ideas, ranging from natural selection through quantum theory to curved spacetime, are introduced with brilliant imagery in this best-selling introduction to modern scientific concepts. Never before have these core ideas of modern civilization been presented in so engaging a manner.

'This book is one of the best panoramic views of nature's extraordinary symmetry, subtlety and mystery currently on offer'
John Cornwell, *Sunday Times*

The Emperor's New Mind: Concerning Computers, Minds, and the Laws of Physics
Roger Penrose

A fascinating roller-coaster ride through the basic principles of physics, cosmology, mathematics, and philosophy, to show that human thinking can never be emulated by a machine.

'Perhaps the most engaging and creative tour of modern physics that has ever been written'
Sunday Times

OXFORD